Asymptotic and Numerical Methods for Partial Differential Equations with Critical Parameters

NATO ASI Series

Advanced Science Institutes Series

A Series presenting the results of activities sponsored by the NATO Science Committee, which aims at the dissemination of advanced scientific and technological knowledge, with a view to strengthening links between scientific communities.

The Series is published by an international board of publishers in conjunction with the NATO Scientific Affairs Division

A Life Sciences **B Physics**	Plenum Publishing Corporation London and New York
C Mathematical and Physical Sciences **D Behavioural and Social Sciences** **E Applied Sciences**	Kluwer Academic Publishers Dordrecht, Boston and London
F Computer and Systems Sciences **G Ecological Sciences** **H Cell Biology** **I Global Environmental Change**	Springer-Verlag Berlin, Heidelberg, New York, London, Paris and Tokyo

NATO-PCO-DATA BASE

The electronic index to the NATO ASI Series provides full bibliographical references (with keywords and/or abstracts) to more than 30000 contributions from international scientists published in all sections of the NATO ASI Series.
Access to the NATO-PCO-DATA BASE is possible in two ways:

– via online FILE 128 (NATO-PCO-DATA BASE) hosted by ESRIN, Via Galileo Galilei, I-00044 Frascati, Italy.

– via CD-ROM "NATO-PCO-DATA BASE" with user-friendly retrieval software in English, French and German (© WTV GmbH and DATAWARE Technologies Inc. 1989).

The CD-ROM can be ordered through any member of the Board of Publishers or through NATO-PCO, Overijse, Belgium.

Series C: Mathematical and Physical Sciences - Vol. 384

Asymptotic and Numerical Methods for Partial Differential Equations with Critical Parameters

edited by

Hans G. Kaper
Argonne National Laboratory,
Argonne, Illinois, U.S.A.

and

Marc Garbey
Université Claude Bernard,
Lyon, France

Technical editor

Gail W. Pieper
Argonne National Laboratory,
Argonne, Illinois, U.S.A.

Springer Science+Business Media, B.V.

Proceedings of the NATO Advanced Research Workshop on
Asymptotic-Induced Numerical Methods for Partial Differential Equations,
Critical Parameters, and Domain Decomposition
Beaune, France
May 25–28, 1992

ISBN 978-94-010-4798-2 ISBN 978-94-011-1810-1 (eBook)
DOI 10.1007/978-94-011-1810-1

Printed on acid-free paper

Preface

This volume contains the proceedings of the NATO Advanced Research Workshop on "Asymptotic-induced Numerical Methods for Partial Differential Equations, Critical Parameters, and Domain Decomposition," held at Beaune (France), May 25–28, 1992.

The purpose of the workshop was to stimulate the integration of asymptotic analysis, domain decomposition methods, and symbolic manipulation tools for the numerical solution of partial differential equations (PDEs) with critical parameters. A workshop on the same topic was held at Argonne National Laboratory in February 1990. (The proceedings were published under the title *Asymptotic Analysis and the Numerical Solution of Partial Differential Equations,* Hans G. Kaper and Marc Garbey, eds., *Lecture Notes in Pure and Applied Mathematics,* Vol. 130, Marcel Dekker, Inc., New York, 1991.) In a sense, the present proceedings represent a progress report on the topic area.

Comparing the two sets of proceedings, we see an increase in the quantity as well as the quality of the contributions. More research is being done in the topic area, and the interest covers serious, nontrivial problems. We are pleased with this outcome and expect to see even more advances in the next few years as the field progresses.

The proceedings of the workshop are organized in six parts. The first three parts give a panoramic view of several fields of research, where asymptotic analysis, numerical analysis, and scientific computing are beginning to be integrated effectively. The last three parts give a more detailed picture of the state of the art in asymptotic analysis, numerical analysis, and computational tools.

Part 1 (Modeling of complex systems with asymptotic-enhanced numerical methods) contains five contributions. In each contribution, significant use is made of asymptotic analysis to model complex physical systems, mostly in fluid dynamics: three-dimensional gas flow around a finite flat plate, tide-induced transport of constituents in shallow water, turbulent flow, transonic flow, some unusual flow problems in paper manufacturing and magnetic recording, and magnetohydrodynamic flow in a tokamak fusion device. In each case, asymptotic methods yield models that form the basis of large-scale numerical simulations and, in addition, provide insight into the underlying physics. In some cases it is actually possible to assess the validity of the asymptotic methods analytically, experimentally, or directly by numerical simulation.

Part 2 (Asymptotic-induced domain decomposition methods) contains three contributions highlighting applications of domain decomposition techniques in multiple-scale problems and singular perturbation problems. Here, the method of matched asymptotic expansions provides useful information not only about the structure of the domain decomposition, but also on the approximation of the PDE inside each subdomain and on the proper implementation of interface conditions. The contributions cover convection-dominated flows, blow-up and quenching phenomena in reaction-diffusion systems, and vortex calculations in superconductivity.

Part 3 (Multiple-scale problems in scientific computing) highlights high-performance computing problems in combustion. Here, we encounter strongly nonlinear problems with moving fronts and internal and boundary layers, which pose formidable challenges to the computational scientists. The two contributions give a good picture of the state of the art in computational combustion; they make extensive use of asymptotic analysis in the derivation of models.

Part 4 (Applied and asymptotic analysis) deals with analytical aspects of differential equations with critical parameters. At critical parameter values, the nature of the solution may change dramatically. The role of analysis is then to bring out these changes and, if possible, suggest consistent ways to simplify the problem description in the various parameter regimes. The contributions in this part of the proceedings do not involve any numerical computation. However, some contributions highlight the source of the difficulties in a direct numerical approach, while others study problems where computations traditionally play an important role in the discovery process. This part will certainly introduce the reader to a number of interesting numerical projects for PDEs with critical parameters.

Part 5 (Symbolic manipulation tools for asymptotic analysis) contains only one contribution: clearly, a demonstration of the fact that symbolic manipulation languages have not made an impact in the field. Nevertheless, the organizers know from experience that symbolic manipulation languages can be used effectively in multiple-scale problems. The tools have to be developed for specific applications, but some level of abstraction is possible, as the contribution in these proceedings illustrates. We certainly expect more progress here.

Part 6 (Numerical methods, algorithms, and computer architectures) contains several contributions that are mostly concerned with the numerical analysis of multilevel methods, domain decomposition, and adaptive methods. They give an idea of the tools available in modern numerical analysis.

The last contribution is concerned with the interplay between the topology of a computer architecture and the efficiency of a given algorithm—an important problem if one lacks a good communications library as in a transputer network.

In summary, the topic theme of the workshop is still very much in the initial stages of development. It is only a matter of time before *asymptotic-induced* or *asymptotic-enhanced* numerical methods will find their rightful place in large-scale scientific computing. We hope that these proceedings will stimulate more researchers to get involved in this area of research.

Many people and organizations contributed to the success of the workshop and the preparation of these proceedings. The organizing committee—consisting of Marc Garbey (Université Claude Bernard, Lyon, France), Hans G. Kaper (Argonne National Laboratory), and F. V. Atkinson (University of Toronto, Toronto, Canada)—worked hard to bring the workshop participants together. Jan Griffin (Argonne National Laboratory) contributed significantly as workshop secretary. Gail Pieper (Argonne National Laboratory) gently pushed the authors to submit their manuscripts on time, did the editing work, gave our fax machine a good workout, and generally did everything to prepare the proceedings for publication.

We thank the NATO Scientific Council for sponsoring this Advanced Research Workshop.

Hans G. Kaper
Marc Garbey

Contents

Contributors

Baranger, Jacques (Prof.)
Université Claude Bernard-Lyon
L.A.N. Bâtiment 101
43 Bd. du 11 Novembre 1918
69622 Villeurbanne Cedex, France

Bayliss, Bernard (Prof.)
Department of Engineering Sciences and Applied Mathematics
Northwestern University
Evanston, Illinois 60208, USA

Bdzil, John B. (Dr.)
Los Alamos National Laboratory
Los Alamos, New Mexico 87545, USA

Bernardi, C. (Dr.)
Laboratoire d'Analyse Numérique
CNRS and Université Pierre et Marie Curie
4 place Jussieu
75252 Paris Cedex 05, France

Boisson, H. C. (Prof.)
IMFT-ENSEEIHT-CNRS UA0005
Avenue Camille Soula
31400 Toulouse, France

Cook, L. Pamela (Prof.)
Department of Mathematical Sciences
University of Delaware 501 Ewing Hall
Newark, Delaware 19716, USA

Desprez, Frederic
LIP/IMAG, CNRS
ENS Lyon
46, Allee d'Italie
69364 Lyon Cedex 07, France

Dijkstra, Douwe (Dr.)
Faculty of Applied Mathematics
University of Twente
P.O. Box 217
7500 AE Enschede, the Netherlands

Doelman, Arjen (Dr.)
Center for Applied Mathematics
Cornell University
Ithaca, New York 14853-6201, USA

Eckhaus, Wiktor (Prof.)
Mathematics Institute
Rijksuniversiteit Utrecht
Budapestlaan 6
Postbus 80.010
3508 TA Utrecht, the Netherlands

El Amri, Harsan (Prof.)
ENS de Fés
BP 34 A
Fés, Morocco

El Ferricha, Mohammed El Alami (Prof.)
ENS de Fés
BP 34 A
Fés, Morocco

Gropp, William D. (Dr.)
Mathematics and Computer Science Division
Argonne National Laboratory
9700 South Cass Avenue
Argonne, Illinois 60439-4844, USA

Herwig, H. (Prof. Dr.-Ing.)
Institut für Thermo- und Fluiddynamik
Ruhr-Universität
4630 Bochum, Germany

Keyes, David E. (Prof.)
Department of Mechanical Engineering
Yale University
New Haven, Connecticut 06520, USA

Kuerten, J. G. M. (Prof.)
Faculty of Applied Mathematics
University of Twente
P.O. Box 217
7500 AE Enschede, the Netherlands

Kwong, Man Kam (Dr.)
Mathematics and Computer Science Division
Argonne National Laboratory
Argonne, IL 60439-4844, USA

Laforgue, Jacques G. (Prof.)
Department of Applied Mathematics FS-20
University of Washington
Seattle, Washington 98195, USA

Maday, Y. (Prof.)
Laboratoire d'Analyse Numérique
CNRS and Université Pierre et Marie Curie
4 place Jussieu
75252 Paris Cedex 05, France

Malmuth, Norman D. (Dr.)
Rockwell International Science Center
1049 Camino Dos Rios
P.O. Box 1085
Thousand Oaks, California 91360, USA

Matkowsky, Bernard J. (Prof.)
Department of Engineering Sciences and Applied Mathematics
Northwestern University
Evanston, Illinois 60208, USA

McCormick, Steve (Prof.)
Computational Mathematics Group
University of Colorado
Campus Box 170, P.O. Box 173364
Denver, Colorado 80217-3364, USA

Mottoni, Pierre de
Matematica
2a Universitá di Roma
Tor Vergata, Italy

Nataf, F. (Dr.)
CMAP, Ecole Polytechnique
91128 Palaiseau Cedex, France

Olagunju, D. O. (Prof.)
Department of Mathematical Sciences
University of Delaware
Newark, Delaware 19716, USA

O'Malley, Robert E., Jr. (Prof.)
Department of Applied Mathematics FS-20
University of Washington
Seattle, Washington 98195, USA

Paris, R. B. (Prof.)
Department of Mathematical and Computer Sciences
Dundee Institute of Technology
Dundee DD1 1HG, UK

Patera, A. T.
Department of Mechanical Engineering
Massachusetts Institute of Technology
77 Massachusetts Avenue
Cambridge, Massachusetts 02139, USA

Perkins, A. Louise (Prof.)
Institute for Naval Oceanography
10003 East Marlin
Bay St. Louis, Mississippi 39520, USA

Riecke, Hermann
Department of Engineering Sciences and Applied Mathematics
Northwestern University
Evanston, Illinois 60208, USA

Rogier, F. (Dr.)
Division Calcul parallèle
ONERA
29 Avenue de la division Leclerc
92322 Châtillon sous Bagneux, France

Salvy, Bruno (Dr.)
Algorithms Project
INRIA Rocquencourt
Domaine de Voluceau B.P. 105
78153 Le Chesnay Cedex, France

Schleiniger, G. F. (Prof.)
Department of Mathematical Sciences
University of Delaware
Newark, Delaware 19716, USA

Schatzman, Michelle (Prof.)
Laboratoire d'Analyse Numérique
Universit/'e Claude Bernard-Lyon
L.A.N., Batiment 101
43 Bd. du 11 Novembre 1918
69622 Villeurbanne Cedex, France

Spiteri, Pierre (Prof.)
IRIT-ENSEEIHT-CNRS UA1399
2 Rue Charles Camichel
31071 Toulouse, France

Stewart, D. Scott (Dr.)
Theoretical and Applied Mechanics
National Center for Supercomputing Applications
5600 Beckman Institute
Drawer 25
405 North Matthews Avenue
Urbana, Illinois 61801, USA

Thompson, Chris P. (Dr.)
Process Studies Department
AEA Petroleum Services
Harwell Laboratory
Didcot, Oxfordshire, OX11 0RA, England

Titi, Edriss S. (Dr.)
Center for Applied Mathematics
Cornell University
Ithaca, New York 14853-6201, USA

Tourancheau, Bernard (Dr.)
LIP/IMAG, CNRS
46 Allee d'Italie
69364 Lyon Cedex 07, France

Van Harten, Aart (Prof.)
Mathematical Institute
Rijksuniversiteit Utrecht
Budapestlaan 6
Utrecht, the Netherlands

Wood, Alastair D. (Prof.)
School of Mathematical Sciences
Dublin City University
Dublin 9, Ireland

Part 1: Modeling of Complex Systems with Asymptotic-enhanced Numerical Methods

ASYMPTOTICS AND NUMERICS FOR LAMINAR FLOW OVER A FINITE FLAT PLATE

D. Dijkstra and J. G. M. Kuerten
Faculty of Applied Mathematics
University of Twente
P.O. Box 217
7500 AE Enschede
the Netherlands

ABSTRACT. A compilation of theoretical results from the literature on the finite flat-plate flow at zero incidence is presented. This includes the Blasius solution, the Triple Deck at the trailing edge, asymptotics in the wake, and properties near the edges of the plate. In addition, new formulas for skin friction at the plate and velocity overshoot at the boundary-layer edge are given. The relevance of the theoretical predictions is shown by a comparison with numerical results calculated with a compressible Navier-Stokes solver at low Mach number and Reynolds numbers between 10^3 and 10^5. This comparison gives clear evidence for the quantitative correctness of the significant upstream influence predicted by the trailing-edge Triple Deck theory.

1. Introduction

The problem of laminar flow over a flat plate at zero incidence has a long history, which started with the Prandtl boundary-layer theory almost a century ago. On the basis of boundary-layer considerations, many theoretical results for the flat plate have been published in the literature.

The best-known approximation of the flow is the Blasius solution of the first-order boundary-layer equations. This solution for large Reynolds number gives an adequate description of the boundary layer over the main part of the plate. Its failure near the edges of the plate has long been recognized.

Goldstein [6] investigated the transition from the Blasius flow to a wake flow near the trailing edge. The full impact of the trailing edge on the flow was elucidated by the Triple Deck model, independently put forward by Stewartson [10] and Messiter [9]. This model is still based on first-order boundary-layer theory, but the difference with the Blasius flow is the presence of a pressure gradient induced locally by the abrupt end of the plate. The novel feature of the Triple Deck formulation is that the pressure is

H. G. Kaper and M. Garbey (eds.),
Asymptotic and Numerical Methods for Partial Differential Equations with Critical Parameters 3–19.

unknown and has to be computed simultaneously with the solution of the trailing-edge problem. An important aspect of the Triple Deck theory is that it allows for upstream influence via the potential flow outside the boundary layer. As a consequence of this upstream influence, the skin friction and drag rise above the Blasius predictions.

Adequate and detailed numerical results for the Triple Deck problem have been obtained by Veldman [12]. Very close to the trailing edge, the Triple Deck model becomes invalid and must be replaced by the full Navier-Stokes equations [5].

Near the leading edge, the flow was investigated by Van de Vooren and Dijkstra [13] and by Yoshizawa [14]. Their results, based on the full Navier-Stokes equations, show an increase in skin friction when compared with the Blasius prediction.

In the present paper, a compilation of the most relevant results from the literature will be given. In addition, new, explicit formulas will be presented for skin friction at the plate and velocity overshoot at the edge of the boundary layer. The adequacy of the theoretical predictions will be illustrated by a comparison with numerical results obtained with a compressible Navier-Stokes solver at a Mach number of 0.3 and three Reynolds numbers in the range $10^3 - 10^5$.

2. Formulation of the Problem

For incompressible, laminar, steady flow in two dimensions the equations in nondimensional form read

$$\frac{\partial u}{\partial x} + \frac{\partial v}{\partial y} = 0, \tag{1}$$

$$u\frac{\partial u}{\partial x} + v\frac{\partial u}{\partial y} = -\frac{\partial p}{\partial x} + \frac{1}{Re}\Delta u, \tag{2}$$

$$u\frac{\partial v}{\partial x} + v\frac{\partial v}{\partial y} = -\frac{\partial p}{\partial y} + \frac{1}{Re}\Delta v, \tag{3}$$

where Δ is the Laplacian.

With the origin at the leading edge, the Cartesian coordinates x and y have been nondimensionalized with the plate length L, the velocity components u and v with the velocity of the incoming flow at infinity U_∞, and the pressure p with ρU_∞^2. Here ρ is the constant density, and the Reynolds number is defined as

$$Re = \frac{U_\infty L}{\nu}, \tag{4}$$

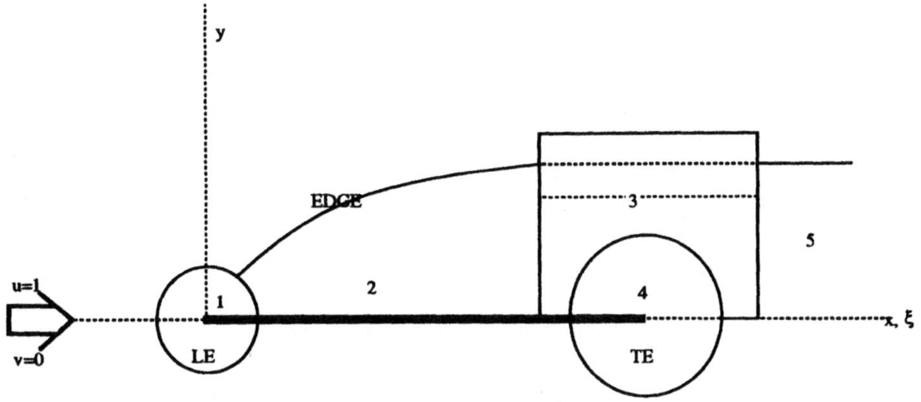

Fig. 1. **Flow regimes over a flat plate. The numbers correspond with the subsections of Section 3. LE: leading edge; TE: trailing edge.**

where ν is the kinematic viscosity. The problem has been outlined schematically in Fig. 1.

Note that the coordinate ξ in Fig. 1 is given by

$$\xi = x - 1. \tag{5}$$

For one side of the plate, the nondimensional local skin friction coefficient is defined by

$$c_f = \frac{\tau_w}{\frac{1}{2}\rho U_\infty^2}. \tag{6}$$

Here τ_w denotes the shear stress at the plate, which is given by

$$\tau_w = \frac{\rho\nu U_\infty}{L}(\frac{\partial u}{\partial y})_w. \tag{7}$$

Combination of (6) and (7) gives

$$c_f = \frac{2}{Re}(\frac{\partial u}{\partial y})_w. \tag{8}$$

The drag D for two sides of the plate is defined as

$$D = 2L\int_0^1 \tau_w dx = \rho L U_\infty^2 \int_0^1 c_f dx, \tag{9}$$

and the corresponding drag coefficient is defined by

$$C_D = \frac{D}{\frac{1}{2}\rho U_\infty^2 L} = 2\int_0^1 c_f dx. \tag{10}$$

3. Flow Regimes and Flow Properties

In Fig. 1 several flow regimes have been outlined. This flow structure develops for large Reynolds number, Re. The limitations of Re used in the present paper are

$$10^3 < Re < 10^5. \tag{11}$$

The lower bound has been chosen in order to keep contact with asymptotics and the upper bound to keep the flow laminar [1].

The flow properties will now be presented for each flow regime given in Fig. 1. Note that for clarity of presentation the number of each regime corresponds directly with the subsection number used in the sequel.

3.1. LEADING-EDGE NAVIER-STOKES

For an adequate description of the flow near the leading edge, the full Navier-Stokes equations (1–3) must be used in Region 1 (Fig. 1). The size of this small region is

$$x, y = \mathcal{O}(Re^{-1}). \tag{12}$$

Accurate computations in this region have been performed by Van de Vooren and Dijkstra [13] and by Yoshizawa [14]. From these papers the following salient property has been selected:

$$c_{f1} \sim \frac{0.755}{\sqrt{xRe}}, \qquad xRe \to 0. \tag{13}$$

The impact of Region 1 on the drag of the plate is of the order of $1/Re$ and thus small, and has hardly any influence on the downstream development of the flow.

3.2. BLASIUS BOUNDARY LAYER

In Fig. 1, Region 2 denotes the boundary layer with size

$$x = \mathcal{O}(1), \qquad y = \mathcal{O}(Re^{-\frac{1}{2}}). \tag{14}$$

The flow in the boundary layer is governed by the classical Prandtl boundary-layer equations with zero pressure gradient. The well-known Blasius solution yields for the velocities in the boundary layer

$$u_2 = f'(\eta), \qquad v_2 = \frac{\eta f'(\eta) - f(\eta)}{2\sqrt{x Re}}, \tag{15}$$

where

$$\eta = y\sqrt{\frac{Re}{x}} \tag{16}$$

and the Blasius function f is given by

$$2f''' + ff'' = 0, \; f(0) = f'(0) = 0, \; f'(\infty) = 1. \tag{17}$$

The numerical solution of this differential equation can be obtained by means of the initial values given in (17) together with

$$f''(0) = \alpha = 0.33205734. \tag{18}$$

Further properties of f are for $\eta \to \infty$

$$f(\eta) \sim \eta - \beta + \mathrm{EXP}, \; \beta = 1.72078765, \tag{19}$$

$$f''(\eta) \sim \gamma e^{-\frac{1}{4}(\eta-\beta)^2}, \; \gamma = 0.23372763, \tag{20}$$

where EXP denotes exponentially small terms. In addition,

$$\eta > 6 \Rightarrow 0 < 1 - f'(\eta) < 10^{-3}. \tag{21}$$

Hence, the 10^{-3} Blasius boundary-layer thickness may be defined by $\eta = 6$, which gives the edge E of the boundary layer by means of Equation (16),

$$y_E = 6\sqrt{\frac{x}{Re}}, \; 0 < x < 1. \tag{22}$$

Beyond y_E the vorticity in the flow may be neglected, which means that one has potential flow. At the edge the velocities as predicted by the Blasius solution are

$$u_2(E) = 1.000, \; v_2(E) = \frac{0.860}{\sqrt{x Re}}. \tag{23}$$

At the plate the normal derivative of u_2 can be obtained from Equations (15), (16), and (18); and this gives the skin friction coefficient (see Equation (8))

$$c_{f2} = \frac{2\alpha}{\sqrt{xRe}} = \frac{0.664}{\sqrt{xRe}}, \quad 0 < x < 1. \tag{24}$$

With Equation (10) the corresponding drag coefficient for two sides of the plate becomes

$$C_{D2} = \frac{8\alpha}{\sqrt{Re}} = \frac{2.656}{\sqrt{Re}}. \tag{25}$$

It should be remarked that the results given above are valid in first-order boundary-layer theory: the pressure gradient vanishes to the order considered. For second-order boundary-layer theory the first step is to calculate the pressure over the displacement body of plate and wake. As far as we know, no explicit formulas are available for this *global* problem. Guided by the work of Veldman [12], we have considered this global problem and constructed an approximate result, which is presented in Section 3.7.

It should also be remarked that the results of boundary-layer theory do not hold in the immediate neighborhood of the edges of the plate. For the leading edge this is immediately clear from Equations (13) and (24). At the trailing edge the situation appears to be much more complicated. This is the subject of the next two subsections.

3.3. TRAILING-EDGE TRIPLE DECK

Region 3 in Fig. 1 denotes the Triple Deck with size

$$\xi = x - 1 = \mathcal{O}(Re^{-\frac{3}{8}}), \; y = \mathcal{O}(Re^{-\frac{3}{8}}). \tag{26}$$

The main feature of the Triple Deck is that it allows for upstream influence via a locally active pressure gradient induced by the abrupt change in boundary conditions. The upstream influence is felt by the skin friction which rises above the Blasius value. As a consequence, the drag of the plate increases significantly.

The Triple Deck theory was developed independently by Stewartson [10] and Messiter [9]. The novel feature of the model is that the boundary condition at the edge of the boundary layer is elliptic: the pressure gradient must be obtained simultaneously with the solution. Numerical results of the problem have been obtained by Veldman [12], Jobe and Burggraf [7], and Melnik and Chow [8]. From these papers the prediction for the skin friction at the trailing edge is found to be

$$\xi = 0, \; c_{f3} = \frac{0.898}{\sqrt{Re}}. \tag{27}$$

This value exceeds the Blasius value (Equation (24)) by 35%.

The drag increase that occurs in the Triple Deck is on the order of $Re^{-\frac{7}{8}}$, and the total theoretical drag coefficient for both sides of the plate is

$$C_{D23} = 2.656Re^{-\frac{1}{2}} + 5.30Re^{-\frac{7}{8}}, \tag{28}$$

where the subscript 23 refers to Regions 2 and 3 in Fig. 1 together.

The wake center-line velocity in the Triple Deck as predicted by Veldman [12] is contained in Table I.

TABLE I
Triple Deck wake center-line velocity

$Re^{\frac{3}{8}}\xi$	$u_3(\xi, 0)Re^{\frac{1}{8}}$
0.037	0.185
0.148	0.298
0.422	0.433
1.000	0.597
2.125	0.805
4.215	1.067
8.000	1.391
14.790	1.777

The behavior for small values of ξ is

$$u_3(\xi, 0) \sim 0.527\xi^{\frac{1}{3}}, \tag{29}$$

and for larger values of $\xi Re^{\frac{3}{8}}$

$$u_3(\xi, 0) \sim 0.772\xi^{\frac{1}{3}}. \tag{30}$$

Note that Equation (30) does not hold as $\xi \to \infty$. The results presented above are also not valid for extremely small values of ξ and must be replaced by results given in the next subsection.

3.4. TRAILING-EDGE NAVIER-STOKES

In Region 4 of Fig. 1 the full Navier-Stokes equations must be used for an adequate description of the flow. The size is

$$(\xi, y) = \mathcal{O}(Re^{-\frac{3}{4}}). \tag{31}$$

The flow in this region has been investigated by Dijkstra [5]. The region has only local impact. The following salient properties are presented here:

$$c_{f4} \sim \frac{0.877}{\sqrt{|\xi|}} Re^{-\frac{7}{8}}, \tag{32}$$

$$u_4(\xi,0) \sim 0.438 Re^{\frac{1}{8}} \sqrt{\xi} \tag{33}$$

valid in the limit

$$|\xi| Re^{\frac{3}{4}} \to 0. \tag{34}$$

3.5. GOLDSTEIN WAKE BOUNDARY LAYER

The wake is Region 5 in Fig. 1. It extends to downstream infinity with vertical size

$$y = \mathcal{O}(Re^{-\frac{1}{2}}). \tag{35}$$

The flow is governed by the first-order boundary-layer equations with zero pressure gradient. The classical solution of these equations in the wake has been given by Goldstein [6] in the form of a coordinate expansion for small values of ξ, that is, the near wake.

The complete wake boundary-layer problem has been solved by Veldman [11]. Numerical values for the wake center-line velocity $u_5(\xi,0)$ can be found in the cited paper. For small, but not too small, values of ξ,

$$u_5(\xi,0) \sim 0.772\xi^{\frac{1}{3}}, \tag{36}$$

which matches the Triple Deck result, Equation (30). In the far wake there holds along the center line

$$u_5(\xi,0) \sim 1 - \frac{0.375}{\sqrt{\xi}}. \tag{37}$$

The velocity profile across the far wake boundary layer is

$$u_5(\xi,y) \sim 1 - \frac{0.375}{\sqrt{\xi}} \exp(-\frac{y^2}{4\xi} Re). \tag{38}$$

3.6. A COMPOSITE RESULT FOR THE SKIN FRICTION

We have constructed a composite expansion formula for the skin friction at the plate,

$$c_{f23} = \frac{0.664}{\sqrt{xRe}} + \frac{1.3}{\sqrt{Re}}(3.6 + |\xi| Re^{\frac{3}{8}})^{-\frac{4}{3}} + \frac{0.30}{Re}|\xi|^{-\frac{2}{3}}, \tag{39}$$

where $\xi = x - 1$. The first term is the Blasius boundary-layer contribution, the second term is an approximation to the Triple Deck contribution, and the third term arises at the interface of Triple Deck and Trailing Edge Navier-Stokes region. Apart from the first term the formula reflects work by Veldman [12] and Dijkstra [5]. Theoretically, it should be valid in the range

$$200 Re^{-1} < x < 1 - Re^{-\frac{3}{4}}. \tag{40}$$

3.7. AN APPROXIMATION OF THE VELOCITY OVERSHOOT

As remarked in the last part of Section 3.2, the effect of second-order boundary-layer theory on u is of the order of $1/\sqrt{Re}$. This effect leads to an overshoot in the velocity $u(E)$, where E denotes the edge of the boundary layer, Equation (22). In order to find a prediction for the overshoot $u(E)-1$, the pressure induced by the displacement body of plate and wake must be obtained. On this basis and with results given by Veldman [12], we have constructed the following approximation of the overshoot:

$$u(E) - 1 = \frac{\beta}{2\pi}\frac{1}{\sqrt{xRe}}\ln(\frac{1+\sqrt{x}}{1-\sqrt{x}}) + 0.50Re^{-1/4}(3.3 + |\xi|Re^{3/8})^{-2/3}, \quad (41)$$

where β is given by (19), $\xi = x - 1$, and $0 < x < 1$. The first term is induced by the displacement body over the plate (Veldman [12]), whereas the second term arises in the Triple Deck. Note the similarity in the formula stated and the formula (39) for the skin friction. Note also that the displacement effect of the far wake (beyond the Triple Deck) has been disregarded.

4. Description of Flow Solver

In this section we describe the numerical method used in the present compressible Navier-Stokes solver. This flow solver is developed as a part of the Dutch ISNaS project, which aims at the construction of an information system for flow simulation based on the Navier-Stokes equations [2]. The compressible flow solver is a joint effort of the National Aerospace Laboratory and the University of Twente, and is aimed especially at two aerodynamic applications, namely, the multielement airfoil and the wing/body combination, both for transonic turbulent flow. A stepwise development has been chosen, in which the first two steps involve inviscid and viscous laminar flow around a single airfoil. One of the test cases for the second step was the viscous laminar flow over a finite flat plate [3].

The discretization of the Navier-Stokes equations follows the method of lines; that is, the spatial discretization is performed first, and subsequently the resulting set of ordinary differential equations is integrated in time, until the steady-state solution is obtained. First, the computational domain is partitioned in quadrilateral cells with the help of a structured boundary-conforming grid. For the spatial discretization a finite-volume method is used in which the integral form of the Navier-Stokes equations is applied to a control volume consisting of a number of grid cells. The basis of the finite-volume method used here is a cell vertex method with overlapping control volumes. In this method the variables are stored at the grid points, and the control volume for each grid point consists of the four neighboring grid cells. The nondissipative flux through the boundaries of a control volume can be accurately approximated by using the trapezoidal rule.

For the calculation of the viscous fluxes an auxilary control volume is used. This control volume is the area between the four neighboring cell centers. The spatial derivatives of the state vector in these cell centers can be approximated with the use of Gauss's theorem on a grid cell. The total viscous flux can then be calculated by using the trapezoidal rule.

The method of central differencing leads to an uncoupling of odd and even grid points and to oscillations near shock waves. Even for viscous flow calculations the presence of viscous dissipation is not sufficient to damp these instabilities outside shear layers. Therefore, nonlinear artificial dissipation is added to the basic numerical scheme. Since no shock waves occur in the present test case, only constant coefficient fourth-order difference terms are incorporated.

At the solid wall boundaries the no-slip condition is used. The density and energy density in the grid points on a solid wall are calculated by solving their respective discrete conservation laws, using the two adjacent cells within the computational domain and their mirror images inside the wall as control volume. The values of the density and energy density in the grid points inside the walls are adjusted such that the adiabatic wall condition is approximated. The boundary conditions at a (subsonic) far-field boundary are based on characteristic theory. However, since the calculations are performed at a low Mach number in order to reduce compressibility effects, the use of this boundary condition at the outflow boundary leads to convergence problems. These are caused by insufficiently suppressed sound waves. Therefore, at the outflow boundary the streamwise velocity component is prescribed according to Equation (38); and the density, normal velocity, and energy density are linearly extrapolated from inside the computational domain.

The large system of ordinary differential equations is integrated in time by using a time-explicit multistage Runge-Kutta method. In the present flow solver a five-stage scheme is incorporated in which the dissipative terms are calculated only at the odd stages. Extra calculation time is saved by applying local time stepping. An improvement of the starting solution is obtained by using "grid sequencing."

The calculations described in this paper are performed on a rectangular grid with 312 cells in the streamwise direction (128 on the plate) and 128 in the normal direction. The grid points in the streamwise direction are clustered around the leading and trailing edge with a smallest distance of 5×10^{-4} to resolve the local small-scale phenomena. The grid spacing in the normal direction is stretched, with the first grid line at a distance of 1.8×10^{-4} plate lengths from the wall. The computational domain extends from $0.5L$ upstream to $25L$ downstream of the plate, and $1L$ above the plate.

5. Results and Discussion

With the compressible Navier-Stokes solver described in Section 4, we have computed results at a Mach number of 0.3 and Reynolds numbers of 2500, 10000, and 40000. The density variations are on the order of 1% and occur mainly at the edges of the plate. We now compare these results with the theoretical predictions given in Section 3.

5.1. VELOCITY COMPONENTS

Figures 2 and 3 show the velocity u at $x = 0.5$ in the boundary layer. The results show an overshoot when compared with $u_2(E)$ at $\eta = 6$, Equation (23). The numerical values of this overshoot are 0.031 and 0.016 for $Re = 2500$ and 10000, respectively. The theoretical approximation (41) yields the values 0.027 and 0.014, respectively, for this overshoot. These values leave some room for the influence of the far wake, which is not contained in Equation (41).

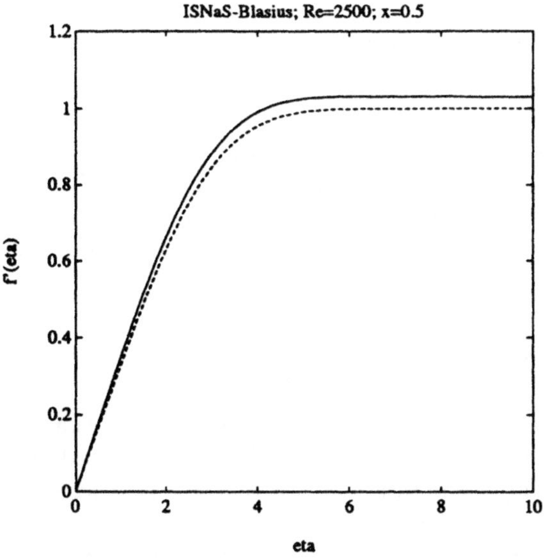

Fig. 2. **Streamwise velocity as a function of η; solid: present solver; dashed: Blasius**

Figure 4 shows the quantity $2v\sqrt{xRe}$ (a scaled normal velocity component) at $x = 0.2676$ in the boundary layer and its value according to Equation (15). It should be remarked that beyond $\eta = 10$ the results of the code slowly decrease to 0 as the boundary of the computational domain is approached.

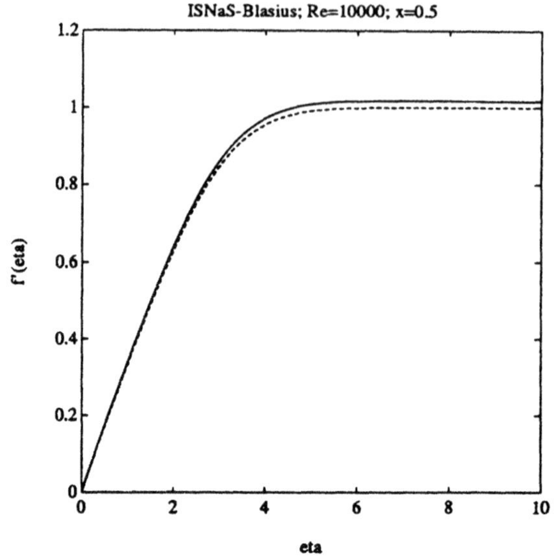

Fig. 3. **Streamwise velocity as a function of η; solid: present solver; dashed: Blasius**

Results for the wake center-line velocity have been given at $Re = 10000$ in Figs. 5 and 6. Note the difference in the scales for ξ. For values of ξ of the order of $Re^{-3/4}$ the results are in good agreement with the predictions of Section 3.4. Apparently the grid is capable of resolving this small-scale flow property.

5.2. SKIN FRICTION

In Fig. 7 the ratio c_f/c_{f2} compares the skin friction with the Blasius result, Equation (24) for $Re = 2500$. Note the rise of 15% at the leading edge, which is in agreement with the prediction (13). Note also that there is a significant rise as the trailing edge is approached.

The adequacy of the composite formula (39) to predict the skin friction over the range (40) can be inferred from Fig. 8. At $Re = 10000$ and 40000 (not shown) the agreement is still better.

5.3. DRAG

In Table II the drag coefficient for two sides of the plate according to Equation (28) is compared with numerical results of the present solver and of Dennis and Dunwoody [4]. It may be concluded that the formula (28) does

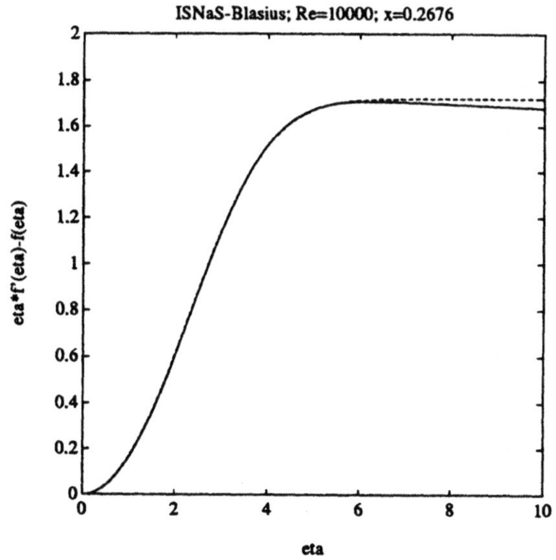

Fig. 4. **Scaled v as a function of η; solid: present solver; dashed: Blasius**

a remarkably good job and is extremely useful for comparison of drag predictions.

TABLE II
Drag coefficient

Re	(28)	[4]	present
1000	0.0966	0.1004	-
2000	0.0662	0.0682	-
2500	0.0588	-	0.0597
5000	0.0406	0.0412	-
10000	0.0282	0.0282	0.0283
40000	0.0138	-	0.0136

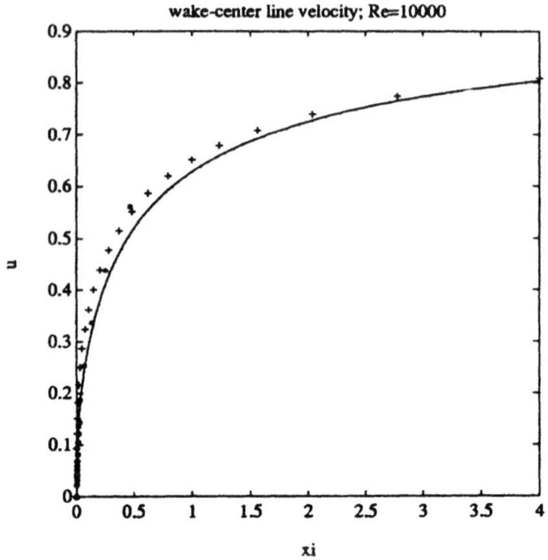

Fig. 5. **Wake center-line velocity; solid line: present solver; o: region 4; *: region 3; +: region 5. The characteristic value of ξ in region 4 is** 0.001.

6. Conclusion

A compilation of results from the literature on finite flat-plate flow has been given, and new formulas for the skin friction and velocity overshoot at the boundary-layer edge have been presented. The confrontation of these theoretical results with a numerical solution of the full Navier-Stokes equations demonstrates the adequacy of the theoretical results. The upstream influence of the trailing edge as predicted by the Triple Deck model is found to be important and in good agreement with the numerical results for the Navier-Stokes equations.

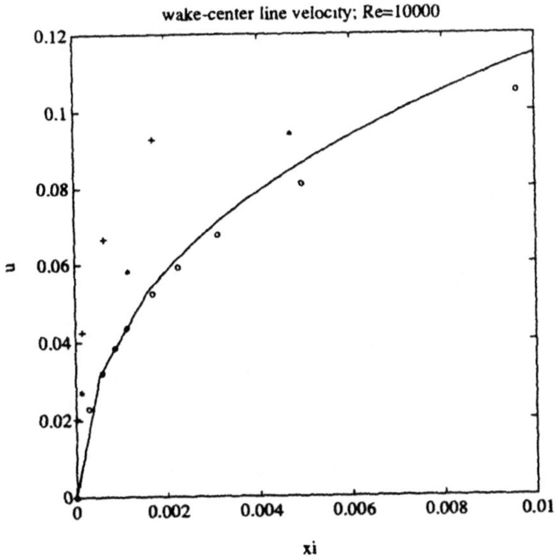

Fig. 6. Wake center-line velocity; solid line: present solver; o: region 4; *: region 3; +: region 5

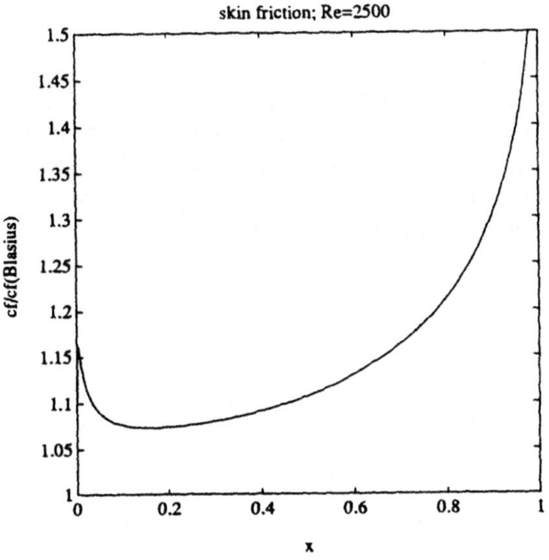

Fig. 7. Ratio of calculated and Blasius skin friction

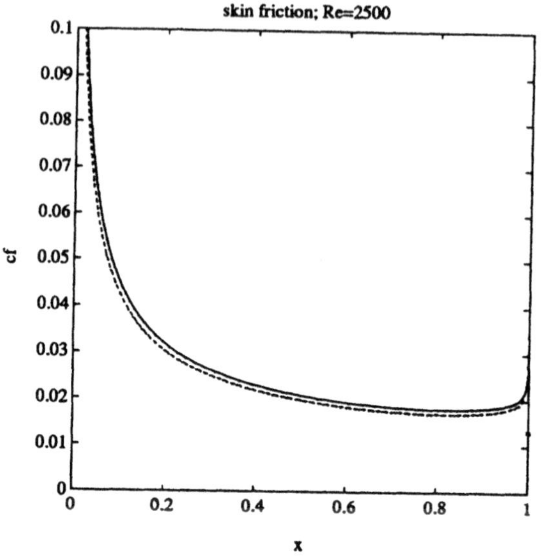

Fig. 8. **Skin friction; solid line: present solver; dashed line: equation (39); *: Blasius value at $x = 1$**

Acknowledgements

We are greatly indebted to Prof. Dr. A. E. P. Veldman for many stimulating discussions. This work is partly supported by the Dutch Ministries of Education and Sciences and of Transport and Public Works. Part of the necessary supercomputer time was sponsored by the National Computing Facilities Foundation, NCF, with financial support from the Netherlands Organisation for Scientific Research, N.W.O.

References

[1] Batchelor, G. K. (1970) *An Introduction to Fluid Mechanics*, Cambridge University Press, Cambridge.

[2] Brandsma, F. J., Vogels, M. E. S., Van der Vooren, J., Dijkstra, D., and Kuerten, J. G. M. (1988) 'Pre-design document of the ISNaS compressible flow simulator', ISNaS 88.04.027.

[3] Brandsma, F. J., and Kuerten, J. G. M. (1990) 'The ISNaS compressible Navier-Stokes solver; First results for single airfoils', in K. W. Morton (ed.), *Twelfth International Conference on Numerical Methods in Fluid Dynamics*, Springer-Verlag, Berlin, 152–156.

[4] Dennis, S. C. R., and Dunwoody, J. (1966) 'The steady flow of a viscous fluid past a flat plate', *J. Fluid Mech.*, **24**, 577–595.

[5] Dijkstra, D. (1974) 'The solution of the Navier-Stokes equations near the trailing edge of a flat plate', Thesis, University of Groningen.

[6] Goldstein, S. (1930) 'Concerning some solutions of the boundary-layer equations in hydrodynamics', *Proc. Cambr. Phil. Soc.*, **26**, 1–30.

[7] Jobe, C. E., and Burggraf, O. R. (1974) 'The numerical solution of the asymptotic equations of trailing edge flow', *Proc. Roy. Soc.*, **A340**, 91–111.

[8] Melnik, R. E., and Chow, R. (1975) 'Asymptotic theory of two dimensional trailing edge flows', Grumman Res. Dept. Report RE 510j.

[9] Messiter, A. F. (1970) 'Boundary layer flow near the trailing edge of a flat plate', *SIAM J. Appl. Math.*, **18**, 241–257.

[10] Stewartson, K. (1969) 'On the flow near the trailing edge of a flat plate', *Mathematika,* **16,** 106–121.

[11] Veldman, A. E. P. (1975) 'A new calculation of the wake of a flat plate', *J. Engg. Math.*, **9**, 65–70.

[12] Veldman, A. E. P. (1976) 'Boundary layer flow past a finite flat plate', Thesis, University of Groningen.

[13] Van de Vooren, A. I., and Dijkstra, D. (1970) 'The Navier-Stokes solution for laminar flow past a semi-infinite flat plate', *J. Engg. Math.*, **4**, 9–27.

[14] Yoshizawa, A. (1970) 'Laminar viscous flow past a semi-infinite flat plate', *J. Phys. Soc. Japan,* **28**, 776–779.

TIME-SCALE DECOUPLING FOR NEARLY PERIODIC ADVECTION-DIFFUSION EQUATIONS

A. van Harten
Mathematical Institute
University of Utrecht
Utrecht
the Netherlands

ABSTRACT. An advection-diffusion problem is considered with an $O(\varepsilon)$ perturbed T-periodic flow and $O(\varepsilon)$ diffusivity, where $\varepsilon > 0$ denotes a small parameter. Such problems arise in tidally induced transport of constituents. An asymptotic approximation of the solution with $O(\varepsilon)$ accuracy on a long $O(1/\varepsilon)$-time scale is constructed by using a rigorous averaging method. The averaged problem on the long time scale is completely decoupled from the $O(1)$ time scale of the nearly periodic flow. As a consequence, numerical computations of the solution of the approximate problem can be based on a much larger time step than for the original problem. Thus a large factor is gained in numerical efficiency. In addition, piling up of discretization errors in the averaged problem is less severe than for numerical computations for the original problem. When this rigorous averaging procedure is used, two aspects of domain decomposition of the space-time cylinder play a role.

1. Introduction

Let us consider constituent transport in shallow water extending in two horizontal directions $X = (x, y) \in \Omega \subset \mathbf{R}^2$. The shallowness will be exploited to integrate variables over depth, so that in the model only variations in horizontal directions have to be taken into account. The depth will be denoted by $h(X)$. The vertically integrated velocity field, $U(X,t)$, induced by the tide is given. In practice, $U(X,t)$ follows by solving the shallow-water equations; cf. [18] and [16]. For tidal motion it is reasonable to assume that the vertically integrated flow, U, has the following structure:

$$U(X,t) = U_0(X,t) + \varepsilon\, U_1(X,t). \tag{1}$$

Here $U_0(X,t)$ is T-periodic in time. This main part, U_0, represents a basic flow pattern with closed periodic orbits. Let us denote the tracks corresponding to this basic flow by $\Phi_0(t; X_0, t_0)$, that is,

$$\frac{d\Phi_0}{dt} = U_0(\Phi_0, t), \tag{2}$$

H. G. Kaper and M. Garbey (eds.),
Asymptotic and Numerical Methods for Partial Differential Equations with Critical Parameters 21–39.

$$\Phi_0 \mid_{t=t_0} = X_0,$$

and $\Phi_0(t+T; X_0, t_0) = \Phi_0(t; X_0, t_0 + T) = \Phi_0(t; X_0, t_0)$ for all t and X_0, t_0. The full tidal flow U is found by superposition of a small perturbation εU_1 to the basic flow U_0. Hence $\varepsilon > 0$ is a small parameter. Note that the small perturbation εU_1 is responsible, in general, for an $O(\varepsilon)$ separation of the trajectories of the full flow after a period T, which determines the Lagrangian residual flow.

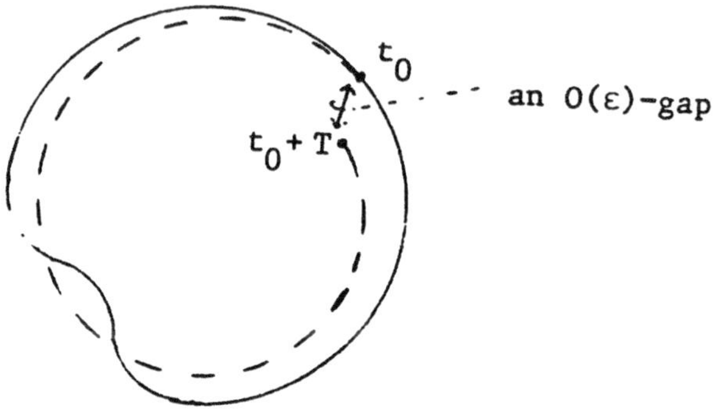

Fig. 1. **Lagrangian residual flow caused by the nonclosedness of the tidal particle tracks over one tidal period**

We shall assume for simplicity that U_0 and U_1 depend C^∞-smooth on X and t.

Let us denote the trajectories generated by U as $\Phi(X_0, t_0, t)$, namely,

$$\frac{d\Phi}{dt} = U(\Phi, t), \tag{3}$$
$$\Phi \mid_{t=t_0} = X_0,$$

with U as in (1). It is well known that solutions of systems of ODEs depend smoothly on parameters contained in the right-hand side; cf. [11]. Hence, it follows that

$$\Phi = \Phi_0 + \varepsilon\, \Phi_1 + O(\varepsilon^2). \tag{4}$$

For future reference it will be convenient to have explicit information available on the dependence of the orbital gap and the Lagrangian residual flow on Φ_0 and εU_1, at least for the leading term in ε.

Let us use $[t]$ for evaluation along the unperturbed trajectory. That is, for $h(X,t)$ we denote $h[t] = h(\Phi_0(t; X_0, t_0), t)$. It is easy to check (cf. [5]) that

$$\Phi_1(t; X_0, t_0) = \frac{\partial \Phi_0}{\partial X_0}(t; X_0, t_0) \int_{t_0}^{t} \frac{\partial \Phi_0^{-1}}{\partial X_0}(\hat{t}; X_0, t_0) \cdot U_1[\hat{t}]\, d\hat{t}. \tag{5}$$

The conclusion is that the Lagrangian residual flow has the following form:

$$U_\ell(X_0,t_0)=\frac{\varepsilon}{T}\int_{t_0}^{t_0+T}\frac{\partial\Phi_0^{-1}}{\partial X_0}(\hat{t};X_0,t_0)\cdot U_1[\hat{t}]\,d\hat{t}+O(\varepsilon^2). \tag{6}$$

This will facilitate recognition of the Lagrangian residual flow later on after averaging.

After this preparation we are ready to formulate the vertically integrated model for constituent transport in shallow water:

$$\frac{\partial C}{\partial t}+U\cdot\nabla C=\frac{\varepsilon}{H}\nabla\cdot\{DH\nabla C\}+\varepsilon f, \tag{7}$$

$$C\mid_{t=0}=\gamma \qquad \frac{\partial C}{\partial n}\mid_{\partial\Omega}=0.$$

Here $\varepsilon D(x,y)$ denotes the small, isotropic diffusion coefficient. The source term is supposed to be of the form $\varepsilon f(X,t,\varepsilon t)$ with f T-periodic in the second coordinate t. Hence the sources are weak and have slowly modulated, periodic behavior in time. The boundary conditions are of Neumann type. For simplicity we consider the case of closed, fixed boundaries, where at $\partial\Omega$ the flow U vanishes:

$$U\mid_{\partial\Omega}\equiv 0. \tag{8}$$

In this paper we shall apply a rigorous averaging technique to the problem (7). The idea is to exploit the fact that after one complete period of the tide, the distribution of the constituent has changed by only an $O(\varepsilon)$ amount. During the tidal cycle at a fixed position, $O(1)$ variations in the concentration will certainly occur, but because of the almost closedness of orbits and the small diffusion, the residual effect after one complete period will be $O(\varepsilon)$ small. Hence, apart from a cyclic behavior over the tidal period, the genuine transport of material by the flow will take place on a long $O(1/\varepsilon)$ time scale. This rigorous averaging method will provide us with an explicit partial differential equation, which will approximately describe the long-time dynamics over $O(1/\varepsilon)$ tidal periods. The coefficients of this PDE are time independent, and they can be expressed explicitly in terms of quantities related to the tidal flow pattern integrated over one tidal period. In this way we obtain a precise formalism to deduce averaged coefficients from the original problem. The averaging procedure as introduced here is rigorous in the sense that validity of the constructed approximation can be demonstrated.

Validity means that the error between the solution of the original problem and the constructed approximation can be estimated. The error turns out to be $O(\varepsilon)$ on a given time interval $[0,T/\varepsilon]$ with $T>0$, $T=O(1)$ measured in the maximum norm. Thus, this type of analysis yields a firm foundation for a discussion of the physical behavior of the solutions.

Before we proceed with the analysis, we first present a few historical notes. Both the theory of averaging and the hydrodynamics of tidal flow have a long history. Already at the end of the eighteenth century, well-known mathematicians such as as Lagrange and Laplace had made contributions to both fields; cf. [13, 14, 15, 20]. Nevertheless, it took quite a while before the theory of averaging reached its maturity in our century (cf. [1, 17]), even for systems of weakly coupled oscillators described by systems of ordinary differential equations.

Applications of this averaging theory to systems of ODEs cover several areas (mechanical vibrations, satellite orbits), but problems related to tidal motion are seldom considered, though there is work of a completely different nature from ours on tidal energy dissipation in binary star systems; cf. [9]. Generalizations of the theory of averaging to PDE problems, including proofs of validity, are of recent origin. In [19, 12] weakly nonlinear strings have been studied by using a Galerkin procedure with the eigenfunctions of the linear string as basis elements. This approach leads to averaging for a countably infinite system of weakly coupled oscillators.

In the second part of his thesis [12], Krol considers averaging for a class of advection-diffusion problems related to tidal flow. Our model (7) does not immediately fit into the class of equations considered in [12], but our approach is certainly inspired by this work. A generalized class containing (7) has been considered in [5].

It is worthwhile to mention that the theory for averaging tidal advection-diffusion equations is rather different from the weakly nonlinear string case. The first problem is of parabolic type, whereas the latter problem is of hyperbolic type, thus making proof of validity a different (but not really a more difficult) job. But, more surprising, for the parabolic case the approximation is found in a (at first sight) completely different way. Nevertheless, there is a strong analogy between the averaging theory in the case of weakly coupled oscillators and the case of tidal advection-diffusion equations. This analogy can conceptually best be explained by thinking in terms of the denumerable set of periodic solutions for the linear string ($\varepsilon = 0$) corresponding to the uncoupled pure modes. In the case of the advection-diffusion problem with $\varepsilon = 0$, there corresponds a periodic orbit to each starting point in the domain. This orbit is found by integrating the unperturbed main tidal flow field without orbital gaps. Hence, the analogy lies in the unperturbed evolution operator describing the dynamics on a set of uncoupled, periodic, closed orbits. The difference is that in the case of the tidal advection-diffusion problems, there is a continuum of periodic, closed orbits labeled with the points of the domain instead of a denumerable set of isolated periodic modes. However, the unperturbed evolution operator is still quite manageable.

Besides the mathematically oriented literature on rigorous averaging, there is a considerable amount of physically oriented literature applying averaging in a heuristic way. For the heuristics of averaged tidal flow we refer to [4]. Often it appears that the heuristic approach fits well with experimental data. By experimental data as well as by heuristic reasoning, one is led to describing the effect of averaged tidal flow in terms of

- a residual flow, and
- a nonisotropic, nonuniform diffusion tensor.

But, though both effects are widely recognized in the heuristic literature, there is no agreement on the matter of calculating these effects rigorously.

As for the nonisotropic, nonuniform averaged diffusion tensor, the situation is even worse. Heuristic proposals for its determination are not available in a general context. Here we mention that for certain special advection-diffusion problems with time-periodic advection, exact solutions can be found, from which the averaged properties can be derived; cf. [2, 21, and 10].

An advantage of our approach is that it works for general realistic problems. It turns out that in a rigorously averaged advection-diffusion equation on the long $O(1/\varepsilon)$ time scale the Lagrangian residual flow plays a role. In addition, we get explicit expressions for corrections resulting from spatial inhomogeneity and for the tidally induced dispersive diffusion coefficients. Another important observation is that the averaging procedure also has advantages from a numerical point of view: after averaging, the time step of the numerical scheme can increase by a large factor compared with a numerical scheme for the original problem.

In the next section we discuss the averaging procedure. In Section 3 several aspects of numerical computations for the averaged problem are considered. We conclude with some results.

2. Averaging of the Advection-Diffusion Model

To start, we introduce the general line of a rigorous averaging procedure. Consider a dynamic phenomenon of the following type:

- almost periodic behavior with a characteristic time scale T (the period), and
- drift in the dynamics with a characteristic time scale NT, $N \gg 1$.

Suppose a model to describe such a phenomenon is given dependent of a parameter $\varepsilon > 0$, where $N = O(1/\varepsilon)$ and $\varepsilon = 0$ corresponds to an unperturbed situation with known purely periodic behavior. Typically, the unperturbed dynamics can then be represented by an evolution operator $E_0(t;p)$ periodic in t with period T. Further, E_0 contains a set of parameters, p, that can be related to the initial conditions in a unique way. Now, in the case $\varepsilon > 0$ a rigorous averaging scheme to describe the drift in the dynamics on an $O(1/\varepsilon)$ time scale consists of several steps:

1. a structure hypothesis for the candidate approximation of the solution;
2. transformation of the original problem into a standard form for averaging;
3. derivation of the averaged problem, which defines the evolution of the approximation on the long time scale;
4. proof of the validity of averaging: error estimates for the approximation on the long time scale; and
5. analysis of the behavior of solutions of the averaged problem.

Usually one looks for an approximation of the form $E_0(t; p(t))$, and the original problem is then converted into a problem for $p(t)$. Note that this is a sort of "variation of constants" procedure, since for $\varepsilon = 0$ p is just a set of time-independent parameters. Hence, it is obvious that for $\varepsilon > 0$, the time variation of $p(t)$ turns out to be $O(\varepsilon)$. Moreover, it is logical that an evolution equation for $p(t)$ will inherit time-periodic "coefficients" from the transformation $E_0(t, p(t))$ which is periodic in t.

Now let the equation for p be of the following form:

$$\frac{dp}{dt} = \varepsilon\ f(t, p), \tag{9}$$

where f is periodic in t. This form will be referred to as a standard form for averaging. It follows that it is reasonable to expect that $p(t)$ can be approximated by a slowly varying function $\bar{p}(\tau)$ with $\tau = \varepsilon t$. Next, an evolution equation for $\bar{p}(\tau)$ is found by averaging the "coefficients" in the evolution equation for p. In other words, Equation (9) implies

$$p(t_0 + T) = p(t_0) + \varepsilon \int_0^T f(t_0 + s, p(t_0 + s))\, ds.$$

In this equation we put $t_0 = \tau/\varepsilon$, and we obtain formally

$$\bar{p}(\tau + \varepsilon T) = \bar{p}(\tau) + \varepsilon \int_0^T f(t_0 + s, \bar{p}(\tau))\, ds + O(\varepsilon^2).$$

Next, the left-hand side is expanded in a Taylor series. Comparing the $O(\varepsilon)$ terms on both sides provides us with the following equation for $\bar{p}$:

$$\frac{d\bar{p}}{d\tau} = \bar{f}(\bar{p}), \tag{10}$$

with $\bar{f}(\bar{p}) = \frac{1}{T} \int_0^T f(t_0 + s, \bar{p})\, ds$.

Finally, the task in proving validity is to show that $p(t) - \bar{p}(\varepsilon t)$ is small in a suitable norm on an interval $0 \leq t \leq L/\varepsilon$.

Now we shall show that the rigorous averaging scheme as introduced can be used in the case of tidally induced advection-diffusion problems of the type (7). Note that then (9), etc., have to be interpreted as evolution equations in a suitably chosen Banach space of space-dependent functions.

To start, we need information on the unperturbed dynamics corresponding with (7). If we set $\varepsilon = 0$, the problem (7) reduces to

$$\frac{\partial C_0}{\partial t} + U_0 \cdot \nabla C_0 = 0, \tag{11}$$

$$C \mid_{t=0} = \gamma.$$

This first-order problem can be solved exactly; its solution is constant along the characteristics, with the orbits as defined in (23). The evolution operator E_0 is given by

$$C_0(X,t) = E_0(t;\gamma) = \gamma(\Phi_0(0;X,t)).$$

Note that if $X = \Phi_0(t;X_0,0)$, then $\Phi_0(0;X,t) = X_0$. In other words, let $P(t)$ represent the map $X_0 \rightarrow \Phi_0(t;X_0,0)$, and let $Q(t)$ represent the map $X \rightarrow \Phi_0(0;X,t)$. Then for all $t : P \circ Q = Q \circ P = I$. As a shorthand notation we shall use $\Psi_0(X,t) = \Phi_0(0;X,t)$ and $\Phi_0(X_0,t)$ instead of $\Phi_0(t;X,0)$ from now on.

Then as a first step in the averaging scheme we make a transformation of coordinates fitted to the structure of the approximation for $\varepsilon > 0$, ε small:

$$C(X,t) = F(\Psi_0(X,t),t), \tag{12}$$

that is, $F(X,t) = C(\Phi_0(X,t),t)$. Note that C and F coincide for special values of t only, namely, $t = nT$ with $n = 0,1,2,\ldots$.

The transformation in (12) brings (7) into a standard form for averaging:

$$\begin{aligned} \tfrac{\partial F}{\partial t}(X,t) &= \tfrac{\partial}{\partial t}\{C(\Phi_0,t)\} \\ &= \left\{\tfrac{\partial C}{\partial t} + U_0 \cdot \Delta C\right\}(\Phi_0,t) \\ &= \varepsilon\left\{(-U_1 + H^{-1}\nabla(DH)) \cdot \nabla C + D\nabla C + f\right\}(\Phi_0,t). \end{aligned}$$

Next one can use (12) to derive expressions for ∇C and ΔC:

$$\begin{aligned} \nabla C(X,t) &= \left(\tfrac{\partial \Psi_0}{\partial X}\right)^t (X,t) \cdot \nabla F(\Psi_0,t) \\ \Delta C(X,t) &= \Delta\Psi_0(X,t) \cdot \nabla F(\Psi_0,t) \\ &\quad + \sum_{i,j} A_{ij}(X,t)\,\frac{\partial^2 F}{\partial X_i \partial X_j}(\Psi_0,t) \end{aligned}$$

with $A = \left(\frac{\partial \Psi_0}{\partial X}\right) \cdot \left(\frac{\partial \Psi_0}{\partial X}\right)^t$.

Thus we find

$$\frac{\partial F}{\partial t} = \varepsilon(-\tilde{U}_1 - \tilde{U}') \cdot \nabla F + \varepsilon \sum_{i,j} \tilde{D}_{ij}\,\frac{\partial^2 F}{\partial X_i \partial X_j} + \varepsilon\,\tilde{f}, \tag{13}$$

$$F\mid_{t=0}=\gamma \qquad \frac{\partial F}{\partial n}=0 \text{ on } \partial\Omega,$$

with

$$\tilde{U}_1=\left\{\frac{\partial\Psi_0}{\partial X}\cdot U_1\right\}(\Phi_0,t),$$

$$\tilde{U}'=-\left\{\frac{\partial\Psi_0}{\partial X}\cdot H^{-1}\nabla(DH)+D\triangle\Psi_0\right\}(\Phi_0,t),$$

$$\tilde{D}=\{DA\}(\Phi_0,t),$$

$$\tilde{f}=f(\Phi_0,t).$$

Note that (13) has an $O(\varepsilon)$ right-hand side with coefficients periodic in t. As a consequence, (13) is in standard form for averaging, indeed.

The averaged problem is obtained simply by averaging the coefficients in (13). This leads us to

$$\frac{\partial\bar{F}}{\partial t}=\varepsilon\left\{(-U_{\ell,0}-\bar{U}')(x)\cdot\nabla\bar{F}+\sum_{ij}\bar{D}_{ij}(x)\frac{\partial^2\bar{F}}{\partial X_i\partial X_j}+\bar{f}(X,\varepsilon t)\right\},\tag{14}$$

$$\bar{F}\mid_{t=0}=\gamma \qquad \frac{\partial\bar{F}}{\partial n}=0 \text{ on } \partial\Omega.$$

Note that

$$\varepsilon\, U_{\ell,0}=\frac{\varepsilon}{T}\int_0^T\left\{\frac{\partial\Psi_0}{\partial X}(\Phi_0(X,\hat{t}),\hat{t})\cdot U_1(\Phi_0(X,\hat{t}),\hat{t})\right\}d\hat{t}.$$

Hence, $\varepsilon\, U_{\ell,0}$ corresponds with the main term of the Lagrangian residual flow as given in (6) with $t_0=0$, since $\frac{\partial\Psi_0}{\partial X}(\Phi_0,t)=\frac{\partial\Phi_0^{-1}}{\partial X}(X,t)$ according to the chain rule.

The other coefficients are given by

$$\varepsilon\,\bar{U}'=-\frac{\varepsilon}{T}\int_0^T\left\{\frac{\partial\Psi_0}{\partial X}\cdot H^{-1}\nabla(DH)+D\triangle\Psi_0\right\}[\hat{t}]\,d\hat{t},\tag{15}$$

$$\varepsilon\,\bar{D}=\frac{\varepsilon}{T}\int_0^T\left\{D\,\frac{\partial\Psi_0}{\partial X}\left(\frac{\partial\Psi_0}{\partial X}\right)^t\right\}[\hat{t}]\,dt,\tag{16}$$

$$\varepsilon\,\bar{f}(X,\tau)=\frac{\varepsilon}{T}\int_0^T\left\{f([\hat{t}],\tau)\right\}d\hat{t}.\tag{17}$$

As before, the notation $[\hat{t}]$ abbreviates $(\Phi_0(X,\hat{t}),\hat{t})$. In each of the averaged coefficients the expression between the braces $\{\ \}$ is integrated along the orbit of the unperturbed flow.

In addition to the Lagrangian residual flow, another advection term $\varepsilon\,\bar{U}'\cdot\nabla\bar{F}$ plays a role. This correction is partly due to nonzero gradients $\nabla(DH)$ stemming from the right-hand side of the original equation (7) by working out the divergence form. Its second part comes from the transformed Laplacian. The expression for $\varepsilon\bar{D}$ in (17) provides us with explicit information on the dispersive action of averaged isotropic diffusion.

Our next point of concern is the calculation of the averaged coefficients. In an actual situation the vertically integrated flow U is found by solving the shallow-water equations. Hence U is known beforehand, but not the splitting $U_0+\varepsilon\,U_1$, as introduced in (1). Now, in principle U_0 can be derived by projecting U on the flow fields with periodic orbits, but this step is not necessary. It is simpler to use in the expressions for the averaged coefficients (15–18) the flow Φ generated by U itself, instead of Φ_0. This causes $O(\varepsilon)$ errors in each of these coefficients, but that is rather harmless.

Define

$$\frac{\partial\bar{F}'}{\partial\tau}=-(U_\ell+\bar{U}'')\cdot\nabla\bar{F}'+\sum D'_{ij}\frac{\partial^2\bar{F}'}{\partial X_i\partial X_j}+\bar{f}'(X,\varepsilon t),\tag{18}$$

$$\bar{F}'\,|_{t=0}=\gamma\qquad\frac{\partial\bar{F}'}{\partial n}=0\text{ on }\partial\Omega,$$

with $\tau=\varepsilon t$ and

$$\varepsilon U_\ell=\frac{1}{T}\,\{\Phi(T;X,0)-X\}\text{ as in Figure 1,}\tag{19}$$

$$\varepsilon\bar{U}''=-\frac{\varepsilon}{T}\int_0^T\left\{\frac{\partial\Psi}{\partial X}H^{-1}\nabla(DH)+D\Delta\Psi\right\}[\hat{t}]'\,d\hat{t},\tag{20}$$

$$\varepsilon\,\bar{D}'=\frac{\varepsilon}{T}\int_0^T\left\{D\frac{\partial\Psi}{\partial X}\frac{\partial\Psi}{\partial X}^t\right\}[\hat{t}]'\,d\hat{t},\tag{21}$$

$$\varepsilon\,\bar{f}'=\frac{\varepsilon}{T}\int_0^T f([\hat{t}]',\tau)\,d\hat{t},\tag{22}$$

where $[t]'$ denotes $(\Phi(X,t),t)$.

The error by changing (14–17) into (18–22) can be estimated as follows.

Lemma 1. *Suppose that U, D, H, and γ and f are smooth with bounded derivatives and that U admits a splitting as in (24) with $U-U_0$ and its derivatives $O(\varepsilon)$ bounded. Then:*

$\Phi - \Phi_0$, $\frac{\partial\Psi}{\partial X} - \frac{\partial\Psi_0}{\partial X}$, $\Delta\Psi - \Delta\Psi_0$ are $O(\varepsilon)$ for $t \in [0, T]$,
$\|U_\ell - U_{\ell,0}\|$, $\|\bar{U}' - \bar{U}''\|$, $\|\bar{D} - \bar{D}'\|$ are $O(\varepsilon)$,
$\|\bar{f}(\cdot, \tau) - \bar{f}'(\cdot, \tau)\| = O(\varepsilon)$ for $0 \leq \tau \leq L$,
and $\|\bar{F}(\cdot, t) - \bar{F}'(\cdot, t)\| = O(\varepsilon)$ for $0 \leq t \leq L/\varepsilon$.

where $\| \quad \|$ *denotes the sup-norm on* $\bar{\Omega}$ *and* $L > 0$ *can be taken any* $O(1)$ *constant.*

Now, we are ready to demonstrate that the averaging procedure is rigorous. Specifically, we estimate the difference $F - \bar{F}'$ with F, the exact solution defined in (13), and $\bar{F}$, the averaged solution defined in (18).

Theorem 1. *Under the same regularity assumptions as before (see Lemma 1), it holds true that for each* $L > 0$, $L = O(1)$,

$$\|F(\cdot, t) - \bar{F}(\cdot, t)\| = O(\varepsilon) \tag{23}$$

for $0 \leq t \leq L/\varepsilon$ *with* $\| \quad \|$ *the sup-norm on* $\bar{\Omega}$.

For the proofs of the lemma and the theorem of the validity of the averaging procedure, we refer to [5]. The proofs are based on the maximum principle for parabolic initial-boundary value problems.

3. On the Numerical Computation of the Solution of the Averaged Problem

Numerically, the solution of the original advection-diffusion problem (7) can be approximated by using a discretization method with a fixed grid size h in the spatial directions and a fixed time step k. Depending on the specific method one uses for a problem with variable coefficients as in (7), one will end with discretization errors of a certain type. Let us investigate how, under reasonable assumptions, in such a direct approach a numerical error of $O(\delta(\varepsilon))$ over a long $O(1/\varepsilon)$ time scale can be obtained.

A first remark is that in this direct approach one has to take

$$h \ll d, \quad k \ll T, \tag{24}$$

where d is a representative length scale for the variation of the coefficients D, H, and U. Of course, (24) has to hold true in order to get small truncation errors. Apart from (24), certain relations between k and h should be satisfied to guarantee sufficient stability of the scheme. Let $\mathbf{E}_n$ denote the maximum error between the solution of (8) and its numerical approximation considered on all gridpoints in space at the n-th time step, that is, at $t = nk$. Then

$$\mathbf{E}_{n+1} = A_n \, \mathbf{E}_n + \mathbf{r}_n. \tag{25}$$

Here A_n denotes the iteration matrix for the discretized evolution, and $\mathbf{r}_n$ denotes the vector of truncation errors. Starting with $\mathbf{E}_0 = 0$, one can explicitly solve the iteration in (25) formally. Let $r(h, k, \varepsilon)$ be an upper estimate of the maximum norm of the truncation error. It is reasonable to suppose that A_n inherits some of the main features of the original advection-diffusion problem, namely,

$$\mathbf{r} \geq \mathbf{0} \Rightarrow A_n\mathbf{r} \geq \mathbf{0}, \quad A_n\mathbf{1} = \mathbf{1}. \tag{26}$$

The first requirement is a positivity property analogous to the maximum principle for the original problem. Second, we require that the constant vector be an exact solution of the homogeneous iteration. This corresponds with the fact that constant functions satisfy the homogeneous advection-diffusion as well as the Neumann boundary conditions. For example, (26) is satisfied unconditionally if one uses a fully implicit, first-order consistent discretization scheme, where as usual the advection term is discretized according to the sign of the x and y component of the advection to the left or right and upward or downward, respectively. In this case one ends up with

$$r(h,k,\varepsilon) = O(k(h+k)).$$

The more common ADI version of a Crank-Nichelson scheme (cf. [18]) has the properties of (26) if

$$\frac{k}{h} \leq \lambda_1, \ \frac{\varepsilon k}{h^2} \leq \lambda_2,$$

where λ_1 is an $O(1)$ constant depending on the advection magnitude only and λ_2 is an $O(1)$ constant depending only on the diffusion magnitude. In this case the truncation errors can be estimated by

$$r(h,k,\varepsilon) = O(k(h^2+k^2)).$$

It is now easy to check with induction that as a consequence of (26)

$$\mathbf{E}_n + n\, r(h,k,\varepsilon)\,\mathbf{1} \geq \mathbf{0},$$

$$\mathbf{E}_n - n\, r(h,k,\varepsilon)\,\mathbf{1} \leq \mathbf{0}.$$

Since n runs from 0 to $[L/(\varepsilon k)]$, it follows that the error in the maximum norm can be estimated by

$$\mid \mathbf{E}_n \mid \leq \frac{L}{\varepsilon k} \cdot r(h,k,\varepsilon). \tag{27}$$

Hence, in order to guarantee an $O(\delta(\varepsilon))$ accuracy on the long time scale, the following condition should be satisfied:

$$r(h,k,\varepsilon) = O(\varepsilon\delta k). \tag{28}$$

If the amount of work per iteration in time is proportional to the number of grid points with proportionality constant $b(\varepsilon, \delta)$, then the computational effort can be estimated as

$$\text{eff} = O((\varepsilon k h^2)^{-1} b). \tag{29}$$

In the fully implicit case $r = O(k(h+k)), \delta = \varepsilon$, this leads to the choice $k = O(\varepsilon^2)$, $h = O(\varepsilon^2)$, and as a consequence an effort $O(\varepsilon^{-7} b)$ is required. In this case $b = (1/\varepsilon)\ln(1/\varepsilon)$ as a result of iterations involved in solving the implicit system of equations. In the ADI case where $r = O(k(h^2 + k^2))$, this leads to a choice $h = O(\varepsilon)$, $k = O(\varepsilon)$ if $\delta = \varepsilon$. The computational effort is now $O(\varepsilon^{-4})$.

It is also worthwhile to mention that the reasoning can be extended to a case where a somewhat weaker condition than positivity holds true, namely,

$$\mathbf{r} \geq 0 \Rightarrow A_n \mathbf{r} \geq -\mu \mid \mathbf{r} \mid \mathbf{1}, \tag{30}$$

where $\mid \mathbf{r} \mid$ denotes the maximum norm of $\mathbf{r}$ and $\mu(h, k, \varepsilon) > 0$. In an analogous way this leads to an estimate

$$\mid \mathbf{E}_n \mid \leq \frac{\exp(\frac{L\mu}{\varepsilon k}) - 1}{1 - \exp(-\mu)} r \tag{31}$$

on the long time scale. This implies an $O(\varepsilon)$ error if, in addition to (28), we require

$$\mu(h, k, \varepsilon) = O(\varepsilon k). \tag{32}$$

It is obvious that the same sort of reasoning yielding (28–29) or (31–32) can immediately be applied to the averaged problem (10). But the difference is that in 1(0) only the slow time scale $\tau = \varepsilon t$ is involved. Discretization leads to $\tau = mk$ with $m = 0, \ldots, [L/k]$. Hence the number of iterations does not contain the factor ε^{-1} which appeared before. Another advantage is that now the coefficients in (10) are time independent.

If we use analogous numerical schemes as before, the truncation error and the error amplification are now characterized by $r(h, k)$ and $1 + \mu(h, k)$, respectively. To obtain an $O(\delta)$ accuracy with $\varepsilon = O(\delta)$ for $0 \leq \tau \leq L$, we use a similar reasoning as before to get the following conditions for the choice of h and k in the averaged problem:

$$\mu(h, k) = O(k), \tag{33}$$

$$r(h, k) = O(\delta k).$$

Note that the conditions are really milder than those of (27), (28). The computational effort for the averaged problem (once the averaged coefficients are known) can be estimated as

$$\text{eff}_{av} = O((kh^2)^{-1} b), \tag{34}$$

with $b(\delta)$ the computational effort per gridpoint per timestep. Note that the numerical schemes mentioned before are mutatis mutandis applicable for the averaged problem. For the ADI version of Crank-Nichelson's method, the requirements for positivity now become

$$\frac{k}{h} \leq \bar{\lambda}_1, \ \frac{k}{h^2} \leq \bar{\lambda}_2,$$

where $\bar{\lambda}_1$, $\bar{\lambda}_2$ depend on the averaged advection and averaged diffusion coefficients, respectively. The order of the errors is the same as before. Now, in the fully implicit case with $\delta = \varepsilon$ we can take $k = O(\varepsilon)$ and $h = O(\varepsilon)$, and this results in an $O(\varepsilon^{-3}\ell n(\frac{1}{\varepsilon}))$ effort. In the ADI case with $\delta = \varepsilon$, we can take $h = O(\sqrt{\varepsilon})$ and $k = O(\varepsilon)$, which results in an $O(\varepsilon^{-2})$ effort. In both cases the reduction in computational effort is really significant. We observe that a choice $k = m\varepsilon T$ with $m = 1, 2, \ldots$ is very natural since it corresponds in terms of the original t time scale with a multiple of the tidal period.

Let us now address the numerical calculation of the averaged coefficients in (10). This amounts to solving a system of ODEs along the orbit through each grid point. For details we refer to [7]. This work has to be done only once, and it is completely parallelizable. In order to determine the averaged coefficients at a given gridpoint with an $O(\varepsilon)$ accuracy, a higher-order ODE solver will typically require an $O(\varepsilon^{-\nu})$ computational effort with $0 < \nu \leq \frac{1}{2}$. Hence, in addition to (34) there is an extra workload

$$\text{eff}_0 = O((\varepsilon^{\nu} h^2)^{-1}) \tag{35}$$

to determine the averaged coefficients with $O(\varepsilon)$ accuracy. The effect of this $O(\varepsilon)$ inaccuracy in the coefficients is similar to an extra truncation error of the order $O(\varepsilon k)$, and it can therefore be absorbed in the right-hand side of (33). Note that the numerical effort to calculate the averaged coefficients given in (35) is small compared with rest of the workload as given in (34). This leads to the conclusion that the averaging procedure is much more efficient for obtaining an $O(\varepsilon)$ approximation on an $O(\frac{1}{\varepsilon})$ time scale than a direct numerical approach. However, one should observe that, because of the large time step, the averaged problem does not provide detailed information within each tidal cycle. Usually, this is no problem whatsoever, since from the outset one is interested in the long time aspects of constituent transport. One should realize that within a tidal cycle or in between two time steps on the long time scale corresponding with $t = nk/\varepsilon = nmT$ and $t = (n+1)k/\varepsilon = (n+1)mT$, the concentration shows an $O(1)$ variation. To make this variation over one tidal period visible with $O(\varepsilon)$ accuracy, say, at the end of the $O(\frac{1}{\varepsilon})$ time interval requires another effort of the same order as calculating the averaged coefficients.

To summarize the numerical averaging procedure, we represent the involved decomposition of the cylinder $\Omega \times [0, L/\varepsilon]$:

Numerical Averaging Procedure	Involved Subsets of the Cylinder
Computationally averaged coefficients	orbits through grid points for $t \in [0, T]$ with time step $O(\varepsilon^{\nu}), \nu \in (0, \frac{1}{2}]$
Computation solution averaged equation	gridpoints for time levels $t = 0,\ mT,\ nmT,\ \ldots,\ L/\varepsilon$
Computational tidal variation at $t = L/\varepsilon$	orbits through gridpoints for $t \in [(L/\varepsilon) - T, T]$ with time-step $O(\varepsilon^{\gamma}), \nu \in (0, \frac{1}{2}]$

4. Behavior of Solutions of the Averaged Problem

First we shall discuss an explicit example; we shall conclude this section with some results on tidally induced dispersion coefficients as calculated in a Dutch coastal water region.

4.1. AN EXPLICIT EXAMPLE

Let us consider the following case:

$$\frac{\partial C}{\partial t} + \alpha \sin(\pi y) \cdot \cos(2\pi t) \frac{\partial C}{\partial x} = \varepsilon \Delta C \tag{36}$$

$$\frac{\partial C}{\partial n} = 0 \text{ for } y = 0 \text{ and } y = 1$$

$$C \mid_{t=0} = \gamma$$

In this case $D = 1, H = 1$.

Here the problem is posed on the strip $0 \leq y \leq 1$, but the analysis goes through without any complication, as one can easily check. Furthermore, $U = U_0 = (\alpha \sin(\pi y) \cos(2\pi t), 0)$, $U_1 = (0, 0)$. The flow is strictly in the x direction with purely periodic orbits:

$$\Phi_0(x_0, y_0, t) = (x_0 + \frac{\alpha}{2\pi} \cdot \sin(\pi y_0) \cdot \sin(2\pi t), y_0)^t. \tag{37}$$

Note that $\Psi_0(x_0, y_0, t) = \Phi_0(x_0, y_0, -t)$. The transformed problem for F with

$$F(x_0, y_0, t) = C(\Phi_0(x_0, y_0, t), t) \tag{38}$$

becomes

$$\frac{\partial F}{\partial t} - \frac{1}{2} \varepsilon \alpha \pi \sin(\pi y) \sin(2\pi t) \frac{\partial F}{\partial x} \tag{39}$$

$$= \varepsilon \left\{ (1 + \frac{1}{4}\alpha^2 \cos^2(\pi y)\sin^2(2\pi t))\frac{\partial^2 F}{\partial x^2} - \alpha \cos(\pi y)\sin(2\pi t)\frac{\partial^2 F}{\partial x \partial y} + \frac{\partial^2 F}{\partial y^2} \right\}$$

$$\frac{\partial F}{\partial n} = 0 \text{ for } y = 0 \text{ and } y = 1$$

$$F\,|_{t=0} = \gamma .$$

Averaging reduces (39) to

$$\frac{\partial \bar{F}}{\partial t} = \varepsilon_{-} \left\{ (1 + \frac{1}{8}\alpha^2 \cos^2(\pi y))\frac{\partial^2 \bar{F}}{\partial x^2} + \frac{\partial^2 \bar{F}}{\partial y^2} \right\} \tag{40}$$

$$\frac{\partial \bar{F}}{\partial n} = 0 \text{ for } y = 0 \text{ and } y = 1$$

$$\bar{F}\,|_{t=0} = \gamma .$$

In the averaged problem the diffusion in the x direction is amplified, and the amplification is stronger, where the gradient of the flow transversal to the x direction is stronger. Starting with an even y independent function $\gamma(x)$ which decreases as $|x|$ increases, the level lines (equi-C curves) will evolve on an $O(1/\varepsilon)$ time scale into a shape as sketched in Fig. 2.

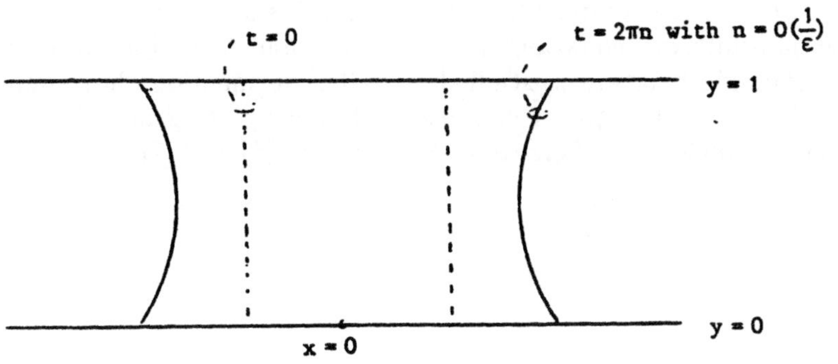

Fig. 2. **Shape of level lines evolving on an** $0(1/\varepsilon)$ **time scale**

A profile with vertical level lines for C at $t = 0$ evolves on the $O(1/\varepsilon)$-time scale into a profile with curved level lines as a result of the variation of the averaged horizontal diffusion in the vertical direction.

4.2. A REALISTIC CASE

The following results have been obtained for (7) with $f = 0$ and constant diffusion coefficient D, by using a numerical solution of the shallow-water equation; cf. [10, 18]. The domain and the flow velocities for various tidal phases are shown in Figs. 3a and 3b. In Fig. 3c some of the particle tracks of the flow field are given. Next, the averaged diffusion coefficients as defined in (18) are calculated. This calculation is done by numerically solving the ODEs that the coefficients of $A = \frac{\partial \Phi}{\partial X}$ have to satisfy:

$$\frac{dA}{dt} = \frac{\partial U}{\partial X}(\Phi(X,t),t) \cdot A \tag{41}$$

$$A(0) = I.$$

Then $\frac{\partial \Psi}{\partial X}(\Phi(X,t),t) = A(t)^{-1}$ and $\bar{D}'$ follow by numerical integration. In Fig. 2d the principal axes of $D^{-1}\bar{D}'$ are shown for a number of points. Strong amplification of diffusion occurs at places with sharp gradients of the current, especially at underwater channels. As in the preceding example, the gradient is roughly perpendicular to the channel, and the longest axis of the averaged diffusion tensor is parallel to the channel. (Note: Figures 3a–d were kindly provided by the Tidal Water Division of Rijkswaterstaat.)

5. Generalizations

To conclude this paper, we mention that the method of averaging introduced here has a wide scope of applicability for problems related to tidal motion. For example, the method can be generalized to quasi-periodic tidal motion or to more complicated advection-diffusion processes. In [8] a case of reaction-diffusion equations is considered, and in [12] a simple problem of morphological nature is studied. Finally, we mention that some of the results derived in this paper can also be found by using models formulated in terms of stochastic differential equations; cf. [6] and [7].

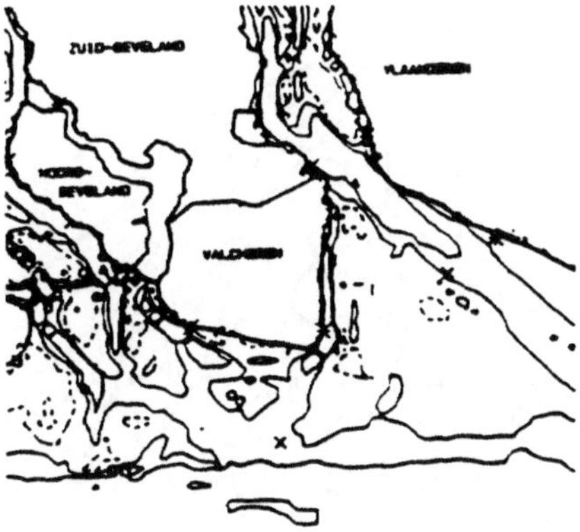

Fig. 3a. The topography of Dutch coastal waters near Walcheren

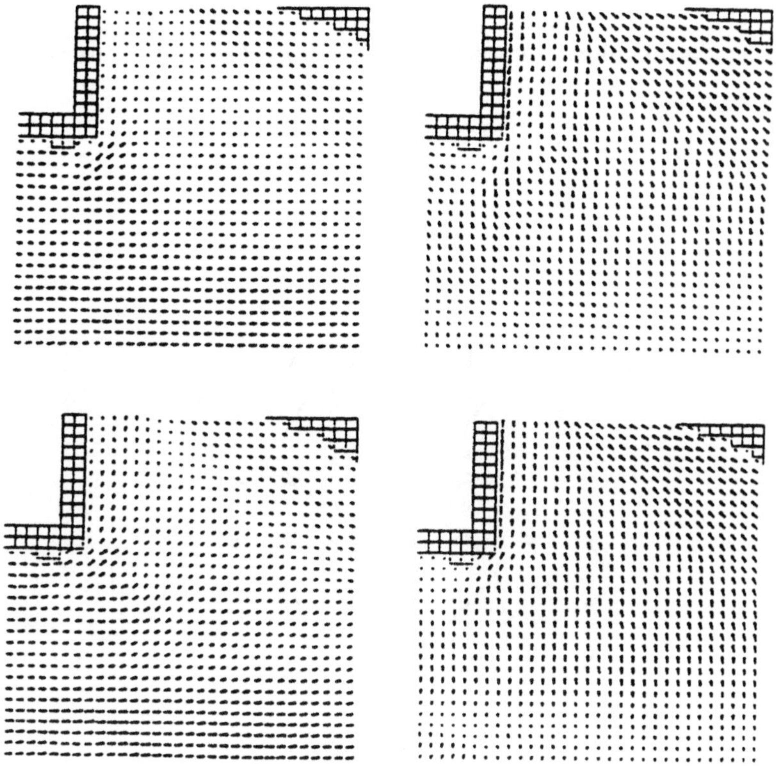

Fig. 3b. Velocities for different tidal phases

Fig. 3c. Some particle tracks induced by the tidal currents

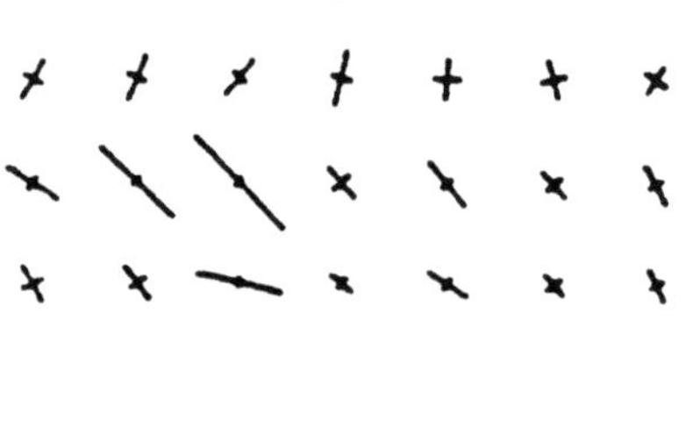

Fig. 3d. Principal axes of the tidally induced averaged diffusion tensor

References

[1] Bogoliubov, N. N., and Mitropolsky, Y. A. (1961) *Methods in the Theory of Nonlinear Oscillations*, Gordon and Breach, New York.

[2] Bowden, K. F. (1965) 'Horizontal mixing in the sea due to a shearing current', *J. Fluid Mech.*, **21**, 113–133.

[3] Eckhaus, W. (1979) *Asymptotic Analysis of Singular Perturbations*, North-Holland, Amsterdam.

[4] Fisher, H. B., List, E. J., Koh, R. C. Y., Imberger, J., and Brooks, N. H. (1979) *Mixing in Inland and Coastal Waters*, Academic Press, New York.

[5] Harten, A. van (1992) 'Analysis of tidally induced transport of constituents', preprint 710, MI, University of Utrecht.

[6] Heemink, A. W. (1991) 'Tidally averaged models for dispersion in shallow water: Averaging random walk models', internal report, Rijkswaterstaat, The Hague.

[7] Heemink, A. W. (1991) 'Tidally averaged models for dispersion in shallow water: Tidally induced dispersion coefficients in Dutch coastal waters', internal report.

[8] Heynekamp, J. J., Krol, M. S., and Verhulst, F. (1991) 'Averaging in non-linear advective transport problems', preprint MI, RUU no. 679.

[9] Hut, P. (1981) *Tidal Evolution in Close Binary Systems for High Eccentricity*, Astronomy & Astrophysics.

[10] Kalkwijk, J. P. Th. 'The modelling of dispersion in 2-D tidal flow over an uneven bed', *Advances in Water Res.*, **10**, 164–172.

[11] Knobloch, H. W., and Kappel, F. (1974) *Gewöhnliche Differentialgleichungen*, Teubner, Stuttgart.

[12] Krol, M. S. (1989) 'On a Galerkin-averaging method for weakly non-linear wave equations', *Math. Appl. Sci.*, **11**, 649–664.

[13] Lagrange (1781) *Mémoire sur la Théorie du Mouvements des Fluïdes*, Oeuvres iv, Serret & Darboux, Paris.

[14] Lagrange (1788) *Mécanique Analytique*, Oeuvres vii, Serret & Darboux, Paris.

[15] Laplace (1799) *Traité de Mécanique Céleste*, Duprat, Paris.

[16] Leendertse, J. J. (1967) 'Aspects of computational model for long period water wave propagation', Rand Corporation, Memorandum RM-5294-PR, Santa Monica, Calif.

[17] Sanders, J. A., and Verhulst, F. (1985) *Averaging Methods in Nonlinear Dynamical Systems*, Springer-Verlag, Berlin.

[18] Stelling, G. S. (1983) 'On the construction on computational methods for shallow water flow problems', *Rijkswaterstaat Communications*, **35**.

[19] van Horssen, W. T. (1988) 'An asymptotic theory for a class of initial boundary value problems for weakly nonlinear wave equations with an application to a model of galloping oscillations of overhead transmission lines', *SIAM J. Appl. Math.*, **48**, 1227–1243.

[20] Verhulst, F. (1984) *Perturbation theory from Lagrange to van der Pol*, Nieuw Archief v. Wiskunde, 428–438.

[21] Young, W. R., Rhines, P. B., and Garrett, C. J. R. (1982) 'Shear-flow dispersion, internal waves and horizontal mixing in the ocean', *Am. Math. Soc.*

TURBULENCE MODELING FROM AN ASYMPTOTIC POINT OF VIEW

H. Herwig
Institut für Thermo- und Fluiddynamik
Ruhr-Universität
4630 Bochum
Germany

ABSTRACT. Turbulence modeling is put into the perspective of a general analysis of modeling in fluid mechanics. It is argued that modeling turbulence necessarily leads to a type of model called the *ad hoc model*, in contrast to a more systematically deduced *asymptotic model*. Based on these considerations, a new way of turbulence modeling is proposed, called indirect turbulence modeling, which may be characterized as an asymptotically motivated ad hoc model. Two examples are given for this kind of turbulence modeling: turbulent Couette flow with heat transfer, and Couette-Poiseuille flow with zero shear stress at the lower wall.

1. Introduction

Ludwig Prandtl, with his boundary layer theory [10], introduced the concept of singular perturbations into fluid mechanics. Though it was based on physical intuition rather than on mathematical rigor, it correctly used all elements of the later-developed mathematical theory of singular perturbations [4, 12]. What Prandtl recognized and what is the basis of singular perturbation theory is a distinct structure of the flow field emerging with increasing Reynolds numbers. A similar observation within the turbulent boundary layer itself was made by Millikan [8], who recognized a two-layer structure in shear flows. Based on these multilayer concepts, the asymptotic theory of turbulent boundary flow was developed about twenty years ago by Yajnik, Fendell, and Mellor (among others); see, for example, [1, 5, 7] for a review of these studies.

In recent years this approach has been less popular, though its full potential has by no means been exhausted [2]. The asymptotic considerations will never replace turbulence modeling completely. However, it may well help to model turbulence in a more rational way than by "ad hoc" models. This point will be illustrated in what follows.

H. G. Kaper and M. Garbey (eds.),
Asymptotic and Numerical Methods for Partial Differential Equations with Critical Parameters 41–52.

2. Turbulence Modeling Seen from a General Perspective of a Theoretical Approach to a Problem

2.1. MODELS AND THEIR "VERIFICATION"

A theoretical approach to a problem is always based on a *theory*, which might be called a *physical/mathematical model*. This model, which is the counterpart to the real system under consideration (and basically exists only in our minds), corresponds to reality in the following sense:

(1) Physical (measurable) quantities can be identified in the real system, like a pressure p. The model states a relation between a finite number of these quantities by means of certain mathematical functions (often by differential equations); that is, certain quantities in reality correspond to certain quantities in the model.

(2) A certain (measurable) behavior of the system can also be observed in the mathematical model, for example, the change of pressure under the change of certain initial or boundary conditions.

These two features are necessary conditions that must be fulfilled in order to use a model for purposes for which it is really needed: to predict the system behavior in a reliable way, that is, to predict a state of the system that cannot be observed by measuring the real system.

How can we make sure that we may trust the model predictions? Actually there would be only one way to do that: If we were able to *verify* the theory. This, however, is impossible: verification would require the existence of a general principle of induction, which does not exist [9]. Being unable to verify a theory, we are nevertheless able to disprove it. Strictly speaking, one counterexample will do. Again and again we try to disprove the theory. During this never-ending process, if it always "fails" (i.e., when again and again we cannot disprove the theory), there evolves an increasing degree of confirmation (Popper in later publications calls it "degree of corroboration"; see [9]). Though this approach seems to be the opposite of what was originally intended, it is the best we can do: that is, a "good" theory has a high degree of confirmation or corroboration.

2.2. MODELS IN FLUID MECHANICS

In light of this process, we can identify different kinds of model that are used in fluid mechanics. First, there is what will be called the *basic model* (or *basic theory*): the Navier-Stokes equations supplemented by the full energy equation. The process of confirmation has been going on for nearly two hundred years and is still under way, in recent years even with new tools (now that direct simulation of turbulent flows is possible).

Aside from this basic model, two categories of model can be identified [6]:

1. *Asymptotic models*: These are asymptotically reduced basic models, that is, the basic model under certain restrictions, such as infinite Reynolds number or zero Prandtl number. In this sense they are special cases of the basic model. For example, laminar boundary layer theory is the theory of the Navier-Stokes equations in the limit $Re^{-1} = 0$. These asymptotic models can claim the same degree of confirmation as the basic model.
2. *Ad hoc models*: This name was introduced by Landahl [6] for models that are not special cases of the basic model in an asymptotic sense. These models are closely related to what Van Dyke calls "irrational approximations"; see [12] for a number of examples in this category. They cannot claim the degree of confirmation of the basic equations and therefore have to undergo the described process to gain a high degree of confirmation.

2.3. TURBULENCE MODELING

After the basic equations (Navier-Stokes plus energy equation) are time-averaged, the system of equations is no longer closed. Closing the system by turbulence modeling results in a new overall model which now is of the "ad-hoc" type. With turbulence modeling, one inevitably leaves the solid ground of the basic model, so that every turbulent theory (except direct simulation, which needs no modeling of turbulence) has to undergo the process that leads to a high degree of confirmation. This is sometimes called, albeit somewhat imprecisely, the "validation of turbulence models."

As was mentioned in the preceding section, one counterexample is sufficient to disprove a theory. For example, based on a Reynolds stress model, such a disproof for high Reynolds number flow can be deduced by an asymptotic analysis of turbulent boundary-layer theory for all models that use the eddy viscosity concept, as has been demonstrated in [3]. They all neglect wall curvature, which can be shown to be a leading-order effect asymptotically; that is, it must be included in a theory that claims the same degree of confirmation as the basic model.

Since the change from the model with a high degree of confirmation to an ad hoc model occurs as soon as a turbulence model is introduced (but not by time-averaging the equations), one might try to delay turbulence modeling as far as possible, that is, to exploit the basic model (or an asymptotic model deduced from the basic model) before introducing a turbulence model. This approach can be achieved by what will be called *indirect turbulence modeling.*

2.4. INDIRECT TURBULENCE MODELING

The conventional way of modeling, known as *direct turbulence modeling,* involves postulating a mathematical model for the unknown Reynolds stress and turbulent heat flux terms in the basic equations. From the now-closed system of equations, the mean velocities and temperature, $\bar{u}$, ..., $\bar{T}$, can be determined.

Indirect turbulence modeling means that $\bar{u}$, ..., $\bar{T}$ is "modeled" (in the sense explained below). The basic equations of the problem thus serve only as a tool to specify the modeling rules but are neither closed nor solved. If, however, the mean quantities modeled according to these rules were inserted into the basic equations, they could be interpreted as equations for the unknown turbulent quantities. In this sense turbulence modeling is performed *indirectly*. Modeling the mean flow quantities $\bar{u}$, ..., $\bar{T}$ is achieved in three steps.

(S1) Asymptotic analysis of the unclosed system of equations

When wall-bounded flow fields are considered, the asymptotic structure near the wall is determined by the fact that turbulent fluctuations are zero at the wall. These fluctuations increase rapidly with distance from the wall and therefore dominate momentum and heat transfer away from the wall. A viscous sublayer thus asymptotically separates the fully turbulent region from the wall, with an overlap zone between them in which matching conditions hold. This near-wall behavior of the flow field is a crucial part of the physics of turbulent flows. In more complicated flows, a multilayered structure will emerge. This analysis will provide

– the solution behavior close to the wall ($y^+ \to 0$) in terms of Taylor series expansions, and
– the solution behavior in overlapping zones between the various layers ($Re \to \infty$) of the flow field in terms of matching conditions.

(S2) Assumption of a general form of solution

A general mathematical form for the mean flow quantities is assumed that

– fulfills all initial, boundary, symmetry, and integral conditions of the flow;
– fulfills all matching conditions ($Re \to \infty$);
– is correct close to the wall ($y^+ \to 0$); and
– has free constants that can be adjusted empirically.

(S3) Determination of the empirical constants

Most of the constants will turn out to be "universal"; that is, they have the same value for different kinds of flow. Some, however, are specific constants for certain kinds of flow (e.g., Couette flow and boundary layers).

In light of the foregoing discussion of modeling, an indirect turbulence model is an ad hoc model. However, the transition from the basic to an ad hoc model occurs in Step (S2) only, not at the beginning. And, for the desired results $\bar{u}$, ..., $\bar{T}$ the asymptotic aspects analyzed in (S1) are still "correct"—however far they may be off the exact solution—since they emerge from the basic equations. To characterize this special feature of the ad hoc model, one might call it an *asymptotically motivated ad hoc model.*

3. Two Examples of Indirect Turbulence Modeling

3.1. COUETTE FLOW

To illustrate the basic ideas of indirect turbulence modeling, we consider first Couette flow, which is the simplest flow possible flow. We include heat transfer, assuming constant (but different) temperatures on both walls (Fig. 1). The basic equations in dimensional form are

$$0 = \frac{d}{dy^*}\left[\rho^*\nu^*\frac{d\overline{u^*}}{dy^*} - \rho^*\overline{u^{*\prime}v^{*\prime}}\right] \tag{1}$$

$$0 = \frac{d}{dy^*}\left[k^*\frac{d\overline{T^*}}{dy^*} - \rho^* c_p^*\overline{T^{*\prime}v^{*\prime}}\right], \tag{2}$$

which, nondimensionalized with

$$H^*, \qquad u_\tau^* := \sqrt{\frac{\tau_W^*}{\rho^*}}, \qquad \text{and} \qquad T_\tau^* := -\frac{q_W^*}{\rho^* c_p^* u_\tau^*}, \tag{3}$$

read

$$0 = \frac{d}{dy}\left[\epsilon\frac{du^+}{dy} - \overline{u'v'}\right] \to \epsilon\frac{du^+}{dy} - \overline{u'v'} = 1 \tag{4}$$

$$0 = \frac{d}{dy}\left[\frac{\epsilon}{Pr}\frac{d\Theta^+}{dy} - \overline{T'v'}\right] \quad - \quad \frac{\epsilon}{Pr}\frac{d\Theta^+}{dy} - \overline{T'v'} = 1. \tag{5}$$

Here $\Theta^+ := \left(\overline{T^*} - T_{Wu}^*\right)/T_\tau^*$. The constants of integration are 1, as a consequence of the boundary condition $\left[\overline{u'v'}\right]_{y=0} = \left[\overline{T'v'}\right]_{y=0} = 0$. Here ϵ is the inverse of Reynolds number Re_τ; that is,

$$\epsilon = \frac{\nu^*}{u_\tau^* H^*} = \frac{1}{Re_\tau}\,. \tag{6}$$

(S1) Asymptotic analysis of the unclosed system of equations
Since $\epsilon \to 0$ for $Re_\tau \to \infty$, ϵ is a perturbation parameter when (4) and (5) are considered as a (singular) perturbation problem. The singular nature is obvious when (4) and (5) are taken as differential equations for $u^+(y)$ and $\Theta^+(y)$, respectively. In the limit $\epsilon = 0$ the highest derivative with respect to y disappears, which in this simple case is the only term with u^+ and Θ^+. This corresponds to the fact that the wall boundary condition $\overline{u'v'} = 0$ for $y = 0$ cannot be fulfilled when (4) degenerates to $\overline{u'v'} = const$ as it does for $\epsilon = 0$.

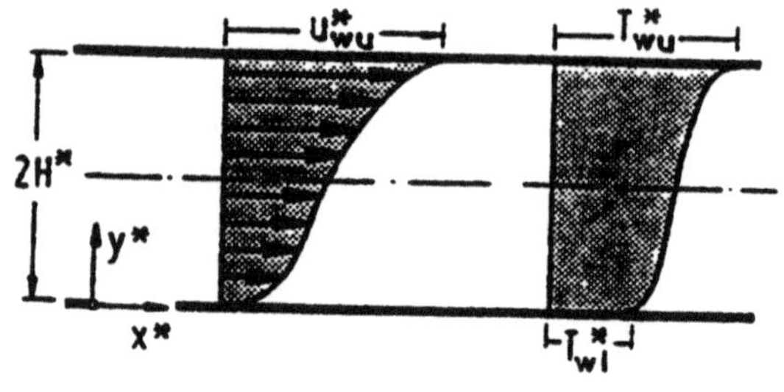

Fig. 1. **Couette flow**

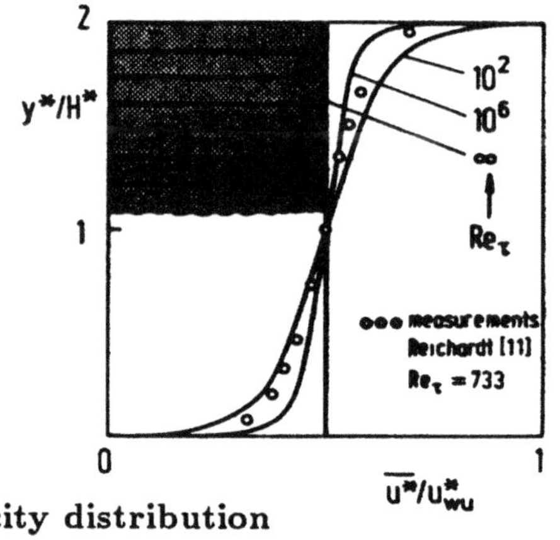

Fig. 2. **Velocity distribution**

Close to the wall, viscous effects must be taken into account since they are the only transport mechanism left when turbulent fluctuations get smaller and smaller for $y \to 0$. The viscous terms are retained when

$$y^+ := \frac{y}{\epsilon} \tag{7}$$

is introduced as a sublayer coordinate, so that in this region the equations are

$$\frac{du^+}{dy^+} - \overline{u'v'} = 1 \tag{8}$$

$$\frac{1}{Pr}\frac{d\Theta^+}{dy^+} - \overline{T'v'} = 1. \tag{9}$$

$$\frac{1}{Pr}\frac{d\Theta^+}{dy^+} - \overline{T'v'} = 1. \tag{9}$$

For $\epsilon \to 0$ these are the equations for the viscous sublayer adjacent to the wall, whereas away from the wall, from (4) and (5) for $\epsilon \to 0$ the equations

$$-\overline{u'v'} = 1 \quad , \quad -\overline{T'v'} = 1 \tag{10}$$

arise (core region). The inner and outer solutions must be matched in an overlap region, namely, for $y^+ \to \infty$ and $y \to 0$, respectively. From this matching procedure we obtain [2]

$$\lim_{y\to 0}\frac{du^+}{dy} = \frac{1}{\kappa y} \quad , \quad \lim_{y^+\to\infty}\frac{du^+}{dy^+} = \frac{1}{\kappa y^+} \tag{11}$$

$$\lim_{y\to 0}\frac{d\Theta^+}{dy} = \frac{1}{\kappa_\Theta y} \quad , \quad \lim_{y^+\to\infty}\frac{d\Theta^+}{dy^+} = \frac{1}{\kappa_\Theta y^+} \tag{12}$$

for $Pr = O(1)$. Here κ and κ_Θ are constants. These conditions state that in the overlap region ($y^+ \to \infty$) velocity and temperature profiles are logarithmic, that is,

$$u^+ = \frac{1}{\kappa}\ln y^+ + C^+$$

$$\text{with} \quad C^+ = \int_0^1 \frac{du^+}{dy^+}dy^+ + \lim_{y^+\to\infty}\int_1^{y^+}\left(\frac{du^+}{dy^+} - \frac{1}{\kappa y^+}\right)dy^+ \tag{13}$$

$$\Theta^+ = \frac{1}{\kappa_\Theta}\ln y^+ + C_\Theta^+(Pr)$$

$$\text{with} \quad C_\Theta^+ = \int_0^1 \frac{d\Theta^+}{dy^+}dy^+ + \lim_{y^+\to\infty}\int_1^{y^+}\left(\frac{d\Theta^+}{dy^+} - \frac{1}{\kappa_\Theta y^+}\right)dy^+. \tag{14}$$

In (12) and (11) $Pr = O(1)$ was assumed. For $Pr \to 0$ and $Pr \to \infty$ a three-layer structure appears instead of the two layers described thus far (for details see [2]).

(S2) Assumption of a general form of solution

From the asymptotic considerations in the preceding step, it is known that there is a two-layer structure (wall region in y^+, core region in y).

– Boundary and symmetry conditions are

$$y^+ = 0 \;:\; u^+ = 0 \;,\; \frac{du^+}{dy^+} = 1 \;;\; \Theta^+ = 0 \;,\; \frac{d\Theta^+}{dy^+} = Pr$$

wall region (15)

$$y = 1 \;:\; \frac{d^2u^+}{dy^2} = 0 \;,\; \frac{d^2\Theta^+}{dy^2} = 0$$

core region (16)

– Matching conditions are

$$y^+ \to \infty : \quad \frac{du^+}{dy^+} = \frac{1}{\kappa y^+} \quad , \quad \frac{d\Theta^+}{dy^+} = \frac{1}{\kappa_\Theta y^+} \qquad \text{wall region} \qquad (17)$$

$$y \to 0 : \quad \frac{du^+}{dy} = \frac{1}{\kappa y} \quad , \quad \frac{d\Theta^+}{dy} = \frac{1}{\kappa_\Theta y} \qquad \text{core region} \qquad (18)$$

– Close to the wall (Taylor series expansion and continuity equation):

$$y^+ \to 0 : \quad \frac{du^+}{dy^+} = 1 - Ay^{+3} + \ldots \qquad (19)$$

$$\frac{d\Theta^+}{dy^+} = Pr\left(1 - A_\Theta y^{+3} + \ldots\right). \qquad (20)$$

These conditions together suggest a mathematical form (which of course is not unique):

$$\text{wall region} \quad : \quad \frac{du^+}{dy^+} = \frac{1}{1 + (A + B)\, y^{+3}} + \frac{By^{+3}}{1 + \kappa By^{+4}} \qquad (21)$$

$$\frac{1}{Pr}\frac{d\Theta^+}{dy^+} = \frac{1}{1 + (A_\Theta + B_\Theta)\, y^{+3}} + \frac{B_\Theta y^{+3}}{1 + \kappa_\Theta Pr B_\Theta y^{+4}} \qquad (22)$$

$$\text{core region} \quad : \quad \frac{du^+}{dy} = \frac{1}{\kappa y} + \frac{1}{\kappa\,(2 - y)} + a_0 \qquad (23)$$

$$\frac{d\Theta^+}{dy} = \frac{\kappa}{\kappa_\Theta}\frac{du^+}{dy}. \qquad (24)$$

Free constants in (21)–(24) are κ, A, B, a_0 for the flow problem and κ_Θ, $A_\Theta(Pr)$, $B_\Theta(Pr)$ for the heat transfer problem.

(S3) Determination of the empirical constants

Most of the above constants are of the "universal" type; that is, they hold for all kinds of flows. The physical reason is that they are all near-wall constants and thus belong to a flow region that is universal for all wall-bounded flows. They are (see [2] for details)

$$\kappa = 0.41 \quad ; \quad A = 6.1 \cdot 10^{-4} \quad ; \quad B = 1.43 \cdot 10^{-3}$$

$$\kappa_\Theta = 0.47 \quad ; \quad A_\Theta = 6.8 \cdot 10^{-4} \cdot Pr \quad ; \quad B_\Theta = \frac{1.5 \cdot 10^{-6}}{Pr}.$$

With these constants, C^+ and C_Θ^+ in (13) and (14), respectively, are $C^+ = 5.0$ and $C_\Theta^+ = 13.7Pr^{\frac{2}{3}} - 7.5$. The only constant remaining that is specific for the Couette flow is the outer region constant $a_0 = 0.41$. In Fig. 2 velocity profiles based on (21) and (23) are shown.

3.2. COUETTE-POISEUILLE FLOW WITH ZERO WALL SHEAR STRESS AT THE LOWER WALL

If a streamwise pressure gradient is superposed on the Couette flow, the shear stress is no longer constant over the cross section but varies linearly with y. For a certain pressure gradient, $dp^*/dx^* = \tau^*_{Wu}/(2H^*) > 0$, where τ^*_{Wu} is the upper-wall shear stress; the wall shear stress will be zero at the lower wall. This special case will be analyzed now, since from it many conclusions can be drawn with respect to separating boundary layers. This example illustrates that the asymptotic analysis by itself, performed in the indirect turbulence modeling approach, will reveal much of the physics of a flow under consideration.

In Fig. 3 the velocity and shear stress profiles are shown. The reference velocity is no longer always u^*_τ according to (3), since the wall shear stress is zero at the lower wall. Instead, from dimensional analysis, the following characteristic velocities can be deduced [2]:

$$\text{core region} \quad : \quad u^*_{\tau u} \tag{25}$$

$$\text{lower wall region} \quad : \quad u^*_l = \left(\frac{\nu^*}{\rho^*} \cdot \frac{dp^*}{dx^*}\right)^{\frac{1}{3}} \tag{26}$$

(S1) Asymptotic analysis of the unclosed equation

For high Reynolds numbers the asymptotic structure is similar to that of the preceding example. There is a core region with no influence of viscosity and two wall layers with viscosity influence.

(a) Close to the upper wall (reference velocity $u^*_{\tau u}$ for core and wall region)

$$\text{core region} \quad : \quad -\overline{u'v'} = \frac{y}{2} \tag{27}$$

$$\text{wall region} \quad : \quad \frac{du^+_u}{dy^+_u} - \left(\overline{u'v'}\right)^+ = 1 - \epsilon\frac{y^+_u}{2} \tag{28}$$

Here $u^+_u = (u^*_{Wu} - \overline{u}^*)/u^*_{\tau u}$ and $y^+_u = (2 - y^*/H^*)/\epsilon$ with $\epsilon = \nu^*/(u^*_{\tau u}H^*)$. The matching conditions are (cf. Equation (11))

$$\lim_{y\to 2} \frac{du^+}{dy} = \frac{1}{\kappa(2-y)} \quad , \quad \lim_{y^+_u\to\infty} \frac{du^+}{dy^+} = \frac{1}{\kappa y^+_u} \tag{29}$$

$$\text{so that} \quad \lim_{y\to\infty} u^+(y) = u^+_{Wu} - \left[\frac{1}{\kappa}\ln\left((2-y)\,Re_{\tau u}\right) + C^+_u\right]. \tag{30}$$

(b) Close to the lower wall (ref. vel. $u^*_{\tau u}$ for the core region, u^*_l for the wall region)

$$\text{core region} \quad : \quad -\overline{u'v'} = \frac{y}{2} \tag{31}$$

$$\text{wall region} \quad : \quad \frac{du^+_l}{dy^+_l} - \left(\overline{u'v'}\right)^+_l = y^+_l \tag{32}$$

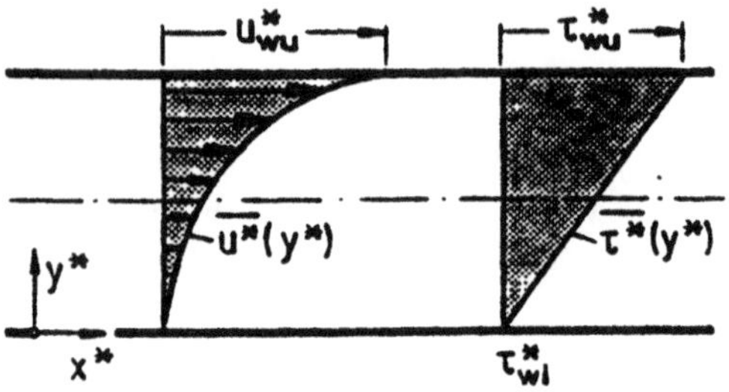

Fig. 3. **Couette-Poiseuille flow** $(\tau_{Wl}^{*} = 0)$

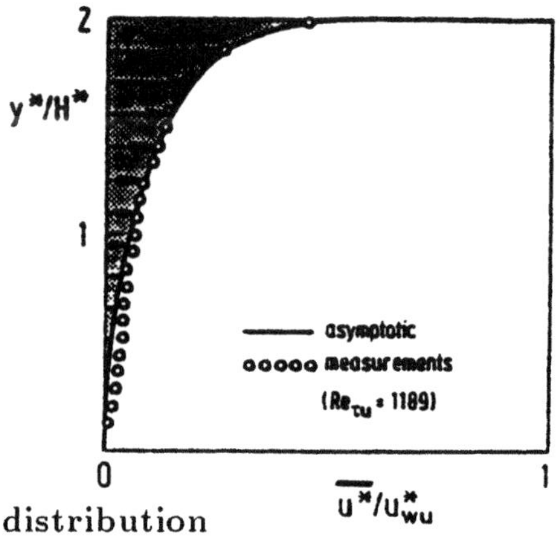

Fig. 4. **Velocity distribution**

Here $u_l^+ = \overline{u}^*/u_l^*$ and $y_l^+ = (y^*/H^*)/\epsilon_l$ with $\epsilon_l = \nu^*/(u_l^* H^*)$.

In the lower-wall region the turbulent shear stress is one order of magnitude smaller than in the core region. Therefore two characteristic velocities, $u_{\tau u}^*$ and u_l^*, were introduced close to the lower wall, with

$$-\left(\overline{u'v'}\right)_l^+ = \frac{\left(-\overline{u^{*\prime} v^{*\prime}}\right) / \left(u_{\tau u}^{*2}/2\right)}{\epsilon_l}. \tag{33}$$

The matching conditions in the overlap zone between the lower-wall layer and the core region are

$$\lim_{y \to 0} y \left[- \frac{\overline{u'v'}}{\epsilon_l} \right]^{-\frac{1}{2}} \frac{du_l^+}{dy} = \lim_{y_l^+ \to \infty} y_l^+ \left[\left(-\overline{u'v'} \right)_l^+ \right]^{-\frac{1}{2}} \frac{du_l^+}{dy_l^+} = \frac{1}{\kappa_0}, \tag{34}$$

from which, together with (30), we obtain

$$\sqrt{\epsilon_l}\lim_{y\to 0}\frac{du_l^+}{dy}=\frac{1}{\kappa_0\sqrt{y}}\qquad ;\qquad \lim_{y_l^+\to\infty}\frac{du_l^+}{dy_l^+}=\frac{1}{\kappa_0\sqrt{y_l^+}}. \tag{35}$$

Integration yields

$$\lim_{y\to 0}u_l^+=\frac{2\sqrt{y}}{\kappa_0\sqrt{\epsilon_l}}+C_l^+\qquad ;\qquad \lim_{y_l^+\to\infty}u_l^+=\frac{2}{\kappa_0}\sqrt{y_l^+}+C_l^+. \tag{36}$$

From this analysis it follows that, for vanishing skin friction, the velocity law in the overlap zone is of square root rather than of logarithmic type (as it is for $\tau_{Wl}^*\neq 0$; cf. Equation (13)). Rewriting in terms of $u^+=\overline{u}^*/u_{\tau u}^*$ with $u_{\tau u}^*=u_l^*\,(2Re_{\tau u})^{\frac{1}{3}}$ yields

$$\lim_{y\to 0}u^+=\frac{\sqrt{2y}}{\kappa_0}+\frac{C_l^+}{(2Re_{\tau u})^{\frac{1}{3}}}. \tag{37}$$

With (37) and (30) the velocity $u^+(y)$ is known for $y\to 0$ and $y\to 2$, respectively. Two universal constants are involved thus far, C_l^+ and C_u^+, which follow from integration over the lower- and upper-wall layer, respectively.

(S2) Assumption of a general form of solution

As demonstrated in the preceding example, a general form for u^+ may be assumed in the two wall layers and the core region. With the known velocity distribution, the velocity on the centerline is

$$u_c^+=C_c+\frac{C_l^+}{(2Re_{\tau u})^{\frac{1}{3}}}\qquad\text{with}\quad C_c=\lim_{y\to 0}\int_y^1\frac{du^+}{dy}dy, \tag{38}$$

and the mean velocity

$$u_m^+=u_c^++C_m\qquad\text{with}\quad C_m=\frac{1}{2}\lim_{y_2\to 2\ ,\ y_1\to 0}\int_{y_1}^{y_2}\left(u^+(y)-u_c^+\right)dy. \tag{39}$$

For determining u_c^+ and u_m^+, only the core region will be modeled by assuming

$$\frac{du^+}{dy}=\frac{1}{\kappa_0\sqrt{2y}}+\frac{1}{\kappa\,(2-y)}-\frac{1}{2\kappa}-\frac{1}{4}\left(\frac{1}{\kappa_0}-\frac{1}{\kappa}\right)y+\frac{1}{\kappa}F(y) \tag{40}$$

$$\text{with}\qquad F(y)=ay+by^2-\frac{1}{2}\left(\frac{a}{2}+b\right)y^3.$$

From this, C_c and C_m in (38) and (39) are functions of a and b only (for details see [2]).

(S3) Determination of the empirical constants

Once again the wall regions are "universal." In the upper wall region $\kappa = 0.41$ holds as in the preceding example. Instead of modeling du^+/dy^+ in the wall regions, the constants C_u^+ and C_l^+ may be extracted from experimental data. At the upper wall it is $C_u^+ = 5.0$ as before.

The lower wall is distinctly different, since here $\tau_{Wl}^* = 0$ holds. From experiments $\kappa_0 = 0.8$ and $C_l^+ = -3.2$. These constants hold for all wall-bounded flows with $\tau_W^* = 0$.

The only constants that are specific for the Coutte-Poiseuille flow with zero shear stress at the lower wall are a and b in (40). From measurements of u_m^+ and u_{Wu}^+ for one $Re_{\tau u}$, $a = -3.87$ and $b = 5.91$ were determined. In Fig. 4 a measured velocity profile that is very close to the case $\tau_{Wl}^* = 0$ is compared to the asymptotic profile according to the indirect turbulence modeling approach. Evidentally, the physics of the flow are modeled correctly.

References

[1] Gersten, K. (1987) 'Some contributions to asymptotic theory for turbulent flows', *Proc. 2nd Int. Symp. on Transport Phenomena in Turbulent Flow*, Tokyo, 201–214.

[2] Gersten, K., and Herwig, H. (1992) *Strömungsmechanik, Impuls-, Wärme- und Stoffübertragung aus asymptotischer Sicht*, Vieweg, Wiesbaden.

[3] Jeken, B. (1992) 'Asymptotische Analyse ebener turbulenter Strömungen an gekrümmten Wänden bei hohen Reynolds-Zahlen mit einem Reynolds-Spannungs-Modell', Dissertation, Ruhr-Universität Bochum, Germany.

[4] Kaplun, S., and Lagerstrom, P. A. (1957) 'Asymptotic expansions of Navier-Stokes solutions for small Reynolds numbers', *J. Math. Mech.*, **6**, 585–593.

[5] Kluwick, A. (1989) 'Interacting turbulent boundary layers', *ZAMM*, **69**, T6.

[6] Landahl, M. T. (1981) 'Mathematical modeling in fluid mechanics', *ZAMM*, **61**, T9–14.

[7] Melnik, R. E. (1987) 'A new asymptotic theory of turbulent boundary layers and the turbulent Goldstein problem', in F. T. Smith and S. N. Brown (eds.), *Boundary Layer Separation*, Springer-Verlag, New York, 217–234.

[8] Millikan, C. B. (1938) 'A critical discussion of turbulent flows in channels and circular tubes', *Proc. 5th Int. Congr. of Applied Mech. Cambridge, Mass.*, 386–392.

[9] Popper, K. R. (1980) *The Logic of Scientific Discovery*, Hutchinson, London.

[10] Prandtl, L. (1904) 'Über Flüssigkeitsbewegung bei sehr kleiner Reibung', in A. Krazer (ed.), *Verh. III Intern. Math. Kongr. Heidelberg*, Teubner, Leipzig, 484–491.

[11] Reichardt, H. (1959) 'Gesetzmäßigkeiten der geradlinigen turbulenten Couette-Strömung', Mitteil. aus dem Max-Planck-Institut für Strömungsforschung, No. 22, Göttingen.

[12] Van Dyke, M. (1975) *Perturbation Methods in Fluid Mechanics*, The Parabolic Press, Stanford.

SOME APPLICATIONS OF COMBINED ASYMPTOTICS AND NUMERICS IN FLUID MECHANICS AND AERODYNAMICS

N. D. Malmuth
Rockwell International Science Center
Thousand Oaks CA 91360

ABSTRACT. Diverse examples are discussed illustrating the power of combining asymptotics and numerical solution procedures. Problems in transonic flow are treated as well as an unusual application occurring in paper manufacturing and magnetic recording. A recurrent theme in these applications is the simplification and clarification of the essential issues by the asymptotics as well as the economization of the computational problem. On the other hand, computation handles nonlinearities inaccessible to the analytics.

1. Introduction

In the 1950s at Caltech and elsewhere, there was a resurgence of interest in perturbation methods, particularly those oriented to singular perturbation problems. As described in [4, 19, and 31], successful applications of the technique have been numerous. With the revolution in high-speed computers, however, attention until recently has been diverted from asymptotic approaches. Indeed, a conviction still held in many circles is that asymptotics are of low relative utility compared with computational procedures.

Nevertheless, strong evidence exists that much is to be gained from an interplay between both asymptotic and numerical viewpoints. On the one hand, the asymptotics illuminates the essential scales and similitudes of the problem. It also provides insights into invariances and physics that are not transparent from a purely numerical solution. Moreover, asymptotic methods give us an idea about parametric dependencies, bring out the essential features of the problem, simplify the formulation (even reducing its dimensionality), and provide information about the smooth, consistent blending of temporal and spatial regions where different effects are important. In fact, success of this matching process often serves as a check of the correctness of the asymptotic procedure itself.

On the other hand, modern computational techniques yield a powerful means of dealing with nonlinearities intractible to analytical methods. Even

H. G. Kaper and M. Garbey (eds.),
Asymptotic and Numerical Methods for Partial Differential Equations with Critical Parameters 53–79.

here., however, new insights and efficient numerical approaches can be derived by exploiting asymptotic ideas. Some of this work is related to older numerical techniques such as "subtracting," or "factoring out," the singularity, which can be represented in many instances as a coordinate rather than parameter expansion. Other ideas currently receiving attention are the use of asymptotic behavior in discretization approximations for derivatives, domain decomposition, and convergence acceleration techniques using solution initialization and defect minimization with asymptotics. Even gridding issues associated with numerical solutions can be resolved by asymptotic methods. An example is the viscous boundary layer of high Reynolds number flow whose length scale is determined from asymptotics and similarity. These suggest the grid clustering necessary to resolve the fine structure of the boundary layer as its thickness and the viscosity become small.

This article surveys examples that illustrate the effectiveness of combined asymptotics and numerics (CAN) methods. The view is primarily the one first mentioned here, namely, allowing the asymptotics to provide an economical formulation that is treated by the numerics. The second viewpoint involving a somewhat more intimate relationship, coined by some authors as "asymptotically induced numerics," will not receive as much emphasis in what follows. It is, however, an active field of research.

Examples to be discussed come from transonic flow and an application of lubrication theory appropriate to foil bearings, paper making, and the production of film and magnetic recording. All of these applications involve thin layers. Some of them contain diverse-length scales. The asymptotics are called "limit-process" asymptotics in the sense that the expansions are increasingly accurate approximations of the solution in a "distinguished limit" as a parameter tends to a limit. Cole and Kevorkian [7] provide formal definitions of these representations, elucidate the basic theoretical ideas, and discuss various applications. All the numerical methods illustrated are of inconsequential to very low computational intensity, illustrating the advantage of asymptotic "preprocessing."

2. Transonic Slender-Body Theory and the Area Rule

In transonic flow, a limiting process can be defined in terms of the freestream Mach number, M_∞, and the body thickness ratio, δ, which measures the slenderness of the body. Useful theories based on this limit can be developed for low-wave drag bodies of practical importance, for example, missiles and fighters such as the ATF. Other slender vehicles such as the hypersonic NASP need to accelerate through the speed of sound to achieve orbital velocity, V_O. Drag at the "transonic pinch point" critically affects attainment of V_O, since accelerations controlled by drag peaks will affect influential parameters such as fuel consumed, payload, and structural and vehicle takeoff weight. Quick response methods are needed to help reach V_O as well as to connect the vehicle shape with its aerodynamic performance. Although computational fluid dynamics (CFD) treats this nonlinear regime, CAN methods offer a lower computational intensity alternative for preliminary aerodynamic design and vehicle sizing, providing quick evaluation of mission and configurational impacts of hundreds of parametric changes and accurate modeling of the nonlinear physics.

Oswatitsch and Keune [28] and others developed the transonic equivalence rule (TER) in which a complicated shape could be replaced by an equivalent body of revolution under mild constraints for purposes of defining the flow field over the more complicated body. This rule was clarified in [6] by a demonstration of how it is embedded in an approximation procedure based on the aforementioned distinguished limit, leading to a model called transonic slender-body theory. Of great practical impact was a TER adjunct called the transonic area rule (TAR), developed in the pioneering work of Whitcomb [33] and Jones [16].

Figure 1 shows a body of thickness ratio δ in a cylindrical polar coordinate system appropriate to limit processes giving the TER. For this system, the flow consists of three regions of different character. The precise asymptotic limits that delineate these zones reveal an "inner" region near the body where crossflow gradients dominate [6]. In the cylindrical coordinate system shown in Fig. 1, if r and x denote dimensionless radial and axial coordinates, respectively, an inner limit is defined in which the crossflow dominated region appears invariantly as $M_\infty \rightarrow 1$. As in TSDT, this limiting behavior is expressed in terms of a transonic similarity parameter K. The inner limit is

$$K = \frac{1 - M_\infty^2}{\delta^2}, \quad r^* = \frac{r}{\delta}, \quad \text{fixed as } \delta \rightarrow 0. \tag{1}$$

If Φ is the velocity potential, an inner expansion for Φ in the limit process (1) gives harmonic behavior in the crossflow plane for a perturbation potential ϕ, where the expansion is

$$\Phi(x, r, \theta; M_\infty, \delta) = U(x + 2\delta^2 \log\delta\ S^*(x) + \delta^2 \phi_2^*(x, r^*; K) + O(\delta^4)). \tag{2}$$

In (2), U is the freestream speed, and $S^*(x)$ is a source distribution to be determined by matching.

Away from the body of Fig. 1, asymmetric effects decay, and nonlinear phenomena become important. This situation occurs on a lateral scale of the steeply inclined Mach waves in supersonic flow zones. This region is defined by the "outer" limit

$$K = \frac{1 - M_\infty^2}{\delta^2}, \quad \tilde{r} = r\delta \quad \text{fixed as } \delta \rightarrow 0. \tag{3}$$

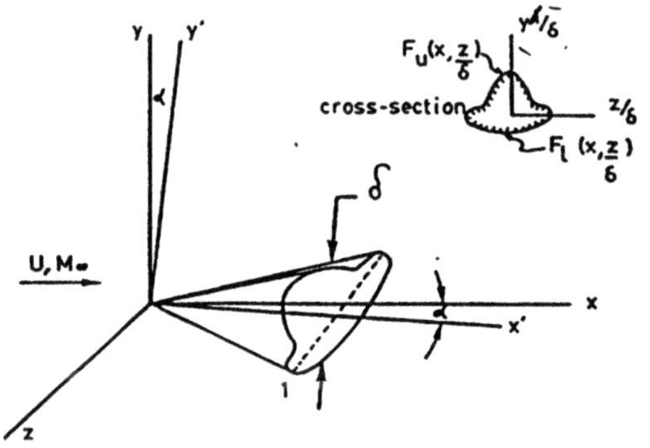

Fig. 1. **Transonic body**

The appropriate asymptotic behavior for Φ in the outer region is

$$\Phi(x, r, \theta; M_\infty, \delta) = U(x + \delta^2 \phi(x, \tilde{r}; K) + O(\delta^4)). \tag{4}$$

Expansions (2) and (4) match only to dominant orders. Reference [6] shows the need for an intermediate expansion to obtain higher-order refinements. Figure 2 shows how the inner and outer expansions are combined according to matching to give a uniformly valid expression for the pressure distribution over an asymmetric body. The nonlinear harmonic boundary value problem for ϕ_2^* is shown in the near field (inner) zone for the body given by

$$r = \delta F(x, \theta). \tag{5}$$

In the coordinates defined in the left-hand side of the figure, c is the reference length, such as the wing chord. The pressure coefficient, C_P, given by Equation (F1) in the figure is partially determined in terms of ϕ_{2x}^* and the squared resultant crossflow speed ${v_B^*}^2 + {w_B^*}^2$ which is related to gradients of ϕ_2^* in (F2).

SC83 24870

NEAR FIELD

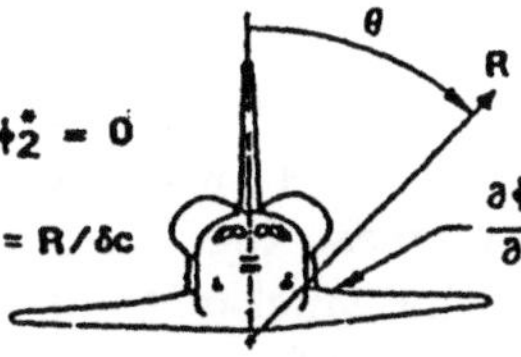

$\Delta\phi_2^* = 0$

$r^* = R/\delta c$

$\frac{\partial\phi_2^*}{\partial n} = \frac{FF_x}{\sqrt{F^2 + F_\theta^2}}$

PRESSURE COEFFICIENT

$$\frac{C_p}{\delta^2} = -\frac{4}{\pi}(\ln\delta)\mathcal{A}''(x) - 2\phi^*_{2_x} - V_B^{*2} - W_B^{*2} - 2g'(x;K) \qquad \text{(F1)}$$

BODY SHAPE

$R = c\delta F(x,\theta)$

CROSS FLOW SPEED

$$V_B^{*2} + W_B^{*2} = |\nabla\phi_2^*|^2 = \left(\frac{\partial\phi_2^*}{\partial r^*}\right)^2 + \frac{1}{r^{*2}}\left(\frac{\partial\phi_2^*}{\partial\theta}\right)^2 \qquad \text{(F2)}$$

FAR FIELD

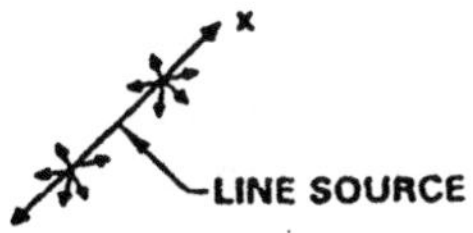

$$(K - (\gamma+1)\phi_x)\phi_{xx} + \frac{1}{\tilde r}(\tilde r\phi_{\tilde r})_{\tilde r} = 0 \qquad \text{(F3)}$$

$$\lim_{\tilde r\to 0}\tilde r\phi_{\tilde r} = \frac{\mathcal{A}'(x)}{2\pi} \qquad \text{(F4)}$$

$$g(x;K) = \lim_{\tilde r\to 0}\left(\phi - \frac{\mathcal{A}'(x)}{2\pi}\log\tilde r\right) \qquad \text{(F5)}$$

$\mathcal{A} = 1/2\oint F^2\,d\theta$

$\tilde r = \delta R/c$

$K = (1 - M_\infty^2)/\delta^2$

Fig. 2. Transonic equivalence rule

The dominant term in (F1) is the second streamwise derivative of the cross-sectional area, $A(x)$. The last remaining component of the pressure $2g'(x)$in (F1) comes from the nonlinear axisymmetric outer potential, ϕ, which is governed by the axisymmetrical Karman transonic small disturbance equation (F3) subject to the boundary condition (F4), which is one of the essential results of dominant order matching. Another is the formula for $g(x;K)$ (F5).

These relations are an excellent illustration of the simplifications afforded by the TER, which reduces the three-dimensional flow problem to an axisymmetric one. By integration, it is evident that the wave drag depends only on the area distribution of the equivalent body of revolution. This is the transonic area rule. By this reduction in dimensionality, the original computational problem is sizably reduced to the numerical treatment of the axisymmetric small disturbance equation (F3) and a boundary integral method for the near-field harmonic boundary value problem. A two-dimensional "panel" approach, which treats each x station body contour by a sequence of straight-line sources around the body contour, works quite well and is a useful method. Modern techniques can increase the speed and accuracy of inverting the full matrix discretization of the panel method singular integral equation. If N is the approximating number of segments, the operation count can be reduced from $O(N^2)$ to $O(N \log N)$ by using these methods.

As an illustration of the power of this approach, Fig. 3 shows a comparison reported in [29] of pressures over a wing body combination as predicted by the TER with experiment. Figures 4 and 5 show comparisons at slightly supersonic Mach numbers for the Shuttle Orbiter. Figure 5 gives pressures on the top centerline of the Orbiter fuselage at $M_\infty=1.05$. These results suggest encouraging elasticity of the theory for not-so-slender shapes. This is another practical bonus of CAN approaches when the underlying asymptotics have such favorable properties. Another application of the TER and TAR is shown in Fig. 6, for the wave drag rise of an advanced fighter configuration. Under certain constraints indicated in [6], the TAR for a body is

$$\frac{D}{\rho U^2} = \int_0^1 A''(x)g(x)\,dx, \tag{6}$$

where D and ρ are the drag and density, respectively. The figure shows the ability of the CAN TAR to give good correlations with experiment, which have been not been reported (to our knowledge) with pure CFD approaches. Appropriate filtering techniques are needed for the numerical evaluation of $A''(x)$. IMSL spline-smoothing routines provide effective tools for this purpose.

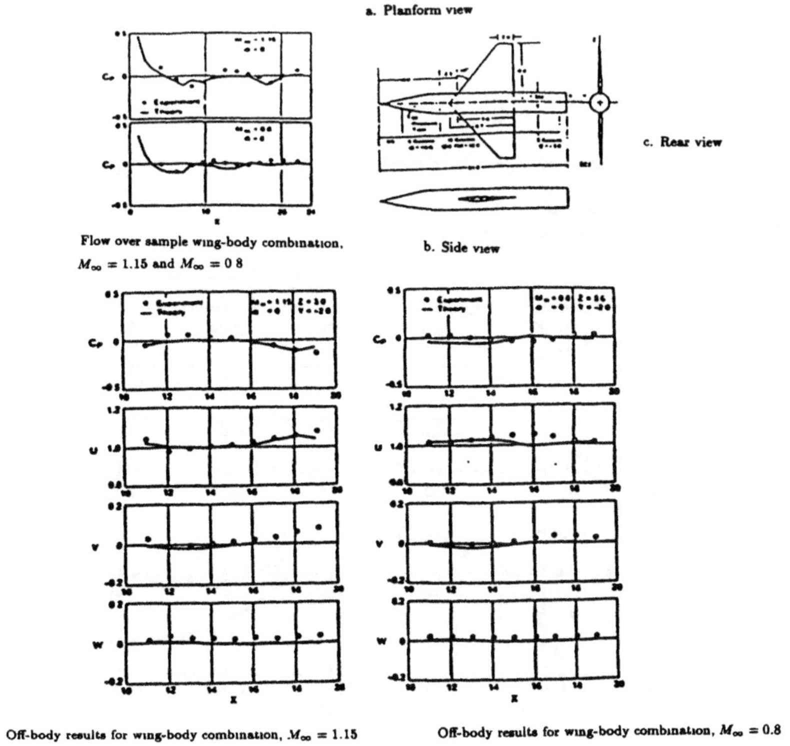

Fig. 3. **Comparison of TER with experiment for generic wing body; M_∞=.8 and 1.15**

3. Shocks over Transonic Bodies of Revolution

An interesting illustration of the power of asymptotics to give surprising insight into the physics of the flow is the application of the slender-body theory just discussed to predict the position of shocks over bodies of revolution. Here, in contrast to planar flows, the theory indicates that the shock position of almost sonic axisymmetric bodies depends only on the properties of the streamwise area progression and occurs at the axial location where there is a point of inflection in that distribution. This property is seen from a simple argument coming from unpublished studies of Malmuth and C. C. Wu. A more detailed analysis [8] has recently been further refined by J. D. Cole.

In the simple argument, if ϕ is the outer perturbation potential, $A(x)$ the cross-sectional area, and x the streamwise Cartesian coordinate, the inner expansion

$$\phi \doteq \frac{A'(x)}{2\pi} \log r + g(x) + \cdots \quad \text{as} \quad r \to 0 \tag{7}$$

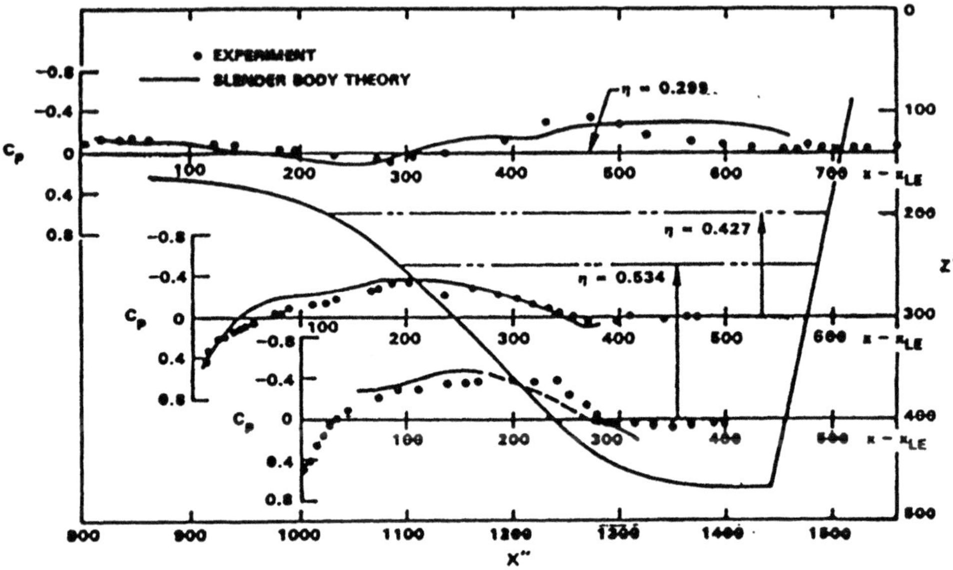

Fig. 4. Comparison of TER with experiment for Shuttle Orbiter wing; M_∞=1.05

holds as well as the Prandtl relation at the foot of the almost normal shock near its intersection with the body

$$\langle \phi_x \rangle = \frac{K}{\gamma + 1} \tag{8}$$

$$\left(K = \frac{1 - M_\infty^2}{\delta^2} = \text{transonic similarity parameter} \right)$$

$$(\langle \ \rangle \equiv \text{average across shock}),$$

where M_∞ is the freestream Mach number and δ is the body thickness ratio.

Differentiating (7) with respect to x and forming the shock average give

$$\langle \phi_x \rangle \doteq \frac{A''(x)}{2\pi} \log r + \langle g(x) \rangle + \cdots \quad \text{as } r \to 0 \cdot \tag{7'}$$

For order-of-magnitude consistency, Equation (8), which is applicable to near-unity Mach numbers, implies

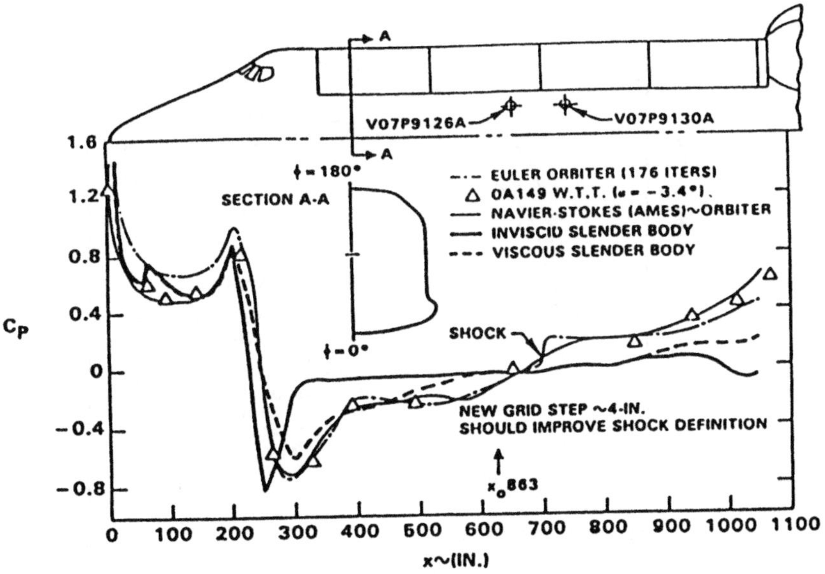

Fig. 5. **Comparison of TER with experiment for Shuttle Orbiter fuselage**

$$A''(x) = 0 \tag{9}$$

in (7′) at the shock's foot. *This is the aforementioned shock position rule.* Shock locations predicted from (9) have been compared with data from [26] for the bodies therein. Consistent with the notation of [26] in which $\xi = \frac{x}{l}$, l=body length, we denote this location as $\xi_{A''=0}$. If the location ξ_{MAX} of the maximum thickness is such that $\xi_{MAX} \leq \frac{l}{2}$, then the equation of the body is

$$H = H_{MAX}C(t - t^n), \tag{10}$$

where

$$t \equiv 1 - \xi,$$

and H is the radius of the body, with n, H_{MAX}, and C as parameters.

If $\xi_{MAX} \geq \frac{l}{2}$, then

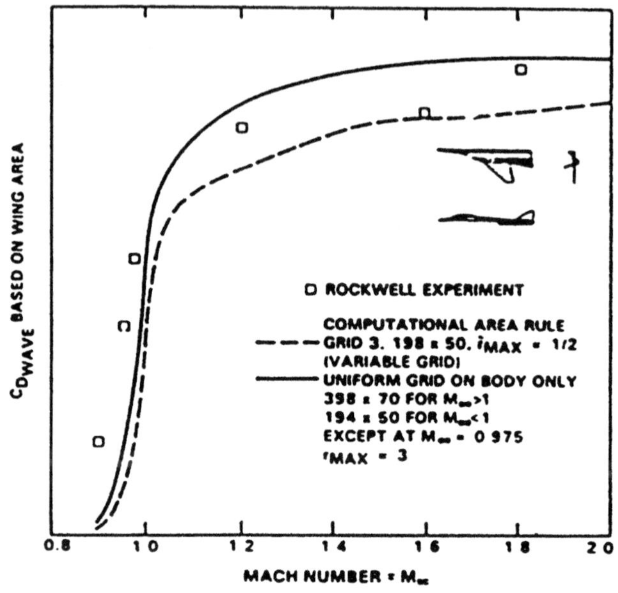

Fig. 6. **Comparison of TAR predicted wave drag rise with experiment**

$$H = H_{MAX} C(\xi - \xi^n). \tag{4}$$

Since $A(x) = \pi H^2$ and $' \equiv \frac{d}{dx}$

$$A'' = 2\pi\left(H'^2 + HH''\right),$$

then

$$A'' = \frac{2\pi C^2 H_{MAX}^2}{l^2}[(1 - n\lambda)^2 - n(n-1)(\lambda - \lambda^2)], \tag{11}$$

where

$$\lambda = \xi^{n-1}.$$

The appropriate expression for $\xi_{A''=0}$ for the case covered by Eq. (9) is therefore

$$t_{A''=0} = \left\{ \frac{b \pm \sqrt{b^2 - 4a}}{2a} \right\}^{\frac{1}{n-1}}, \tag{12}$$

where

$$a = 2n^2 - n, \quad b = n^2 + n.$$

For the case covered by Equation (10), t in Equation (12) is replaced by ξ. These formulas were checked against values obtained from Fig. 4(e) in [26]. The positive branch is taken for the equation (9) case and the negative for that of Equation (10). Note that the branch other than that for the shock-body intersection corresponds to the body intersection of the upstream limb of the sonic line.

In the table on page 4 in [26], values of n are given for the various bodies tested. Figure 7 compares the shock location, ξ_{SHOCK}, obtained from our theoretical result

$$\xi_{SHOCK} = \xi_{A''=0} \tag{13}$$

with experiment.

The comparison in Fig. 7 is affected by sting-induced viscous phenomena occurring on the rearward part of the bodies tested, which are shown in the figure. Despite the fact that the test Reynolds numbers are on the order of 40×10^6, strong diffusion of the shock appears at its foot, presumably a result of shock-boundary layer interaction and shock-induced separation. Krupp and Murman [17] and Malmuth et al. [25] have computed this case with a TSDT code. In Fig. 18 of [17], the M_∞=.999 case is apparently labeled as $M_\infty = 1$. The computations of [25] were for M_∞=.99. Krupp and Murman used certain empirical scalings in the definition of K, r stretching, and the pressure coefficient C_P when they did these calculations. As far as the shock position is concerned, both computations produce results that are nearly identical. They are shown by the solid square in Fig. 7. The *inviscid* result agrees better with the theory than the experiments, in accordance with expectations. Note also that the data in [26] were not corrected for wind-tunnel wall interference. Figure 8 shows isomachs discussed in [25] that were used to define the shock position, ξ_{SHOCK}, in Fig. 7.

4. Transonic Wind-Tunnel Wall Interference

Another example of transonic slender-body theory is wind-tunnel wall interference. As in the preceding applications, there is a need to rapidly estimate the effects of the walls on the model to extrapolate the measured aerodynamic characteristics to the free field (i.e., the free-flight configuration of the

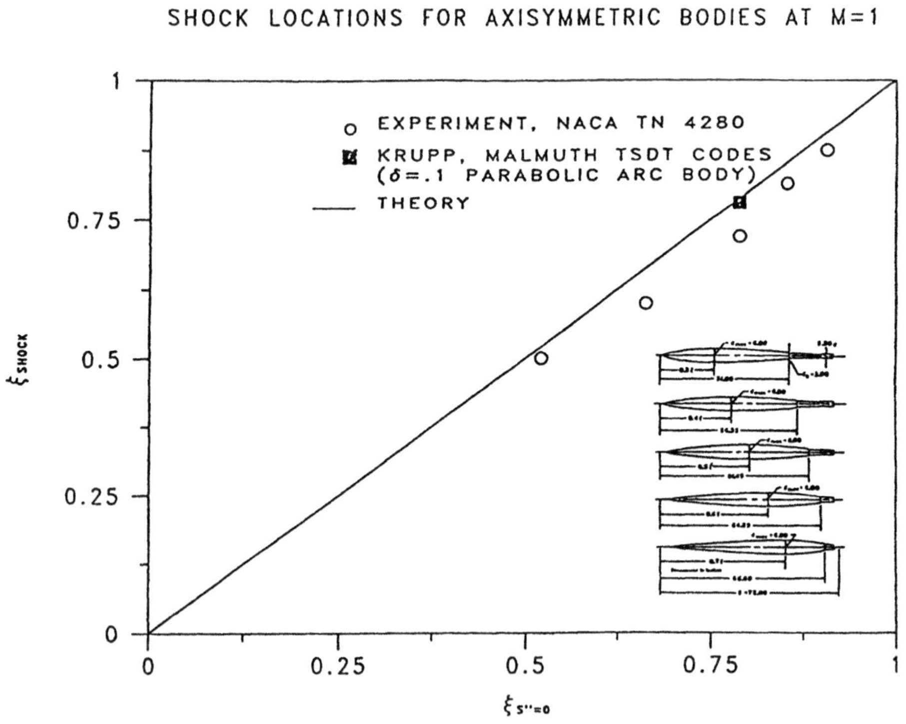

Fig. 7. Comparison of shock locations from Equation (9) with data of [26]

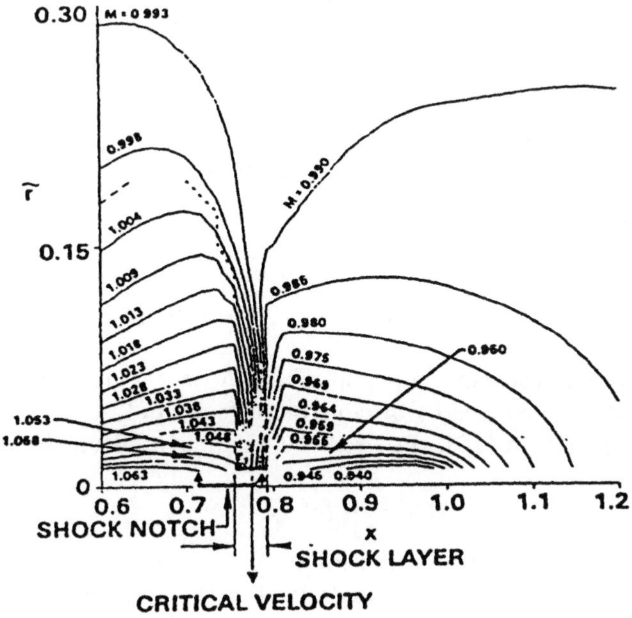

Fig. 8. Isomachs for M=.99 flow over parabolic arc body; δ =.1

walls absent). The same duality exists as discussed previously. Pure CFD models have been used. Since large parametric sets are tested, a need exists for an inexpensive complement to these simulations to assess interference quickly. CAN approaches are a promising candidate to fill this requirement.

References [25] and [24] describe CAN approaches to treat slender airplane configurations and high-aspect ratio wings, as well as procedures in which the asymptotics can be integrated with experimental measurements to assess wall interference. For both the slender-body and high-aspect ratio cases, the wall interference is obtained by a systematic asymptotic expansion procedure. Each is represented by a secondary approximation within a Karman-Guderley (KG) TSDT framework.

In the slender case, if the airplane model is represented as the surface

$$r = \delta F(x, \theta) \quad , \tag{14}$$

within the coordinate system indicated in Fig. 9, with δ = the characteristic thickness ratio and overbars representing dimensional quantities, the asymptotic expansion of the velocity potential Φ in terms of the freestream speed, U, is

$$\frac{\Phi}{U} = x + \delta^2 \phi(x, \tilde{r}, \theta; K, H, A) + \cdots \quad , \tag{15}$$

which holds for the KG outer limit,

$$x, \tilde{r} = \delta r, \theta, K = (1 - M_\infty^2)/\delta^2, H = h\delta/c, A = \alpha/\delta \text{ fixed as } \delta \to 0, \tag{16}$$

where M_∞ is the freestream Mach number, K is the KG similarity parameter, H is the scaled height of control surface, and A is the incidence parameter. For Equation (16), the pressure formula valid on the interface,

$$C_P = -2\delta^2 \phi_x, \tag{17}$$

leads to a (primary) KG formulation for ϕ.

Secondary expansions for ϕ can be derived in the limit of $H \to \infty$. Three asymptotic decks can be identified for this limit. Shown schematically in Fig. 9 is an inner deck where cross-flow gradients dominate, a central region which is weakly perturbed by the walls or interface deviations from free field conditions, and a wall layer in which the model appears as a multipole in Prandtl-Glauert flow for "classical" Dirichlet and Neumann conditions associated with free jet and solid tunnel walls. Nonclassical conditions associated with slotted and porous walls can be handled by assuming that pressures are specified on a cylindrical "interface" surface near the walls. Matching procedures (detailed in [25] and [22]) provide the effect of reflections of the

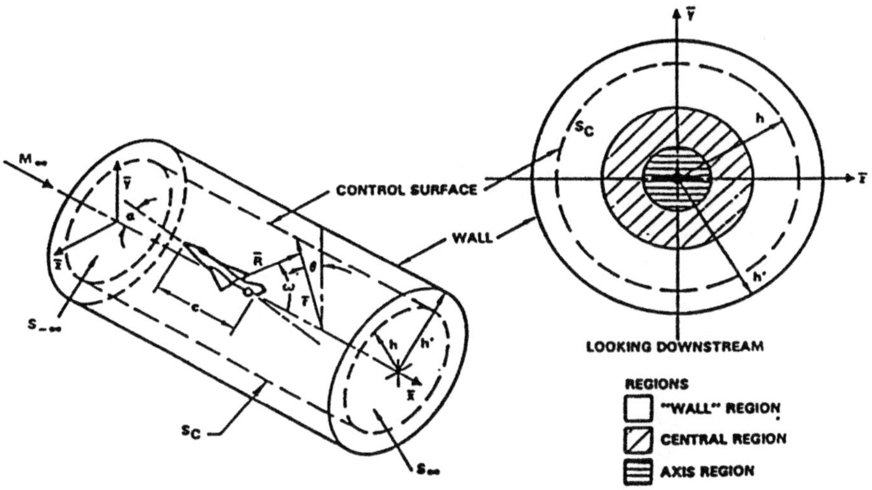

Fig. 9. **Confined slender airplane model**

multipole in the interface for the classical case and give far-field boundary conditions for a linearly perturbed KG problem for the interference. These are generalized in [25] to treat the nonclassical case as well. The boundary conditions as $\tilde{r} \to 0$ are obtained from matching with the inner deck or axis layer. From these considerations, the boundary value problem for the interference potential ϕ_1 representing the wall-induced perturbations on the ϕ_0 free field flow in the central region is as follows:

$$[K_0^* - (\gamma + 1)\phi_{0_x}]\phi_{1_{xx}} - (\gamma + 1)\phi_{0_x}\phi_{1_x} + \frac{1}{\tilde{r}}(\tilde{r}\phi_{1_{\tilde{r}}})_{\tilde{r}} = -K_1^*\phi_{0_{xx}} \qquad (18a)$$

$$\phi_{1_{\tilde{r}}}(x, 0) = 0 \qquad (18b)$$

$$\phi_1 \simeq \alpha_0 R^2 P_2(\cos\omega) + \alpha_1 R\cos\omega + \alpha_2 \quad \text{as } R \to \infty, \qquad (18c)$$

where $P_2(\cos\omega)$ is a Legendre polynomial, $R^2 = \frac{x^2}{K_0^*} + \tilde{r}^2$,

$$\alpha_0 = \mathcal{C}_0 + \mathcal{D}_0 \tag{18d}$$

$$\alpha_1 = B_0\alpha_0 \tag{18e}$$

$$\alpha_2 = \alpha_0 C \tag{18f}$$

$$4\pi B_0 = -S(1) + \int_0^1 S(x)dx + \pi(\gamma+1)\int_{-\infty}^{\infty} dx \int_0^{\infty} r^2\phi_{0_x}^2 d\tilde{r}, \tag{18g}$$

$K_0^* = \frac{1-M_\infty^2}{\delta^2}$ is the free-field similarity parameter, and K_1^* is the perturbation of this parameter that can be adjusted for tunnel Mach number corrections. The constant C is the strength of the quadrupole part of the ϕ_0 far field, and $C_p(x,\theta;H) = f_x$, with $S(x) = \frac{A'(x)}{2\pi}$, where $A(x)$ is the streamwise area progression. The 0th order ϕ_0 free field "base" flow solves a KG problem for which $S(x)$ is the source strength of the equivalent body of revolution (EBR) on the x axis. The quantities $\mathcal{C}_0$ and $\mathcal{D}_0$ in Equation (18d) are given by

$$\mathcal{C}_0 = -\frac{1}{\pi}\int_0^{\infty}\frac{k^2dk}{I_0(k)}\int_0^{\infty} f(X)\cos kX\,dX \tag{18h}$$

$$\mathcal{D}_0 = \frac{-S(1)}{2\pi^2\sqrt{K_0^*}}\int_0^{\infty}\frac{k^2K_0(k)}{I_0(k)}dk \quad . \tag{18i}$$

Here, ω is the scaled analogue of the polar angle defined in Fig. 9; that is, $\omega = \cos^{-1} X/R$, with $X = x/\sqrt{K_0^*}$, and K_0 and I_0 are modified Bessel functions. In the numerics, the third term in Equation (18c) is neglected.

These equations are appropriate to pressure interface (f) data that have axial streamwise symmetry about $x = 0$. More general interface distributions are treated in [25]. For the free=jet case, $\mathcal{C}_0 = 0$ in Equation (18e). Solid-wall conditions are modeled by making $\alpha_0 = b_0' = b_0S(1)/\sqrt{K_0^*}$, and $\alpha_1 = 8\pi B_0 b_0$, with $b_0 = .063409$.

In addition to the boundary conditions given by Eqs. (18b) and (18c), the problem (18) is subject to the appropriate perturbation of the shock conditions. These are satisfied on the 0th-order flow shocks. Since the interference flow ϕ_1 (1st order) boundary value problem is sensitive only to $A(x)$ (area rule for interference), it can be used to rapidly assess interference for slender but realistic three-dimensional configurations such as missiles and fighters, provided that the angles of attack are moderate enough to warrant the application of this theory. References [9] and [5] provide other asymptotic frameworks to handle higher incidences.

Equation (18a) and its high aspect ratio analogue are "variational" equations of the interference flow. They are linear perturbations on the KG base flow with discontinuous coefficients associated with the shocks. Near the shock, they can be modeled by a generalized Lavrentef–Bitsadze equation. The variational equations are of mixed type; accordingly, modifications of the classical successive overrelaxation (SLOR)-type sensitive differencing algorithms originally developed in [27] were used to treat them. The procedures used are generalizations employed by Small [30]. Because the coefficients of the second derivatives of ϕ_1 are frozen, the CPU time is an order of magnitude less than that for a KG solver. In contrast to the KG base flow, the shock is fitted rather than captured, and the fact that the Rankine–Hugoniot jump perturbations can be linearized on the 0th-order captured surface motivates the use of an inscribing "notch" on which the perturbation jumps serve as internal boundary conditions for the interference perturbation potential ϕ_0. Special methods used to couple the jumps along the notch with the SLOR method are described in [25]. For high-aspect ratio shapes, an added technique is required, since there is an additional coupling of the shock notch-induced perturbations with the circulation of the interference flow, Γ_1. This motivated the application of convergence acceleration methods developed by Hafez and Cheng [15]. Another feature of the numerical method involves subtracting the far-field singularity given by Equation (18c). This allows homogeneous Dirichlet conditions to be imposed at the outer boundaries of the computational domain and the dependent variables to be regularized by avoiding their scaling with the size of the domain. Convergence considerations associated with the size of the domain are thereby simplified. This regularization procedure introduces forcing terms in the variational equations.

For the slender-body case, an important accuracy issue is the discretizations of the boundary conditions at $\tilde{r}=0$. Another feature of the numerical methods used in both our KG and variational equation solvers is the use of the axis layer asymptotics in the boundary operators. We believe that this gives enhanced accuracy. This is an excellent example of the *asymptotically induced numerics* approach described in the introduction of this paper. A typical validation for an $H = O(1)$ simulation and our 0th-order code is shown in Fig. 10. The data is for the B–3 body tested by Couch et al. [10] at $M_\infty = .99$. In spite of the potent stagnation singularities for this blunt-nose case, agreement with the data is excellent. Validations of this kind have provided encouraging evidence of the accuracy of the ϕ_0 solver.

In regard to the interference flow, Fig. 11 gives the Mach number dependence of the interference pressure distribution over a parabolic arc body confined by solid walls. For this part of the discussion, the parameter K signifies K_0^* introduced in Equation (18a). Because the shock moves as a result of the change of the Mach number in the tunnel as compared to the free field, there is a spike in these distributions in the vicinity of the free-field shock location. Predictably, the intensity of this blip decreases, while its width increases as the Mach number decreases. The qualitative variation

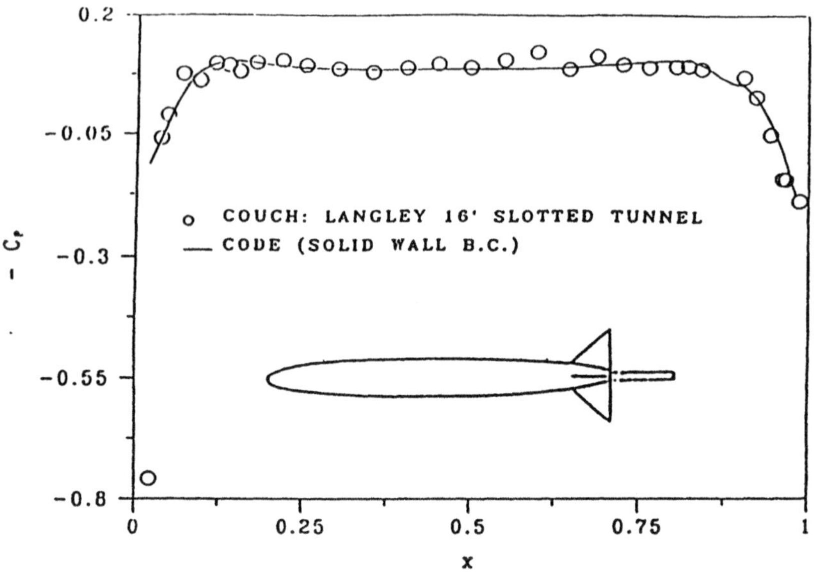

Fig. 10. Comparison of theory with [10]

of the interference pressure away from the location resembles that studied by Malmuth for incompressible flow [21] by analytical methods, where the $H \to \infty$ theory was validated against an exact finite H solution. The interference mechanism is associated with a doublet reflection from the walls, which gives a constant level in the large H limit, since there is an averaging effect as a result of the large distance of the wall from the model.

The associated effect on the total surface pressures is shown in Fig. 12, which indicates the shock movement and the antisymmetry about the critical pressure level. This behavior is in accord with Prandtl's relation for normal shocks, which was assumed to apply at the virtual intersection of the shock and the body. Natural outgrowths of the asymptotic representations that lead to the previously discussed boundary value problems are the similitudes of the interference pressure and drag. Matching of the three decks determines "gauge" functions in the expansions that appropriately order the terms. The ordinates and the abscissas of Fig. 11 and a plot of the interference drag given in Fig. 13 indicate this scaling, which is consistent with that determined by Goethert [13] using nonasymptotic procedures. An interference thrust that increases with Mach number is evident in Fig. 13. Although the walls provide an increase in Mach number and drag at the lower transonic Mach numbers because of their constrictive effect on the free-field streamlines, this tendency is offset at the higher Mach numbers, presumably because of the post-shock compressions on the forward-facing parts of the afterbody. These increase

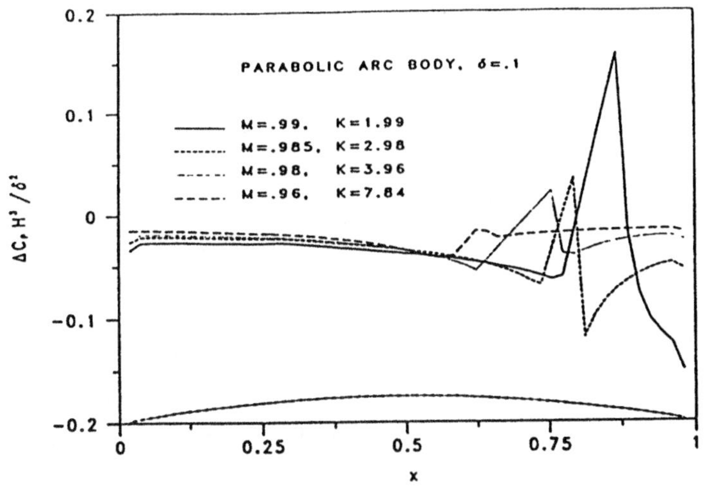

Fig. 11. K **dependence of reduced interference pressures for parabolic arc body**

in intensity and move rearward with increasing M_∞.

5. Web Levitation and Foil Bearings

A common occurrence in industrial processes and lubrication technology is the transport of single and multiple continuous sheets (called webs) through rollers and over other guiding, folding, forming, and cutting elements. The applications are diverse, including film and tape manufacture as well as paper production and tape transport devices used in the computing and entertainment industries (associated with magnetic recording and tape drives). Commercial and defense aspects relate to yet another application involving foil bearings in which the web has flexural rigidity. The configuration of the foil moving over the solid surface is directly analogous to the aforementioned web transport applications in which the web usually has unlimited flexibility. A typical foil-bearing configuration is shown in Fig. 14.

To augment lubrication, foil bearings frequently use *external pressurization* to inject fluid in the region of potential contact between sliding surfaces. Slots, holes, and grooves on the cylinder C shown in Fig. 1 can be used to convect the lubricating fluid in the region to be lubricated. A typical example of gas as the lubricating fluid is the gas bearing. Although there is a rich literature [12, 3, 11, 20, 2, 34, 1, 14, 32, 18] on these applications, theoretical

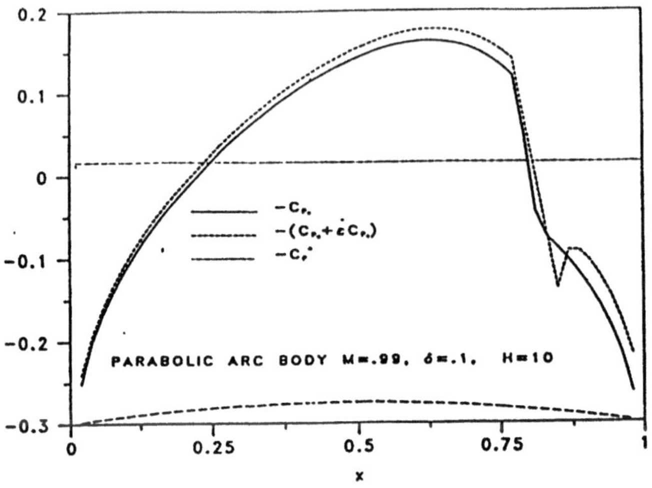

Fig. 12. **Comparison of** 0th**-order and total** C_P, $H = 10+$, **parabolic body;** $M_\infty = .99$, $\delta = .1$, $K = 1.99$

treatment of the underlying fluid mechanics seems to be limited to lubrication theory. Eschel and Elrod [11] treated foil bearings without external pressurization. Singular perturbations and matching can be applied for the externally pressurized case. Our approach applies large-tension approximations for the first time to treat arbitrary variations of pressure and velocity on the slotted cylinder bearing surface using this consistent approximation scheme.

Figure 14 shows the overall physical flow system to be considered. The web in this figure is assumed to be traveling at the velocity U perpendicular to the symmetry axis of the bearing cylinder C. Figure 15 shows a cross section of the web flow over the cylinder C, in which external pressurization is applied to the gap between the web and C from the plenum P through the slot system shown. A blowup of the gap region and an overall view of the major flows involved in this application are shown in Fig. 16. It is important to understand the fluid-dynamical processes associated with gasdynamic lubrication by levitation in this kind of foil bearing. Clearly, diverse-length scales interact. Regions I–IV shown in Fig. 16 are coupled. Here Region I is the thin-gap lubrication zone, and Region II is the outer region in which

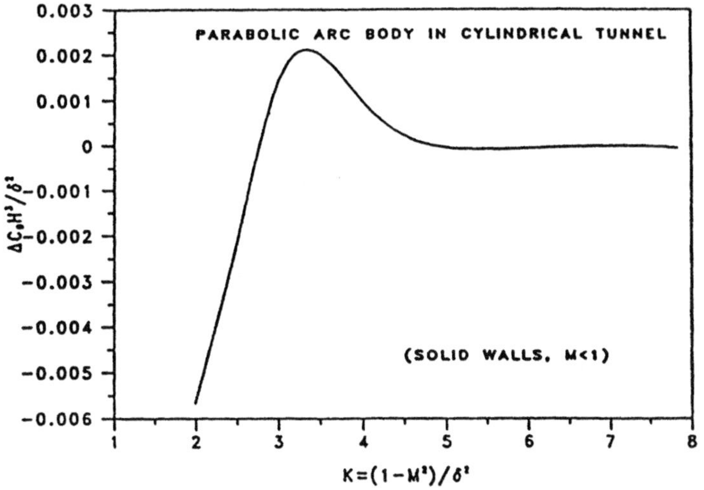

Fig. 13. **Normalized interference drag $\frac{\Delta C_D H^3}{\delta^3}$ as a function of the transonic similarity parameter K**

circulations are induced by the moving web. Other circulations and eddies can be produced in Region IV, and there is a strong interaction between Region I and the plenum (Region III).

As previously indicated, the global flow structure in the narrow-gap Region I in Fig. 16 can be approximated by lubrication theory. If $\mathbf{q} = \overline{q}_r \mathbf{1}_r + \overline{q}_\theta \mathbf{1}_\theta$, asymptotic expansions can be used to define a consistent approximation scheme to solve the elastohydrodynamic interaction of the web with the flow field. If the Re based on an $O(1)$ length equals ϵ^{-1}, then if $\overline{p}$ is the dimensional pressure, $\rho =$ density, and $p = \frac{\overline{p}}{\rho U^2}$, the expansions

$$q_\theta(x, \tilde{y}; \epsilon) = u_0(x, \tilde{y}) + \cdots \tag{19a}$$

$$q_r(x, \tilde{y}; \epsilon) = \epsilon v_0(x, \tilde{y}) + \cdots \tag{19b}$$

$$p(x, \tilde{y}; \epsilon) = \frac{p_0(x, \tilde{y})}{\epsilon} + \cdots \tag{19c}$$

for the distinguished limit

$$x, \tilde{y} = y/\epsilon \text{ fixed as } \epsilon = \frac{1}{Re} \to 0, \tag{19d}$$

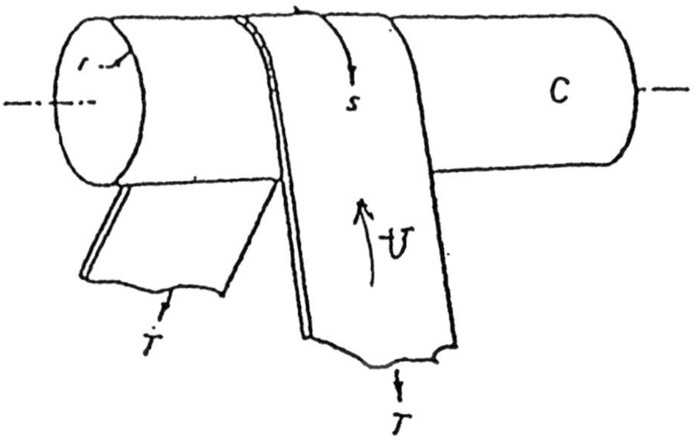

Fig. 14. Schematic of foil bearing

using the coordinate system shown in Fig. 16, when substituted into the incompressible Navier-Stokes equations, lead to the lubrication theory equations. Here, the stretched coordinate $\tilde{y}$ is used to keep the gap region on view in the limit (19d).

Boundary conditions for these equations are related to appropriate matching with the wall boundary conditions on the cylinder surface and the local slot or hole flow fields, as well as dynamical surface traction boundary conditions on the web.

A model has been constructed allowing for arbitrary distributions of normal and tangential velocity components as well as pressure on the ventilated surface. Details of the analysis are given in [23]. Actually, the connection between the slot density and geometry with these distributions needs to be established. This connection is important in determining power requirements and the economics discussed earlier.

Presumably, asymptotic matching can be used to connect the hole or slot geometry and the "inner" slot flow with the outer thin-gap (Region I) lubrication flow. Region I, in turn, could provide outer boundary conditions

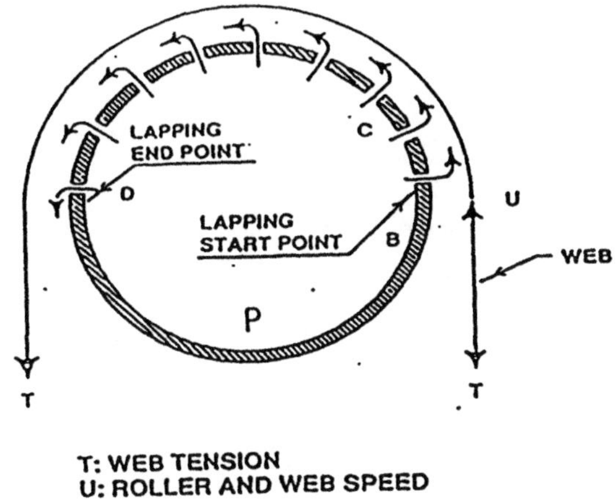

Fig. 15. **Schematic of airbar**

for the local slot flow problem.

The solution of the Region I flow problem discussed in [23] is

$$u_0 = u_{0_w}(x) + p'_{0_w}(x)\frac{\tilde{y}^2}{2} + \left\{\frac{1 - u_{0_w}(x)}{f(x)} - \frac{p'_{0_w}(x)f(x)}{2}\right\}\tilde{y} \tag{20a}$$

$$v_0 = v_{0_w}(x) - u'_{0_w}(x)\tilde{y} - p''_{0_w}\frac{\tilde{y}^3}{6}$$
$$+\left\{\frac{u'_{0_w}(x)}{f(x)} + \frac{p''_{0_w}(x)f(x)}{2} + \left[\frac{1 - u_{0_w}(x)}{f^2(x)} + \frac{p'_{0_w}(x)f'(x)}{2}\right]\right\}\frac{\tilde{y}^2}{2} \tag{20b}$$

$$p_0 = p_{0_w}(x), \tag{20c}$$

where subscript w quantities refer to the arbitrary ventilated cylinder surface distributions mentioned previously. If we denote the wall deflection $\delta(x) = \epsilon f(x)$ in the coordinate system shown in Fig. 16, use of the surface traction and no-slip boundary condition leads to a first-order nonlinear differential equation for $f(x)$, which can be integrated to give the cubic

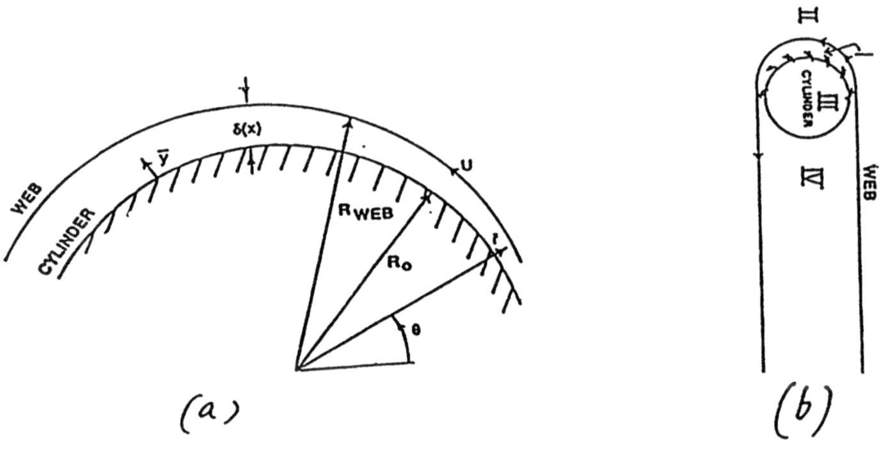

Fig. 16. **Narrow-gap flow Region I in relation to others**

$$Af + Bf^3 - C = 0, \tag{21}$$

where A, B, and C are functions of the wall distribution and a scale height f_0 determined by matching with end layers near the web entry and exit wrap-around points and an integral conservation law. A typical solution of the cubic (21) is plotted in Fig. 17, which shows the interplay between the blowing interface pressures and uniformity of levitation. Variations such as these have demonstrated the profound effect of the ambient decay of pressure on levitation uniformity, which interacts with lubrication effectiveness.

6. Conclusions

Various examples have been discussed that illustrate the power of combined asymptotics and numerics (CAN) for nonlinear problems. Although the transonic equivalence and area rule have been known for many years, their combination with modern computational techniques has been untapped until recently. This combination is quite useful for preliminary design and under-

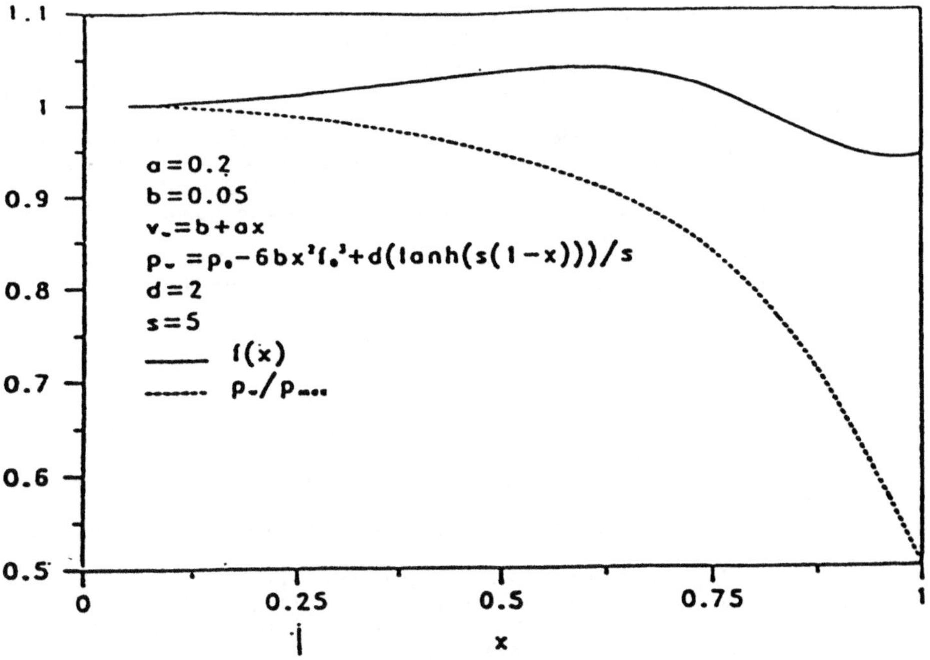

Fig. 17. **Web height variation,** $x = \frac{\theta}{\theta_{MAX}}$, $f_0(0) = 1$, $u_{0_W} = 0$, $f'(0) \neq 0$

standing the basic physics of the flow. Extensions of transonic slender-body theory to CAN can test the applicability of the shock location rule for bodies of revolution to asymmetric bodies. Examples relating to wind-tunnel wall interference and foil bearings provide useful insights and generalizations.

Acknowledgements

A portion of the efforts described here were supported by NASA Contract NASA9–14000 as well as AEDC and AFOSR U.S. Air Force Contracts F40600–84–C–0010 and F46920–92–C–0006DEF, respectively.

This article is printed with the permission of the Society for Industrial and Applied Mathematics from *Some Recent Developments in the Asymptotic Theory of Transonic Aerodynamics*, edited by L. Pamela Cook, copyright 1993 by the Society for Industrial and Applied Mathematics, Philadelphia.

References

[1] Barnum, T. B., and Elrod, H. G., Jr. (1971) 'A theoretical study of the dynamic

behavior of foil bearings', *Trans. ASME J. Lubrication Technology,* January, 133–142.
[2] Barlow, E. J. (1968) 'Derivation of governing equations for self-acting foil bearings', *Trans. ASME J. Lubrication Technology,* January, 334–340.
[3] Block, H., and vanRossum, J. J. (1953) 'The foil bearing—A new departure in hydrodynamic lubrication', *Lubrication Engineering,* December, 316–320.
[4] Cole, J. D. (1968) *Perturbation Methods in Applied Mathematics,* Blaisdell Publishing Company, New York.
[5] Cole, J. D., and Cook, L. P. (1988) 'Some asymptotic problems of transonic flow theory', presented at the Symposium Transsonicum III, Göttingen, West Germany, May.
[6] Cole, J., and Cook, P. (1986) *Transonic Aerodynamics,* North Holland.
[7] Cole, J., and Kevorkian, J. (1980) *Perturbation Methods in Applied Mathematics,* Springer-Verlag, New York.
[8] Cole, J. D., and Malmuth, N. (1989) 'Shock wave location on a slender transonic body of revolution', *Mechanics Research Communications,* **10**, 6, November-December, 353–335.
[9] Cole, J. D., and Malmuth, N. D. (1988) 'Wave drag due to lift of transonic airplanes', Transonic Symposium: Theory, Application and Experiment, NASA CP-3020, 293–308.
[10] Couch, L. M., and Brooks, C. W., Jr. (1973) 'Effect of blockage ratio on drag and pressure distributions for bodies of revolution at transonic speeds', NASA TN D-7331.
[11] Eshel, A., and Elrod, H. G., Jr. (1965) 'The theory of the infinitely wide, perfectly flexible, self-acting foil bearing', *Trans. ASME J. Lubrication Technology,* December, 831–836.
[12] Fuller, D. (1956) *Theory and Practice of Lubrication for Engineers,* John Wiley, New York.
[13] Goethert, B. H. (1961) *Transonic Wind Tunnel Testing,* Pergamon, New York.
[14] Gross, W., et al. (1980) *Fluid Film Lubrication,* John Wiley, New York.
[15] Hafez, M. M., and Cheng, H. K. (1975) 'Convergence acceleration of relaxation solutions for transonic computations', AIAA Paper 75-71.
[16] Jones, R. T. (1956) 'Theory of wing-body drag at supersonic speeds', NACA Rpt. 1318, September.
[17] Krupp, J. A., and Murman, E. M. (1972) 'Computation of transonic flows past lifting airfoils and slender bodies', *AIAA J.,* **10**, no. 7, July, 880–886.
[18] Lacey, C. A., and Talke, F. E. (1990) 'A tightly coupled numerical foil bearing solution', *IEEE Trans. Magnetics,* **26**, 6, November, 3039–3043.
[19] Lagerstrom, P. A. (1988) *Matched Asymptotic Expansions, Ideas and Techniques,* Springer-Verlag, New York.
[20] Licht, L. (1968) 'An experimental study of elastohydrodynamic lubrication of foil bearings', *Trans. ASME J. Lubrication Technology,* July, 199–220.
[21] Malmuth, N. D. (1987) 'An asymptotic theory of wind tunnel wall interference on subsonic slender bodies', *J. Fluid Mechanics,* **177**, 19–35.
[22] Malmuth, N. D., and Cole, J. D. (1984) 'Study of asymptotic theory of transonic wind tunnel interference', Final Report for Period May 30, 1982, through August 30, 1983, Contract No. F40600-82-C-0005, Arnold Engineering Development Center/DOS Report AEDC-TR-84-8, Tullahoma, Tennessee, May.
[23] Malmuth, N., Hobson, D., and Bivins, L. (1990) 'Aerodynamics of web handling', Rockwell Science Center Internal Report SC-071.90, September 19.
[24] Malmuth, N. D., Jafroudi, H., Wu, C., Mclachlan, R., and Cole, J. (1991) 'Asymptotic methods applied to transonic wall interference', AIAA Paper 91-1712, presented at the AIAA 22nd Fluid and Plasma Dynamics Conference, Honolulu, Hawaii, June 22–26.
[25] Malmuth, N. D., Wu, C. C., Jafroudi, H., Mclachlan, R., Cole, J. D., and Sahu, R. (1991) 'Asymptotic theory of wind tunnel wall interference', AEDC-TR-91-24, Final Report for Contract F40600-84-C-0010, December 1991.
[26] McDevitt, J., and Taylor, R. (1958) 'Pressure distributions at transonic speeds for slender bodies having various axial locations of of maximum diameter', NACA Tech-

nical Note TN 4280, July.

[27] Murman, E. M., and Cole, J. D. (1971) 'Calculation of plane transonic flows', *AIAA J.* **9**, January, 114–121.

[28] Oswatitisch, K., and Keune, F. (1955) 'Ein Aquivalensatz fur nichtfangstelle Fluge keiner Spannwelte in schallnaher Strömung', *Zeitschrift für Flugwissenshaften*, **3**, no. 2, 29–46.

[29] Rajagopal, K., Malmuth, N., and Lick, W. (1988) 'Application of transonic slender body theory to various shapes', AIAA Paper 88-0005, presented at the 26th Aerospace Sciences Meeting, Reno, Nevada, January 11–14.

[30] Small, R. D. (1978) 'Studies in transonic flow VI, calculation of a transonic lifting line theory', UCLA Report UCLA-ENG-7836, April.

[31] Van Dyke, M. D. (1975) *Perturbation Methods in Fluid Mechanics*, Parabolic Press, Stanford.

[32] Vogel, S. M., and Groom, J. L. (1974) 'White light interferometry of elastohydrodynamic lubrication of foil bearings', *IBM J. Res. Dev.*, November, 521–528.

[33] Whitcomb, R. T. (1952) 'A study of zero lift design rise characteristics of wing-body combinations near the speed of sound', NACA Report 1273.

[34] Wildmann, M. (1969) 'Foil bearings', *Trans. ASME, J. Lubrication Technology*, January, 37–44.

ASYMPTOTIC-ASSISTED NUMERICS FOR SPECTRAL PERTURBATION IN MHD INSTABILITY

R. B. Paris
Department of Mathematical and Computer Sciences
Dundee Institute of Technology
Dundee DD1 1HG
UK

and

A. D. Wood
School of Mathematical Sciences
Dublin City University
Dublin 9
Ireland

ABSTRACT. Within the setting of a singular perturbation problem arising in MHD instability theory, we discuss how a formal asymptotic analysis may be employed to show why a numerical method is breaking down and to find ways to correct it. We also remark how rigorous large variable asymptotics may be used to verify the position of the eigenvalue loci for large values of the flow parameter.

1. Introduction

The spectral perturbation problem arises from a computational study of the effects of equilibrium plasma flow, fluid viscosity, and a gravitational acceleration on the onset of resistive instabilities in magnetohydrodynamics. Such flows occur in many Tokamak machines, where modes predicted to be stable in the absence of flow are found to be unstable when the flow is switched on. It is a good example of the use of asymptotic techniques to provide insight into the character of the solution, to identify dominant balances, and to locate regions of validity of the numerical method.

Although we are not primarily concerned with the physical modelling aspects, we offer a brief description of the physical background in Section 2 in order to aid the interpretation of the parameter values in Section 5. In Section 3 we reduce the governing partial differential equations to a second-order, singular boundary value problem for an ordinary differential equation

H. G. Kaper and M. Garbey (eds.),
Asymptotic and Numerical Methods for Partial Differential Equations with Critical Parameters 81–90.

on each of the positive and negative half-lines. The problem contains an eigenvalue parameter P, corresponding to the growth rate of the mode, which appears not only linearly in the equation but also in a jump condition for the solution across the origin.

This eigenvalue problem has been solved explicitly in terms of special functions related to the parabolic cylinder function [2] in the special case when the flow parameter R, which will play the part of the perturbation parameter, is zero. The position of the unperturbed eigenvalues is thus known. When R is very large, asymptotic solutions have been given in [1] and [4]. For intermediate values of R, a numerical method must be used. In Section 4 this is found to break down for very small values of R when Re$P < 0$. We use the method of dominant balance to explain why this happens and to estimate a key parameter determining the length of the interval over which the numerical method is applied. This method permits us to bring our plots of the eigenvalue loci $P(R)$ closer to the origin in the complex P-plane.

We conclude with a discussion of the perturbation of the eigenvalue as both the flow parameter R and viscosity parameter N are increased from zero. We present the marginal stability curves and make some physical conclusions in Section 5.

2. Derivation of the Governing Equations

In a separate paper [3] we give the full physical details of the effects of equilibrium flow on the resistive tearing-mode instability. We offer here only a broad outline of the derivation of the eigenvalue problem in the resistive boundary layer. Furth, Killeen, and Rosenbluth [2] in 1963 studied resistive instabilities in a plane current layer with zero-order flow ignored. In that an d subsequent papers the problem was simplified by scaling variables in such a way that resist ive diffusion across the boundary layer occurs on a much longer time scale than the growth of the res istive modes. In the present work, which is based on the papers of Paris and Sy [4] and Bondeson and Persson [1], we consider the effect of a sheared plasma flow in the current layer along the equilibrium magnetic field. and order the variables in such a way that the equilibrium flow significantly modifies the linear growth of the modes. This ordering is relevant to Tokamak plasmas and has the advantage that the effect of flow need be considered only in the resistive boundary layer. The external region of infinite conductivity is not modified.

We start from the set of resistive MHD equations for an incompressible plasma, which we linearize and manipulate into a pair of coupled equations in the velocity and magnetic field. By assuming a complex exponential variation in all perturbed quantities and applying the standard transformation to dimensionless variables of [2], we arrive at a pair of coupled ordinary differential equations of second order in the scaled components of velocity and magnetic field. In the high conductivity limit as the magnetic Reynolds number tends to infinity, a boundary layer appears, centered on the resonant plane. The dynamic and dissipative effects are concentrated in this narrow boundary layer, and we in turn concentrate attention on it. All we require from the outer region of infinite conductivity is the discontinuity in the logarithmic derivative of the magnetic field perturbation across the resonant plane. W We take this as a given quantity Δ' dependent on the global structure of the problem.

In the boundary layer, after various ordering arguments and a further rescaling described in [1, 4], we arrive at the problem

$$(P + iR\theta)\frac{d^2H}{d\theta^2} - \theta^2 H + \frac{G}{F'^2}\frac{H - iR}{P + iR\theta} - N\frac{d^4H}{d\theta^4} = P\theta - \left(\frac{F''}{F'}\right)\Delta \tag{1}$$

$$\Delta' = \Delta \int_{-\infty}^{\infty} (P + \theta H)d\theta, \tag{2}$$

where θ is the stretched variable across the boundary layer, and, following [1], we have written the complex variable $\Delta' - i\pi F''/F'$ in polar form:

$$\Delta' - \frac{i\pi F''}{F'} := \Delta e^{i\chi}, \quad \Delta > 0, \chi \in (-\pi, \pi]. \tag{3}$$

The new, scaled variables P, R, H, G, and N are related to the physical quantities growth rate, flow velocity, velocity perturbation, gravitational forces and viscosity. For the tearing mode we take the gravitational term $G = 0$. The location of the boundary layer is determined by $F = 0$, where F is the component of the equilibrium magnetic field in the direction of the wave vector and F' and F'' denote the numbers $F'(0)$ and $F''(0)$. Equation (2) is the matching condition between the outer region and the resistive boundary layer.

3. The Eigenvalue Problem

We shall denote by L the formal differential operator

$$Lh := \frac{d^2h}{dk^2} + R\frac{d}{dk}(k^2h) - (k^2P + k^4N)h.$$

If we use the Fourier transform $h(k) = \int_{-\infty}^{\infty} H(\theta)e^{-ik\theta}d\theta$, (1) becomes

$$Lh = 2\pi iP\delta'(k) - \left(\frac{F''}{\Delta F'}\right)2\pi\delta(k).$$

When we replace these delta functions by the corresponding jump conditions, we obtain a pair of boundary value problems on $(-\infty, 0)$ and $(0, \infty)$, namely,

$$Lh = 0, \quad -\infty < k < 0 \quad \text{and} \quad 0 < k < \infty \tag{4}$$

$$h(k) \to 0 \quad \text{as} \quad k \to \pm\infty$$

$$h(0+) - h(0-) = 2\pi i P \tag{5}$$

$$h'(0+) - h'(0-) = -2\pi \frac{F''}{\Delta F'}. \tag{6}$$

On applying the Fourier transform to condition (2), we find that

$$\Delta' = \frac{1}{2} i\Delta\{h(0+) + h(0-)\}.$$

Also, from (3) and (6),

$$h'(0\pm) = -i \exp[\pm i\chi].$$

Substitution in (5) yields the relation for the eigenvalue P

$$2\pi P = \frac{h(0-)}{h'(0-)} e^{-i\chi} - \frac{h(0+)}{h'(0+)} e^{i\chi}. \tag{7}$$

If we set $h(k) = y(k)\exp[-\frac{1}{6}k^3 R]$, and observe that under this change of dependent variable the ratios $h'(0)/h(0)$ and $y'(0)/y(0)$ remain equal, we obtain the problems

$$y'' + [kR - k^2 P - k^4(N + \frac{1}{4}R^2)]y = 0, \quad -\infty < k < 0, \quad 0 < k < \infty, \tag{8}$$

$$y(0) = 1, \quad y(\pm\infty) = 0, \tag{9}$$

where the eigenvalues P are the zeros of the function

$$f(P) := 2\pi P + \frac{e^{i\chi}}{y'(0+, P)} - \frac{e^{-i\chi}}{y'(0-, P)}. \tag{10}$$

Notice that the only quantity from the differential equation (8) required in the eigenvalue relation (10) is the slope of the solution $y(k, P)$ on either side of the origin .

4. Numerical and Asymptotic Methods for Perturbation

The number χ is given, through (3), by the global equilibrium structure, but the flow parameter R arising from neutral beam injection and viscosity parameter N ma y be varied. We study the locus of the eigenvalue $P \equiv P(R, N, \chi)$ in $\mathbf{C}$ as R and N increase from zero for values of χ/π between 0 and 1. The paths obtained for $-1 < \chi/\pi < 0$ are simply complex conjugates. We consider first the inviscid limit $N = 0$. When $R = 0$ it is known from [2] that the eigenvalue $P(\chi)$ lies on the nonnegative real axis for

$0 \le \chi/\pi \le 0.5$ and for $0.5 < \chi/\pi \le 1$ we obtain a pair of complex conjugate eigenvalues located on the rays $\arg P = \pm 4\pi/5$. As $R \to \infty$, the location of the eigenvalue is known from the asymptotics of [1] and [4] to be

$$P \sim \left(\frac{Re^{i\chi}}{2\pi}\right)^{\frac{1}{2}} + \frac{3^{-\frac{2}{3}}\Gamma\left(\frac{1}{3}\right)}{2\pi} R^{-\frac{1}{3}} \cos\chi .$$

For intermediate values of R we adopt the following numerical scheme. Fix R, χ and choose $k_\infty > 0$ and a tolerance $\varepsilon > 0$. Set $P = P_0$, our first guess at the eigenvalue. For $n = 0, 1, 2, \cdots$, until $|P_{n+1} - P_n| < \varepsilon$, solve (by finite differences) the differential equation (8) with $P = P_n$ and $y(0) = 1, y(k_\infty) = 0$. Compute $y'(0\pm)$, and solve the eigenvalue relation (10) (by a generalised Newton-Raphson method) to obtain a new estimate $P = P_{n+1}$.

We check the validity of our choice of k_∞ by solving the problem again with the condition $y(2k_\infty) = 0$ and comparing the solutions. For most values of P and R it is sufficient to take $k_\infty = 3$. When $R \ll 1$ and Re $P < 0$ computational difficulties arise, and it is not possible to follow the eigenvalue locus close to the origin. Numerical plots of the eigenfunctions show a domain of growing oscillations on the k axis until algebraic decay eventually sets in. The amplitude of these oscillations can become very large, and they may persist over a considerable length of the k axis. It is clear that, unless we choose k_∞ beyond the interval of oscillation, the condition $y(k_\infty) = 0$ will be unrealistic and will yield a poor estimate of $h'(0)$ for substitution in the eigenvalue relation (10).

The presence and extent of this domain of oscillation may be confirmed by a formal asymptotic analysis. We shall work with the original equation (4) in $h(k)$ with $N = 0$. Recall that h is related to y by $h(k) = y(k)\exp\left(-\frac{1}{6}k^3 R\right)$. When Re $P > 0$, a straightforward application of the Liouville-Green substitution $h(k) = \exp[S(k)]$ gives $h(k) \sim \exp\left[-\frac{1}{3}Rk^3\right]$ as $k \to +\infty$ and $h(k) \sim \frac{1}{k}\exp[Pk/R]$ as $k \to -\infty$. For Re $P < 0$, and $k > 0$, the same method yields exactly the same dominant behavior as $k \to +\infty$. When $k < 0$, we proceed differently. If we assume $h'' \gg k^2 Ph$ as $k \to -\infty$, (4) becomes $h' + Rk^2 h \sim$ constant, which may be integrated to yield

$$h(k) \sim ce^{-\xi(k)} \int_{-\infty}^{\xi(k)} \frac{e^{\xi(k)}}{Rk^2(\xi)}\, d\xi,$$

where c is a constant and $\xi(k) := \frac{1}{3}Rk^3$. Using the standard result

$$\int_{-\infty}^{-X} \frac{e^x}{f(x)} dx \sim \frac{e^{-X}}{f(x)} \quad \text{as} \quad X \to +\infty$$

we obtain $h(k) \sim c/Rk^2$ as $k \to -\infty$.

But for Re $P < 0$, $k < \infty$ and $R \ll 1$, it is clear from the unperturbed equation that the corresponding behavior will be of parabolic cylinder type. Thus we may postulate the existence of a number $k_0 > 0$ such that, for constants A and B,

$$h(k) \sim \frac{A}{\sqrt{k}} \exp\left[\frac{1}{2}\sqrt{P}k^2\right] + \frac{B}{\sqrt{k}} \exp\left[-\frac{1}{2}\sqrt{P}k^2\right], \qquad 0 < -k < k_0,$$

which is oscillatory because Re $P < 0$. Beyond k_0, we recover the previous behavior

$$h(k) \sim \frac{c}{Rk^2}, \qquad -k > k_0.$$

It remains to locate k_0. At $k = k_0$, the second and third terms in (4) must balance, leading to

$$\sqrt{P}Rk_0^3 \approx Pk_0^2 h$$

or

$$k_0 \approx \frac{\sqrt{|P|}}{R}.$$

For small R and P with small negative real part, we therefore choose $k_\infty > k_0$, which leads us to take $k_\infty = 50$, greatly increasing the computation. For very small values of R it was found easier to solve the boundary layer equations (1),(2) by a direct numerical method, rather than applying the Fourier transform, although the latter was always more efficient for reasonably large values of R, when good agreement with the asymptotics was obtained.

5. Conclusions

We observe from Fig. 1 that, for $0 \leq \chi/\pi \leq 0.5$, the grow th rate Re P increases with R and when $0.5 < \chi/\pi < 1$ new eigenvalues emerge from $P = 0$, initially with Re$P < 0$ (stable) but moving over into Re$P > 0$ (unstable) as R increases. This discontinuous behavior of the spectrum is not unexpected, as the differential equation (8) is singularly perturbed for small R. We also see that the stable eigenvalues originally on the ray arg $P = 4\pi/5$ simply move towards and out along the negative real axis with increasing R. The physical interpretati on of Fig. 1 is that shear flow along the field lines destabilizes the tearing mode, with important consequences for neutral injection in Tokamaks.

As soon as we introduce viscosity effects, the spectrum undergoes another qualitative change. The loci emerging from $P = 0$ for $0.5 < \chi/\pi < 1$ are no longer present. These originate instead from the eigenvalues that lie on the ray arg$P = 4\pi/5$ when $R = N = 0$, although this ray is itself distorted as

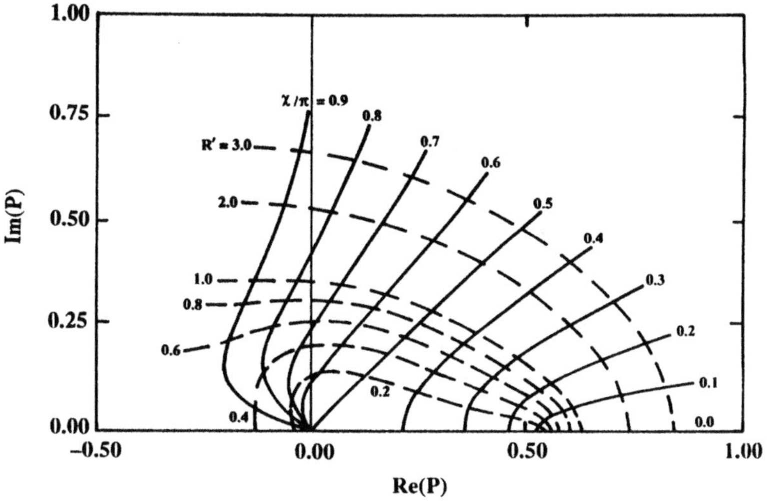

Fig. 1(a)

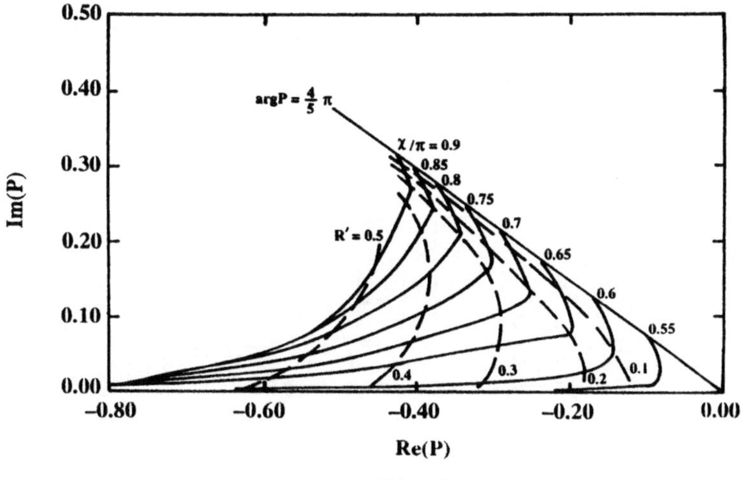

Fig. 1(b)

Fig. 1. **Loci of the eigenvalue P originating on the non-negative real axis as R increases from 0 to 3 for** $0 \leq \chi/\pi \leq 0.9$. (N=0, G=0) are shown in (a), the unstable branch. In (b) we show the stable branch of loci of the eigenvalue P originating on the ray $\arg P = 4\pi/5$ as R increases from 0 to 0.6 for $0.55 \leq \chi/\pi \leq 0.9$ ($N = 0, G = 0$). We indicate the level curves of R by dashed lines.

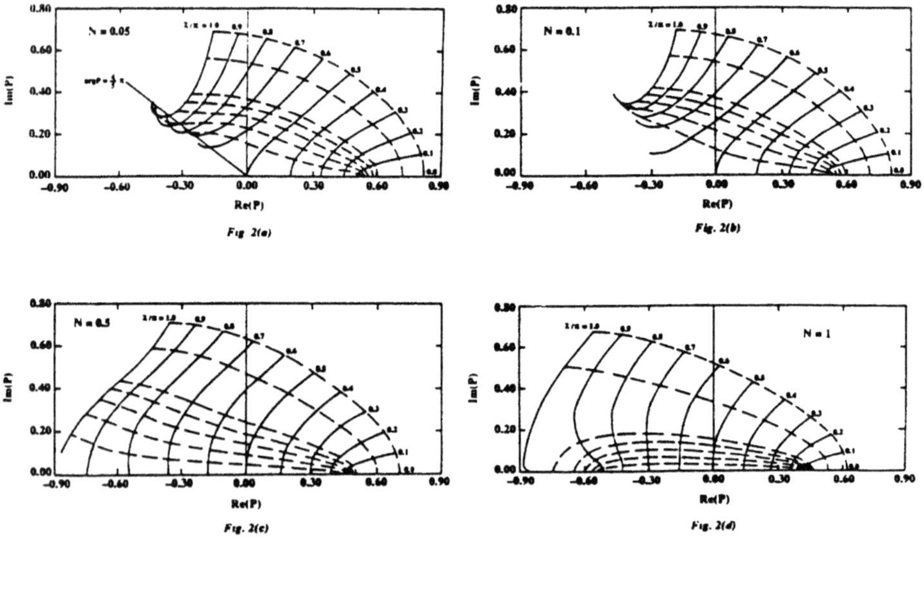

Fig. 2. **Loci of the eigenvalue P as R increases from 0 to 3 for the viscosity parameter $N = 0.05, 0.1, 0.5, 1.0$ with $0 \leq \chi/\pi \leq 1$ $(G = 0)$**

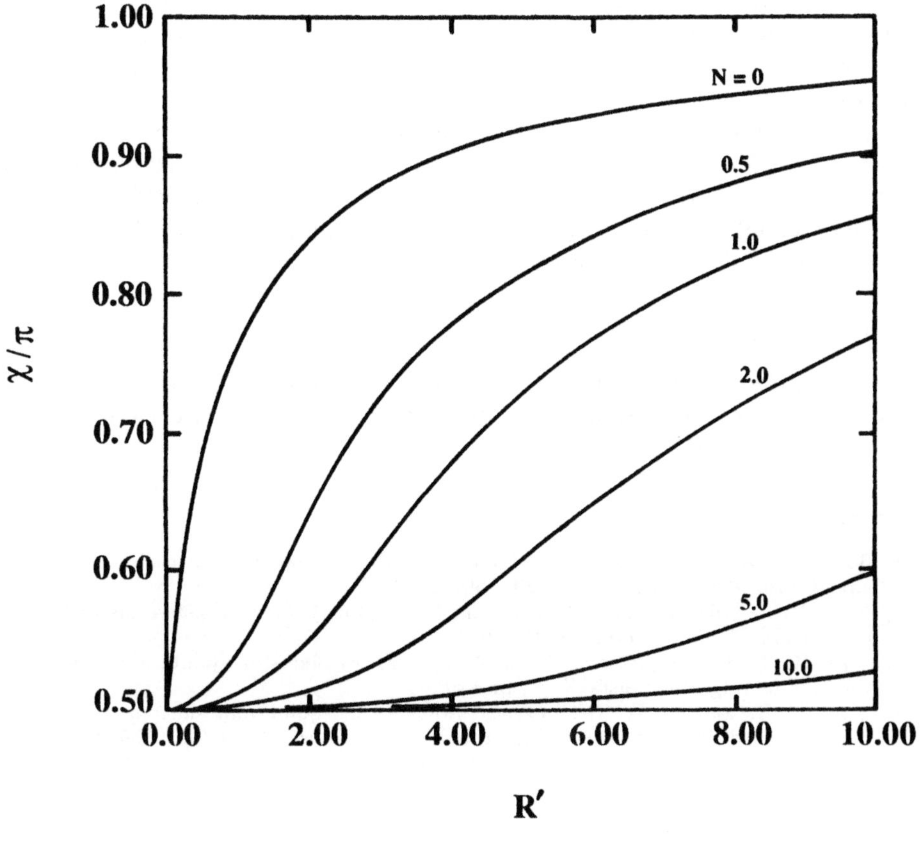

Fig. 3. The marginal stability curves of χ/π against the flow parameter R for different values of the viscosity parameter N

N increases, being pulled down gradually on to the negative real axis. This is shown in Fig. 2, which has the not unexpected physical interpretation that increasing viscosity improves stability, although it remains true for any viscosity that the tearing mode is destabilized by sufficiently increasing flow.

The value of χ for marginal stability ($\mathrm{Re}P = 0$) corresponds to where the new eigenvalue loci cross over from the left-hand to the right-hand P-plane. In Fig. 3 we show the variation of the marginal value of χ against the flow parameter R for different values of N. We observe that in the inviscid limit $N = 0$ the instability condition $|\chi| < \pi/2$ in the absence of flow is replaced by $|\chi| < \pi$ for sufficiently large R. But as N increases, the range of unstable χ for a given value of R is progressively reduced until the condition effectively becomes $|\chi| < \pi/2$ again. This viscous limit corresponds to the neglect of the inertial term k^2P in Equation (4). For small or moderate values of N, this neglect results in an underestimate of the unstable range of χ values caused by the dominance of the inertial term over the viscous term near $k = 0$. Dominant balance arguments similar to those used earlier show that this neighborhood of zero is defined by $|k| \leq (|P|/N)^{1/2}$. Retention of this inertial term in (4) gives the smooth transition shown between the inviscid and viscous dominated limits.

Acknowledgements

We acknowledge the assistance of S. Stewart at Dublin City University with the numerical computation. This work was supported by EC Fusion Contract 401-89-11 FUA IRL.

References

[1] Bondeson, A,. and Persson, M. (1986) 'Resistive tearing modes in the presence of equilibrium flows', *Phys. Fluids*, **29**, 2997–3007.
[2] Furth, H. P., Killeen, J., and Rosenbluth, M. N. (1963) 'Finite-resistive instabilities of a sheet pinch', *Phys. Fluids*, **6**, 459–481.
[3] Paris, R. B., Stewart, S., and Wood, A. D. (1992) 'Effects of equilibrium flow on the resistive tearing mode instability', preprint.
[4] Paris, R. B., and Sy, W. N-C. (1983) 'Influence of equilibrium shear flow along the magnetic field on the resistive tearing instability', *Phys. Fluids*, **26**, 2966–2975.

Part 2: Asymptotic-induced Domain Decomposition Methods

DOMAIN DECOMPOSITION AS A MECHANISM FOR USING ASYMPTOTIC METHODS

William D. Gropp
Mathematics and Computer Science Division
Argonne National Laboratory
Argonne IL 60439

and

David E. Keyes
Department of Mechanical Engineering
Yale University
New Haven CT 06520

ABSTRACT. This paper summarizes some recent advances in numerical analysis for PDEs, particularly those in algebraic domain decomposition techniques, and demonstrates how such methods may be combined with asymptotic methods to provide robust and effective solvers.

1. Introduction

The use of asymptotic methods in numerical schemes is complicated by the fact that many numerical methods are based on a global data structure, while asymptotic methods often address the behavior of a problem in a small area, such as a boundary layer. This paper discusses how some recent numerical techniques provide a good framework for implementing asymptotic-induced numerical methods, and gives an example of a method that combines robust numerics with some relatively simple asymptotics.

We can ask why asymptotic methods and numerical analysis have not mixed well in the past. One answer is that the emphasis in numerical analysis is on methods that can attack a range of problems, while asymptotics is typically used to attack specific problems. Asymptotic methods must also adapt to the features of the solution they are approximating. Some of this is a result of the available tools to express ideas, particularly those from asymptotics, within a computer program. Weak tools tend to lead to weak methods, whether they be weak mathematical or computer tools. This paper

H. G. Kaper and M. Garbey (eds.),
Asymptotic and Numerical Methods for Partial Differential Equations with Critical Parameters 93–105.

reviews some recent developments in computer science and numerical analysis that provide significant expressive power to computational scientists, particularly those using asymptotic methods.

Our approach is to use a basis of approximation functions that depends on the characteristics of the problem. The choice of approximation basis is made on a subdomain-by-subdomain basis, where the subdomains are somewhat arbitrarily specified.

The art of creating a numerical method for the approximate solution of a PDE involves a number of steps. One of the most critical is the choice of approximation. General numerical methods tend to pick a simple family of functions, such as low-order polynomials or trigonometric functions, on some mesh or set of elements. From a convergence standpoint, these choices may seem fine, since it can be proved that in exact arithmetic, as the size of the basis tends to infinity, the approximation error tends to zero. Unfortunately, we can use neither exact arithmetic nor an infinite number of terms. Further, for PDEs with critical parameters, these choices are often poor. Asymptotics may be used to guide the selection of a better set of basis functions and set of elements.

In fact, two extreme positions exist. In one, an enormous number of basis functions is used. For a simple, one-dimensional PDE with a boundary layer of size ϵ, this might require $\mathcal{O}(\epsilon^{-1})$ elements. At the other extreme, a carefully derived asymptotic solution may require $\mathcal{O}(1)$ elements, for example, an inner and an outer solution. This paper advocates a middle ground: use as much asymptotics as is relatively easy to apply, as long as it can improve the useful accuracy or efficiency of the problem. Note that in existing codes, there may be significant constraints on the ability to modify the method; here, asymptotics may help by providing a replacement set of basis functions that is more appropriate for a particular problem.

This approach is related to that of subtracting a known singularity from the problem, but has the advantage that the magnitude of the term being "subtracted" need not be known; the system solves for it. The approach is particularly suitable for problems where much of the computation is in the outer solution, or where the inner solution has a relatively simple structure. Problems that require highly accurate resolution of the inner solution, such as computational combustion, need a different approach.

This choice of approximation basis is combined with a robust numerical method for solving the resulting system of equations. The method used here is an algebraic domain decomposition method; recent progress on these methods is reviewed.

Finally, a simple advection/diffusion problem is solved. This problem was chosen mostly to allow us to compare with some limiting cases. The problem is a good null-test in the sense that any method for this class of problem should be able to solve this one. The partitioning of the algebraic problem can be chosen to correspond to subdomains for the asymptotic method. In turn, the asymptotic method is relied upon only to aid in the construction of a preconditioner for the full discrete problem. Instead of trying to exactly match the inner and outer solutions, we use a method accurate for both solutions in some small part of the domain, and a method accurate for the "smooth" part of the solution in the outer region. In the overall system matrix, all of the terms of the PDE are discretized; hence this method depends upon the asymptotics only for rapid convergence, not for accuracy, and it remains robust even when the "small" parameter is only moderately small.

2. Recent Trends in Scientific Computing

Significant advances have occurred in the technology that is available to solve problems numerically. These include advances in software technology (ways to express algorithms) and algorithm technology (ways to solve problems). In addition, changes in computer technology (mostly hardware) alter some of the decision points in determining which algorithms and approaches are best.

2.1. SOFTWARE TECHNOLOGY

Advances in computer science have made (more) practical the creation of object-oriented libraries that provide efficient methods that are much more closely connected with their mathematical counterparts. A significant example is the application of linear transformations and their inverses. Fast, efficient methods for general, sparse matrices have been created. These permit the simple and efficient use of, for example, general functions in a Galerkin approximation.

Another example is dynamic data structures. These allow general decompositions to be represented easily. For example, general and dynamic meshes can be manipulated relatively easily by using recursive coding techniques. The overhead associated with these grids can be reduced by using regular tree-subdivision techniques (such as quad-trees and oct-trees).

These two advances make it relatively easy to implement a method that uses a range of basis functions and domain decompositions to attack a problem.

2.2. COMPUTER TECHNOLOGY

Years ago, the relative cost of floating-point operations was much greater than practically any other computer operation. This is no longer true. The cost of memory accesses is often far greater than the cost of a floating-point operation (often a factor of ten or more for unordered accesses). However, the amount of memory that is available has grown enormously in the same time. In fact, the tradeoff is between small, fast memory or large, slow memory. These two changes mean that the best way to solve a problem may have changed. One example is that matrix-vector and matrix-matrix operations, because they minimize the amount of memory references needed, can be much faster than vector-vector operations.

2.3. ALGORITHM TECHNOLOGY

Many algorithms now have time and space complexities that are linear or nearly linear in the number of degrees of freedom. These algorithms include multigrid and new Schwarz methods for solving linear systems, the fast multipole method for finding potential fields (central force problems), and wavelets (for some integral equations). These methods require less structure in the problem than the previous generation of methods, such as ADI (alternating direction implicit method), FFTs, and locally one-dimensional methods. These advances require algorithm designers to re-examine various decisions. For example, in time-evolution problems, particularly on general domains, explicit methods have been favored over implicit methods by many because of the far greater cost of solving a linear system. With the newer solution techniques (such as the additive Schwarz method), this cost is much lower and, further, does not grow (per degree of freedom) as the size of the problem grows. If using a large time-step is feasible (for accuracy reasons), then the implicit method may now be a better choice.

3. Framework

This section presents an example of use of the ideas expressed above. Some simple asymptotics will be used to improve the numerical solution of a simple advection-diffusion problem with two different kinds of boundary layers. The underlying numerical method is an algebraic domain decomposition technique; it provides a robust method for solving the linear system that our asymptotics generates.

3.1. ALGEBRAIC DOMAIN DECOMPOSITION

In recent years, there has been a resurgence of interest in algebraic domain decomposition methods, that is, methods for solving systems of linear equations, particularly those arising from discretizations of PDEs, that exploit various kinds of domain decomposition. The proceedings of the SIAM conferences on domain decomposition [2, 3, 4, 6, 7] describe the development of these methods. The methods were designed to be robust techniques for solving elliptic PDEs. Recently, optimal versions of these, many based on variants of Schwarz's original method, have been developed [9, 5]. Optimal here means that it takes an amount of work proportional to the number of degrees of freedom in the discretization to solve the linear system, independent of the number of degrees of freedom. Versions for parallel computers that are as optimal as possible have also been developed. Briefly, these methods construct a preconditioner for an Krylov-space iterative method (such as conjugate gradient or GMRES) consisting of (a) subdomain partitioning with local solvers, (b) some method for interactions between neighboring subdomains, and (c) a global, coarse grid solver (this is the key element needed for optimal convergence of the iterative method).

To make this slightly more concrete, we consider the following problem:

$$Ax = b,$$

solved for x by using a preconditioned iterative method, where A comes from the discretization of a PDE on the unit square. It is well known that it is important to choose preconditioner B such that

$$AB^{-1} \approx I$$

(more precisely, one that is spectrally equivalent to A) and such that $B^{-1}v$ is relatively inexpensive to compute.

Consider the domain in Fig. 1. Let A be partitioned as

$$A = \begin{pmatrix} A_I & A_{IB} & A_{IC} \\ A_{BI} & A_B & A_{BC} \\ A_{CI} & A_{CB} & A_C \end{pmatrix},$$

where

$$\begin{aligned} A_I &= \text{Block diagonal subdomain} \\ A_{IB} &= \text{Coupling to interior from interfaces} \\ A_{IC} &= \text{Coupling to interior from vertices} \\ A_B &= \text{Block diagonal tile interface} \\ A_{BC} &= \text{Coupling to interface from vertices} \\ A_C &= \text{Diagonal vertex term.} \end{aligned}$$

We can construct a preconditioner in block triangular form as

$$B = \begin{pmatrix} \tilde{A}_I & \tilde{A}_{IB} & \tilde{A}_{IC} \\ 0 & \tilde{A}_B & \tilde{A}_{BC} \\ 0 & 0 & \tilde{A}_C \end{pmatrix}.$$

The use of tildes emphasizes that the entries in B are related to those in A but may be different. In particular, the appropriate choice for $\tilde{A}_C$ is often a discretization of the PDE on the mesh of subdomain vertices.

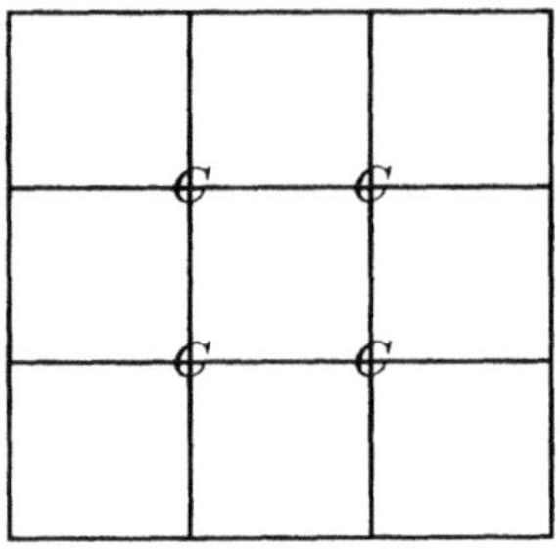

Fig. 1. **Sample decomposition.** C are cross points, and lines are tile-tile boundaries.

This preconditioner involves only one set of subdomain solves per iteration. More sophisticated versions can be written as products of block-triangular matrices. Some of these methods involve the use of overlapping subdomains and bear some resemblance to the classical Schwarz alternating procedure. However, they also possess a coarse-grid term (similar to the $\tilde{A}_C$ here) and can be shown to be optimal for a wide variety of problems.

3.2. EFFICIENCY OF DOMAIN DECOMPOSITION PRECONDITIONERS

As an example of the effectiveness of these methods, we present an example from [1]. The problem

$$\begin{aligned}\mathcal{L}u \;=\; & -((1+\tfrac{1}{2}\sin(50\pi x)u_x)_x \\ & -((1+\tfrac{1}{2}\sin(50\pi x)\sin(50\pi y)u_y)_y \\ & +20\sin(10\pi x)\cos(10\pi y)u_x \\ & -20\cos(10\pi x)\sin(10\pi y)u_y - 70u\end{aligned}$$

is solved on the unit square with Dirichlet boundary conditions. Table I shows the performance of two optimal methods (MSM and ASM) and three incomplete factorization methods (ILU(0), ILU(1), and ILU(2)). Note that classical Schwarz (MSR) fails for this problem.

Note that the number of iterations is small and nearly independent of mesh size for the Schwarz methods. The number of iterations is also nearly independent of the number of subdomains, which makes these methods appropriate for parallel computers.

TABLE I
Iteration counts for solving the variable-coefficient, nonsymmetric indefinite problem

	$H = 1/4$			$H = 1/16$	
$h =$	1/32	1/64	1/128	1/64	1/128
MSR	∞	∞	∞	∞	∞
MSM	15	15	14	10	10
ASM	33	35	35	19	18
	H=1				
ILU(0)	44	78	312		
ILU(1)	28	44	99		
ILU(2)	22	36	76		

3.3. ADDING ASYMPTOTIC INFORMATION

Asymptotic information may be added in several ways, including
- choice of grid,
- choice of approximating functions, and
- choice of problem to model.

We will consider the first two of these in our example. The third may also be handled in the context of domain decomposition, but chosing the different problems on a subdomain by subdomain basis. For example, in a problem involving combustion, on some subdomains where no combustion takes place, but which cannot be omitted from the coupled chemical hydrodynamics model because of the absence of boundary conditions on a more compact domain, the problem being modeled could be restricted to simple fluid flow.

4. A Simple Family of Model Problems

We consider a family of singular perturbation problems parameterized by a small parameter $\epsilon(> 0)$, a Dirichlet-Neumann boundary condition blending parameter $\theta(0 \leq \theta \leq 1)$, and a general interior forcing function f:

$$\left\{ \begin{aligned} -\epsilon(u_{xx} + u_{yy}) + u_y &= \text{f in } \Omega \\ u &= 1 \text{ on } \Gamma_1, \\ u_x &= 0 \text{ on } \Gamma_2, \\ u &= 0 \text{ on } \Gamma_3, \\ \theta u + (1-\theta)u_x &= 0 \text{ on } \Gamma_4, \end{aligned} \right. \tag{1}$$

where $\Omega = [0,1] \times [0,1]$ and the Γ_k, $k = 1, \ldots, 4$ are its bounding segments, enumerated counterclockwise, beginning at the base (see Fig. 2). Physically, u represents a passive scalar (temperature provides a convenient example)

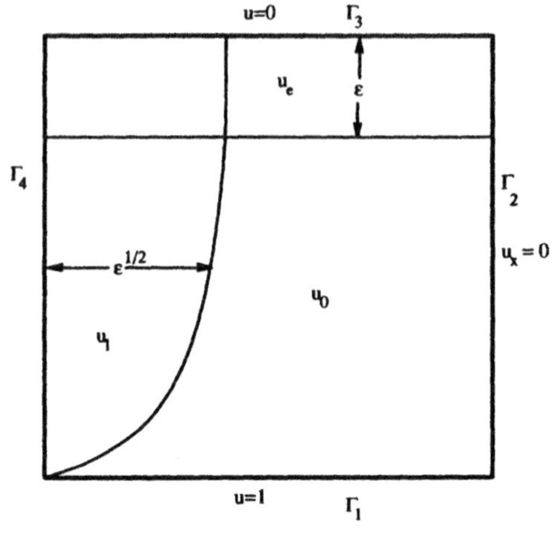

Fig. 2. **The domain Ω and boundary conditions**

that is convected in the positive-y direction with unit velocity while it is diffused isotropically with diffusivity ϵ. A nonzero f represents internal sources, or sinks.

Throughout most of the discussion we shall put $f = 0$ because of linearity, but we point out that our procedure requires only that f be smooth.

As ϵ approaches zero, the differential equation reduces to $u_y = 0$, or simple vertical advection of the upstream boundary condition given along Γ_1, except, possibly, for thin regions of the domain near the boundaries in which the second derivatives of the solution with respect to x or y are $\mathcal{O}(\epsilon^{-p})$ in size, for certain $p > 0$. These regions arise when the boundary conditions are incompatible with the solution to the reduced differential equation. For the system (1) with $f = 0$, the reduced solution, denoted u_0, is simply $u_0(x, y) = 1$. It is clear that a boundary layer must exist at the exit boundary Γ_3, where the condition $u = 0$ imposes a sudden quenching. When the parameter θ is zero, the exit layer is the only boundary layer in the problem. When $\theta > 0$, a layer also appears along Γ_4. Exploiting linearity, we write the solution as the sum of the reduced, lateral, and exit contributions, $u(x, y) \simeq u_0(x, y) + u_l(x, y) + u_e(x, y)$. Asymptotic analysis then yields that

$$u_l(x, y) = -\operatorname{erfc}\left(\frac{x}{2\sqrt{\epsilon y}}\right) + e^{-\beta x + \beta^2 \epsilon y} \operatorname{erfc}\left(\frac{x}{2\sqrt{\epsilon y}} - \beta\sqrt{\epsilon y}\right) \tag{2}$$

and

$$u_e(x, y) = -(1 - u_l(x, y))e^{-(1-y)/\epsilon}, \tag{3}$$

where $\beta \equiv \theta/(1-\theta)$.

It is evident that $u_l(x,y) \to 0$ as $\theta \to 0$; in other words, the problem has no structure at all in x when both of the lateral boundary conditions are Neumann. This provides a convenient exact (one-dimensional) solution for evaluation of the numerics in a limiting case, namely,

$$u(y) = \frac{1 - e^{-(1-y)/\epsilon}}{1 - e^{-1/\epsilon}}. \tag{4}$$

Up to four regions of Ω with distinctly different solution behavior can be identified (see Fig. 2). The bulk of the domain (for small ϵ) is approximately governed by the hyperbolic reduced equation $u_y = f$, which can be integrated along the characteristics. The lateral layer adjacent to Γ_4 has a thickness of $\mathcal{O}(\sqrt{\epsilon})$ and is governed by the parabolic equation $-\epsilon u_{xx} + u_y = 0$. The exit layer adjacent to Γ_2 has a thickness of $\mathcal{O}(\epsilon)$ and is governed by the one-dimensional elliptic problem $-\epsilon u_{yy} + u_y = 0$ Finally, in the region in which the lateral and exit layers intersect, all three terms of the original PDE are important and must be retained. The different forms of the governing equation in different regions can be exploited in preconditioning the overall PDE.

4.1. APPROXIMATION

Our discretization to this problem uses several approaches. The first is mesh stretching (see Fig. 3).

We use the known functional dependencies in the layers to move the mesh points so that more points are in the layers and so that they are well placed in some sense. The second approach is to change the basis of approximation functions. For example, in the top layer, we let u be approximated locally by

$$u(y) = a + be^{(y-1)/\epsilon}$$

instead of the usual second-order polynomial. Thus, the approximation for the second derivative becomes

$$-u_{yy} \approx a_{-1}U(x_i, y_{j-1}) + a_0 U(x_i, y_j) + a_1 U(x_i, y_{j+1}),$$

where

$$\begin{aligned}
a_{-1} &= \frac{1}{\epsilon^2}\left[\frac{1}{\exp(-(y_j - y_{j-1})/\epsilon) - 1 + \frac{y_j - y_{j-1}}{y_{j+1} - y_j}\left(\exp((y_{j+1} - y_j)/\epsilon) - 1\right)}\right] \\
a_1 &= \frac{y_j - y_{j-1}}{y_{j+1} - y_j} a_{-1} \\
a_0 &= -(a_{-1} + a_1).
\end{aligned}$$

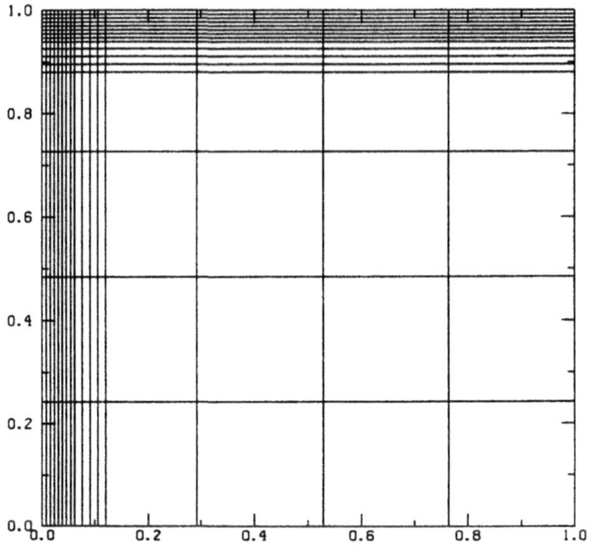

Fig. 3. **Sample mesh** (16×16 **for** $\epsilon = 10^{-3}$)

A similar formula is used for the u_y. Note that as ϵ becomes large, this reduces to the usual centered difference approximation.

Thewith third approach is local mesh refinement. This is used in the upper-left corner where we have no asymptotic solution.

4.2. RESULTS

Using our method, we have found the solution for a number of test cases, of which we present two samples. The cases presented here are for $\epsilon = 10^{-6}$. The first case has an exact solution and demonstrates the advantage of choosing appropriate representation functions. The second case has no exact solution. The linear systems were solved with the algebraic domain decomposition technique described in [8].

In the case $\theta = 0$, the exact solution is given by Equation (4). Using our discretization (stretched functions), we obtain the computed solution to machine precision. For comparison, the error by a conventional finite difference approach is shown in Fig. 4.

In the case $\theta = 1$, there is no exact solution to compare with. We show in Figure 5 the difference between the asymptotic approximation and the computed solution. There are two important features to note here. First,

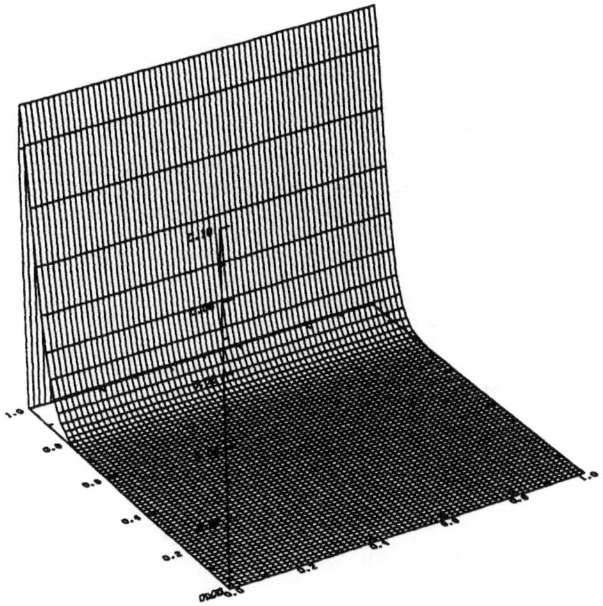

Fig. 4. **Error for finite differences (with mesh refinement)**

the boundary layer "birth" region at the origin is not well represented by the asymptotic solution. This fact is not surprising, since we did not use the asymptotic form given by Equation (2) to handle that case. It is in fact a strong endorsement of using locally modified difference schemes that handle such behavior. The second feature to note is the shallow trough; this may represent an approximation to the second term in the asymptotic expansion (or, more properly, all but the first term).

5. Conclusions

We have discussed how recent developments in scientific computing are creating new opportunities for combining asymptotic and numerical methods. We have given an example for the model convection-diffusion problem that demonstrates one way in which these methods may be combined. It is important to note that asymptotic methods have their limits as well. Using a good numeric method to back up the asymptotic method gives a much more useful and robust method.

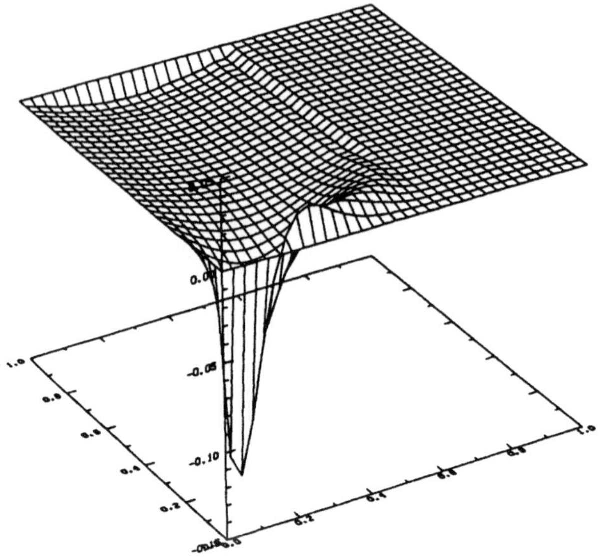

Fig. 5. **Difference between asymptotic approximation and computed solution for $\theta = 1$**

Acknowledgements

This work was supported by the Applied Mathematical Sciences subprogram of the Office of Energy Research, U. S. Department of Energy, under Contract W-31-109-Eng-38.

References

[1] Cai, X.-C., Gropp, W. D., and Keyes, D. E. (1992) 'A comparison of some domain decomposition algorithms for nonsymmetric elliptic problems', in T. F. Chan, D. E. Keyes, G. A. Meurant, J. S. Scroggs, and R. G. Voigt (eds.), *Fifth International Symposium on Domain Decomposition Methods for Partial Differential Equations,* SIAM, Philadelphia, to appear.

[2] Chan, T., Glowinski, R., Périaux, J., and Widlund, O., eds. (1989) *Domain Decomposition Methods,* SIAM, Philadelphia.

[3] Chan, T., Glowinski, R., Périaux, P., and Widlund, O., eds. (1990) *Third International Symposium on Domain Decomposition Methods for Partial Differential Equations,* SIAM, Philadelphia.

[4] Chan, T. F., Keyes, D. E., Meurant, G. A., Scroggs, J. S., and Voigt, R. G., eds. (1992) *Fifth Conference on Domain Decomposition Methods for Partial Differential Equations,* SIAM, Philadelphia, to appear.

[5] Drya, M., and Widlund, O. B. (1991) 'Additive Schwartz methods for elliptic finite element problems in three dimensions', in T. F. Chan, D. E. Keyes, G. A. Meurant, J.

S. Scroggs, and R. G. Voigt (eds.) *Fifth Conference on Domain Decomposition Methods for Partial Differential Equations,* SIAM, Philadelphia, to appear.

[6] Glowinski, R., Golub, G. H., Meurant, G. A., and Périaux, J., eds. (1987) *First International Symposium on Domain Decomposition Methods for Partial Differential Equations,* SIAM, Philadelphia.

[7] Glowinski, R., Kuznetsov, Y. A., Meurant, G. A., Périaux, J., and Widlund, O., eds. (1991) *Fourth International Symposium on Domain Decomposition Methods for Partial Differential Equations,* SIAM, Philadelphia.

[8] Gropp, W. D., and Keyes, D. E. (1992) 'Domain decomposition with local mesh refinement', *SIAM J. Sci. Stat. Comp.,* to appear.

[9] Smith, B. F. (1989) 'An optimal domain decomposition preconditioner for the finite element solution of linear elasticity problems', Tech. Report 482, Courant Institute of Mathematical Sciences, NYU.

DOMAIN DECOMPOSITION: A BLOWUP PROBLEM AND THE GINZBURG–LANDAU EQUATIONS

Man Kam Kwong
Mathematics and Computer Science Division
Argonne National Laboratory
Argonne IL 60439-4801

ABSTRACT. In two of our recent projects, the technique of domain decomposition plays a crucial role in the numerical integration of the partial differential equations. In our problems, the local behavior varies because of the nonlinear and singular nature of the equations. Attempts to introduce different length scales did not lead to improvement. We found that different forms of the equations have to be used over different subdomains. The first project arises in an effort to determine the profile at the moment of blowup or quenching of a nonlinear heat equation. The second project arises in the investigation of the Ginzburg-Landau equation in the theory of superconductivity.

1. Blowup and Quenching

1.1. PROBLEM FORMULATION

Let Ω be a domain in R^n with boundary $\partial\Omega$, and let $p > 1$ and $\lambda > 0$ be constants. We are interested in blowup phenomena for the nonlinear heat problem

$$u_t(x,t) = \Delta u(x,t) + \lambda u^p(x,t), \quad x \in \Omega, t > 0, \tag{1}$$

$$u(x,t) = \alpha, \quad x \in \partial\Omega, t > 0, \tag{2}$$

and the related problem

$$u_t(x,t) = \Delta u(x,t) + \lambda e^{u(x,t)}, \quad x \in \Omega, t > 0, \tag{3}$$

$$u(x,t) = \beta, x \in \Omega, t > 0, \lambda > 0. \tag{4}$$

With a suitable scaling, we can reduce α to 1 in (2) and β to 0 in (4). In an appropriate sense, (3) is the limit form of (1) as $p \to \infty$.

It is well known that if the initial profile $u(x,0)$ and/or λ is sufficiently large, the solutions of the initial-boundary value problems will blow up at

H. G. Kaper and M. Garbey (eds.),
Asymptotic and Numerical Methods for Partial Differential Equations with Critical Parameters 107–122.

some finite time. More specifically, there exists a finite $T > 0$, such that the solution $u(x,t)$ exists for all $0 < t < T$, but

$$\lim_{t \to T} \max \{u(x,t) : x \in \Omega\} = \infty. \tag{5}$$

The phenomenon of blowup has been studied extensively. Criteria on the nonlinear source function and initial profile that lead to blowup are known. Estimates on the blowup time have been derived, and many properties of the set of blowup points have been established. There are so many papers on the subject that it is impossible to cite them all; a substantial number of references can be found in the monograph by Bebernes and Eberly [2]. The survey paper by Friedman and McLeod [13] discusses blowup phenomena for higher-order parabolic equations and hyperbolic equations. Blowup has also been observed in solutions of the nonlinear Schrödinger equation; see, for example, Landman et al. [21].

A related phenomenon is exhibited by the solutions of the equation

$$u_t(x,t) = \Delta u(x,t) + \frac{\lambda}{(1 - u(x,t))^p}, \quad x \in \Omega, t > 0, 0 < p < 1, \tag{6}$$

subject to the same boundary conditions (2). In this case, if the initial profile and/or λ is sufficiently large, then there exists a finite time T such that that solution $u(x,t)$ exists for all $0 < t < T$ but

$$\lim_{t \to T} \max \{u(x,t) : x \in \Omega\} = 1. \tag{7}$$

Here, the solution $u(x,t)$ remains bounded, but a singularity in the derivatives of $u(x,t)$ occurs at T, as the second term on the right-hand side becomes infinite. This phenomenon is called quenching. A survey of the subject together with a useful reference list has been given by Levine [22].

1.2. BEHAVIOR NEAR BLOWUP

There is a great deal of interest in the asymptotic behavior of the solution of (1) and (3) near the blowup time, stimulated by the classical work of Giga and Kohn [15]. Papers by Bressan [5, 6], Filippas and Kohn [11], and Herrero and Velázquez [16] are only a small sample of the more recent work on the subject. It is interesting to note that similar work on the quenching problem is lacking. Below, we shall describe a way to unify the two subjects through a single equation (9). This suggests a means to extend known results on the phenomenon of blowup to cover the phenomenon of quenching.

1.3. THE REGULARIZING TRANSFORM

The motivation for our recent work stems from the search for a reliable numerical method to compute the heat profile near the blowup/quenching time T, as well as the limiting profile at time T, the so-called blowup or quenching profile. Obviously, the unboundedness of $u(x,t)$ or its derivatives renders any regular numerical scheme for solving parabolic equations useless once t is near T. Numerical estimations for the blowup or quenching time have been carried out, usually by solving the parabolic equation until the solution or its derivatives become larger than some arbitrary bound. The only systematic numerical study of the profile in the vicinity of the blowup time that we are aware of is the rescaling algorithm proposed by Berger and Kohn [4], based on the asymptotic results obtained in Giga and Kohn [15]. No similar work has been done on the quenching problem yet.

Recently, Bellout [3] proposed a new approach based on a change of variable, which is inspired by known asymptotic results for the solutions. The method avoids the costly rescaling. Following Bellout, we use the transformation

$$v = u^{1-p} \quad (v = e^{-u}) \tag{8}$$

to change (1) ((3)) to the equation

$$v_t = \Delta v - \lambda - \mu \frac{|\nabla v|^2}{v}, \tag{9}$$

where $\mu = p/(p-1) > 1$ for (1) and $\mu = 1$ for (3). One can easily normalize the equation with a scaling in the space variable to reduce λ to 1. In the one-dimensional experiments we describe below, we retain λ in the equation and, instead, choose the scaling to reduce the domain to the unit interval.

We discovered that by using the change of variable

$$v = (1-u)^{p+1}, \tag{10}$$

(6) was transformed into an equation of the form (9), where $\mu \in (0,1)$. We have thus found a unified approach to treat all three equations (1), (3), and (6).

Blowup/quenching in $u(x,t)$ now corresponds to the vanishing of $v(x,t)$. The new variable $v(x,t)$ remains bounded up to the blowup/quenching time. However, the previous difficulty of having to deal with unbounded functions is replaced by having to deal with the *potential* singularity of the term $|\nabla v|^2/v$ as $v \to 0$. Bellout believes that this new difficulty is only apparent and conjectures that this term remains bounded uniformly up to the blowup time. The idea is that near where v is a minimum, ∇v is also small, so the ratio $|\nabla v|^2/v$ remains bounded. It can be shown that the conjecture is equivalent to the assertion that Δv remains uniformly bounded. If the conjecture is true, then (9) is a regular parabolic equation, and classical methods of integration can be applied with no difficulty up to the blowup time.

In an attempt towards settling the conjecture, Bellout proved in [3] a useful regularity result in the one-dimensional case for (1) with $p > 5$, which implies that v_x is bounded for $p > 5$. Using a refined argument of Bellout's proof, we are able to extend his regularity result to all $p > 1$ as well as to cover the case of quenching. Furthermore, we are able to show that $v_x^2/(v|\ln v|^\gamma)$ is bounded for some $\gamma > 0$. The proofs will be given elsewhere. We have also established a rather long list of assertions that are equivalent to the conjecture. The conjecture is, however, still open.

The partial regularity results are nevertheless important because they imply that the differential equation satisfied by v^θ for some suitable power $\theta > 1$ has regular coefficients. Thus, v^θ can be solved numerically as a regular parabolic problem with classical algorithms, and the challenge to determine the blowup profile can essentially be considered answered. In practice, however, it is much more convenient to solve the differential equation for v than the differential equation for v^θ. It is also an intellectual challenge to be able to resolve the conjecture.

1.4. NUMERICAL SIMULATION

We have performed many numerical experiments, and the results always supported the conjecture that the term $|\nabla v|^2/v$ is uniformly bounded up to the blowup time. Most of the experiments were done using MATLAB, version 4.0. An interesting problem arises when μ is increased beyond 4. That is when domain decomposition comes in.

We have treated both one-dimensional and two-dimensional problems. For the sake of simplicity, we confine our discussion below to the one-dimensional case, with $\Omega = (0,1)$. Instead of dealing with Equation (9) directly, we solve for $U(x,t) = 1 - v(x,t)$. This means that we have to deal only with the simpler homogeneous boundary conditions $U(0,t) = U(1,t) = 0$. Furthermore, the evolution of $U(x,t)$ now resembles more that of the original variable $u(x,t)$; $U(x,t)$ increases towards 1 as $u(x,t)$ blows up, whereas $v(x,t)$ *decreases*.

We use the method of lines, discretizing only the space variable x, to obtain a matrix ordinary differential equation in t,

$$\frac{dU}{dt} = LU + \lambda + \mu\frac{U_x^2}{1-U}, \tag{11}$$

where U is a column vector representing the values of $U(x,t)$ at the chosen node points, L is the usual finite-difference tridiagonal matrix representing the second derivative, and U_x is some suitable finite-difference approximation of the first derivative. The fraction in the last term is to be understood in the componentwise sense. The experiments show that U develops some sort of a boundary layer in which U_x assumes very large values. To improve accuracy, we use a five-point numerical differentiation formula to approximate U_x.

In our initial experiments, we used a uniform grid partitioning $[0,1]$ into 49 or 99 grid points. We integrate the system of Equations (11) over time intervals of size $k = t_{n+1} - t_n$. As is well known, the system is stiff and requires special techniques such as the Crank-Nicholson and other implicit methods. In our situation, we exploit the fact that the coefficient matrix L is a constant. Over each time interval $[t_n, t_{n+1}]$, we approximate the nonhomogeneous part of (11), $\lambda + \mu U_x^2/(1-U)$, by a linear function in t, namely, $A + Bt$, where A and B are vectors given by

$$A = \lambda + \mu \left. \frac{U_x^2}{1-U} \right|_{t=t_n}, \qquad A + Bk = \lambda + \mu \left. \frac{U_x^2}{1-U} \right|_{t=t_{n+1}}. \tag{12}$$

The solution $U(t_{n+1})$ at the new time step is estimated by solving the approximated form of (11) exactly:

$$U(t_{n+1}) = U(t_n) + K_1(LU(t_n) + A) + K_2 B, \tag{13}$$

where K_1 and K_2 are constant matrices given by

$$K_1 = L^{-1}(L^{Lk} - I), \tag{14}$$

$$K_2 = L^{-2}(e^{Lk} - I - Lk), \tag{15}$$

where I denotes the identity matrix. This formula, however, cannot be used directly since only A, not B, is determinable from $U(t_n)$ by using (12). We use the usual predictor-corrector technique: we take an initial guess $B = 0$ and use (13) to predict the value $U(t_{n+1})$, which is then substituted into the second formula in (12) to give a corrected value for B. This step may be iterated to give successively better estimates of B, but in practice we find that one cycle is usually sufficient. Note that the matrices K_1 and K_2 can be computed easily once the spectral decomposition of L is known. Moreover, they need to be computed only once for each choice of k.

Our method of integration is unconditionally stable for any value of k, and thus the choice of k is restricted only by the error introduced in the approximation of the nonhomogeneous part of the equation. The criterion we used to determine a good choice of k is to compare the difference in the results obtained by using k and $2k$, respectively. A smaller k is usually chosen to repeat the last few time steps before the blowup time if higher accuracy is desired near the final moment.

Our numerical scheme works efficiently and accurately for a wide range of the parameter λ and for $\mu < 3$. For illustration we include the graphs from three typical runs. We used 99 grid points and the time step $k = 10^{-4}$.

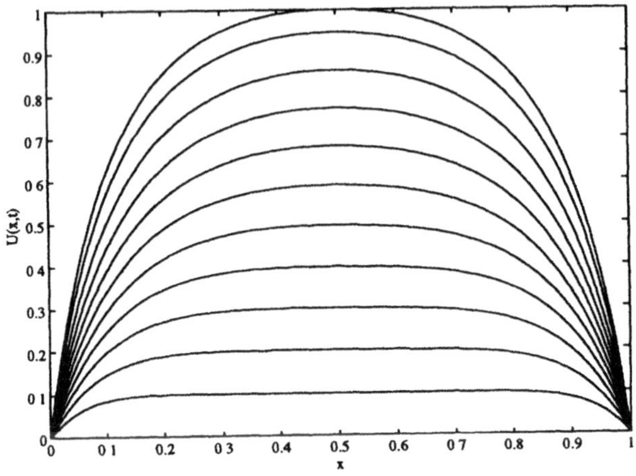

Fig. 1. $\lambda = 20, \mu = 1, u(x,0) \equiv 0$

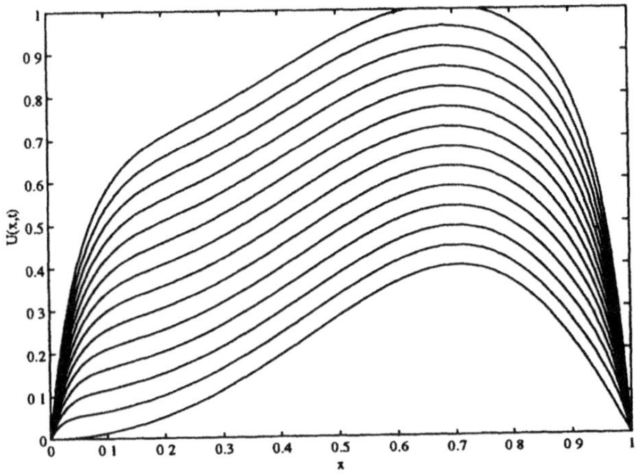

Fig. 2. $\lambda = 100, \mu = 1, u(x,0) \equiv 0.4\sin(\pi x^2)$

The figures show the profiles $U(x, t_m)$ at selected time t_m that are uniformly apart except for the last one which is some moment before blowup. Figure 1 traces the evolution from a trivial initial profile. Figure 2 is an example of a nonsymmetric initial profile. Figure 3 is an example of a solution blowing up at more than one point.

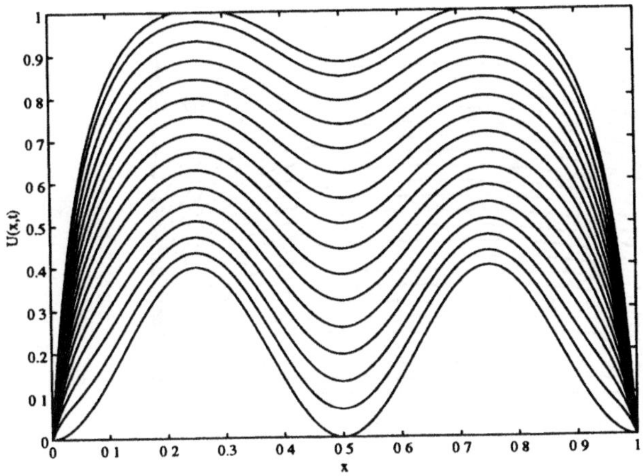

Fig. 3. $\lambda = 100, \mu = 1, u(x,0) \equiv 0.4\sin^2(2\pi x)$

The method of rescaling is a never-ending task; a new rescaling is needed once the rescaled solution grows sufficiently large. Our method, on the other hand, allows one to accurately compute the blowup time by extrapolating $\max\{U(x,t_n), x \in (0,1)\}$ for various t_n, to find out when it will reach $U = 1$. The blowup profile can then be accurately computed by integrating (11) using $T - t_n$ as the final time step size.

1.5. DOMAIN DECOMPOSITION TECHNIQUE FOR LARGE μ

The numerical method described in Section 1.4 fails when μ is large and the technique of domain decomposition has been used to overcome the difficulty.

When the parameter μ is increased, the profile becomes flatter near the center of the interval. This produces, moments before blowup, two "boundary layers" near the endpoints, in which the solution shoots up sharply from 0 to almost 1 within a short distance. For $\mu = 4$, near the blowup time, $U(x,t)$ developed two "horns" as shown in Fig. 4. In the next time step, the solution blew up at two points. The same phenomenon prevails for larger values of μ. If one had blindly trusted computers, one could have declared the discovery of a new kind of blowup behavior. However, it has been rigorously proved (see, for example, Friedman and McLeod [13]) that if the initial profile is symmetrical and has only one local maximum, at the midpoint of the interval, then all subsequent profiles behave in the same way. In particular, it is impossible for two "horns" to develop and the solution can blow up only at a single point, namely, the midpoint of the interval.

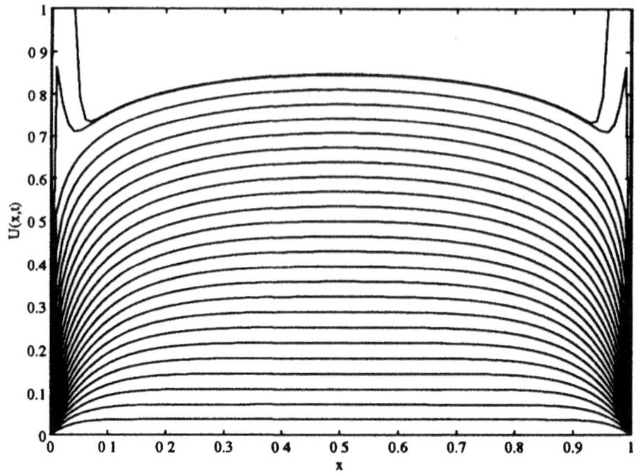

Fig. 4. **Instability sets in:** $\lambda = 100, \mu = 5, u(x,0) \equiv 0$

Our first reaction was to repeat the experiments with a significant reduction in the time step. We soon found that the plan did not work. Our next attempt was to use a higher-order approximation for the nonhomogeneous term. Not only did that lead to a much more complicated computation, but it did nothing to prevent the formation of the "horns." Our next strategy was to try to scale down the boundary layer by multiplying $U(x,t)$ by a factor such as $x(1-x)$, hoping that the final profiles would take less of the shape of a plateau. The transformed differential equation is complicated, and that discouraged us from attempting to modify the code to handle the new variable.

For a while, multiple-length scale seemed to be a reasonable explanation of the difficulty. The rapid increase of $U(x,t)$ within a short distance must mean that we need many more grid points within the boundary layer in order to represent the variation adequately. We restructured our programs to handle nonuniform grids. The first attempt was to have three subintervals, the two boundary layers and the middle section, each having a uniform subgrid. In one example we used 40 grid points between 0 and 0.1, 40 grid points between 0.9 and 1, but only 20 between 0.1 and 0.9. The results were encouraging. The appearance of the "horns" was delayed by a few time steps, and their locations were pushed outward towards the endpoints. Yet all attempts to completely eliminate the "horns" failed. We then overhauled the programs to deal with very general nonuniform grids. We used grids that became progressively finer towards the endpoints: we started with 20 uniformly spaced grid points in $[0, 1]$, subdivided the first and last intervals into 10 subintervals, and repeatedly subdivided the resulting first and last intervals. Our program includes a subroutine to compute the coefficients for a nonuniform four-point finite difference formula. To our dismay, no matter how finely we subdivide the first and last intervals, "horns" always occur at the second or third grid point next to the endpoints.

We are puzzled by the persistence of the instability. We believe that the explanation lies in the fact that the term v_x^2/v is of the same order of magnitude as the leading term v_{xx}. The stability of the numerical scheme is upset by the presence of the nonlinear term in the same way that a large time step in the Euler method for solving parabolic equations can lead to instability. It would be interesting to see a rigorous proof.

We finally succeeded in circumventing the instability problem by using the technique of domain decomposition. We discovered that within the boundary layers the original form of the equation, namely, (1), could be solved without any instability problem. Theory guarantees that if we stay away from the blowup points (there are only a finite number of them), the original solution $u(x,t)$ remains uniformly bounded up to T, even though it may be large.

Our algorithm is as follows. Identify two suitable subdomains that contain the boundary layers. In our experiments, we used [0, 0.1] and [0.9,1]. Based on the value of U at t_n, compute the solution in [0.1, 0.9] at the new time step t_{n+1} by integrating (11) over the entire domain [0, 1] using the method described above and discarding the part in the boundary layers. Solve the original equation (1) over each of the boundary layers, requiring that the solution be continuous across the common boundary point of the subdomains. The programs have been tested for values of μ as high as 20.

In theory, the requirement for the matching of the solution at a single common boundary point between two subintervals is not sufficient to guarantee a smooth matching, for there is the possibility of a jump in the first derivative. The proper algorithm should use an iterative scheme to match the solution in the common boundary points of overlapping subintervals. In all our experiments, however, the simple matching procedure suffices.

2. The Ginzburg-Landau Equations

2.1. PROBLEM FORMULATION

Our second example concerns the Ginzburg-Landau equations in the theory of superconductivity. Part of the work reported here is done in collaboration with H. G. Kaper. More details will appear in a forthcoming joint paper [18]. For the physical background, we refer to the many excellent references cited in our earlier work [18, 19, 20] and also in the papers by Du et al. [8, 9].

We are interested in finding a complex scalar ψ and a three-dimension real-valued vector $\mathbf{A}$, both functions of position in a given domain $\Omega \in R^3$, that minimize the Helmholtz free-energy functional:

$$G(\psi, \mathbf{A}) = \int_\Omega \left(-|\psi|^2 + \frac{1}{2}|\psi|^4 + |(\nabla - i\mathbf{A})\psi|^2 + \kappa^2 \, |\nabla \times \mathbf{A}|^2 \right) d\Omega, \quad (16)$$

where $i = \sqrt{-1}$ and κ is a characteristic constant of the superconducting material. The Euler-Lagrange equations for this variational problem are called the Ginzburg-Landau equations.

$$(\nabla - i\mathbf{A})^2\psi + \left(1 - |\psi|^2\right)\psi = 0, \quad (\kappa^2\nabla \times -|\psi|^2)\mathbf{A} + \mathrm{Im}(\psi^*\nabla\psi) = 0.(17)$$

Of particular interest is the quantity $|\psi|$ which represents the amount of superconducting activity going on at the pertinent location. For type-II superconductors, characterized by the fact that $\kappa > 1/\sqrt{2}$, Abrikosov [1] showed analytically, by using singular perturbation techniques, the existence of solutions that exhibited a lattice pattern.

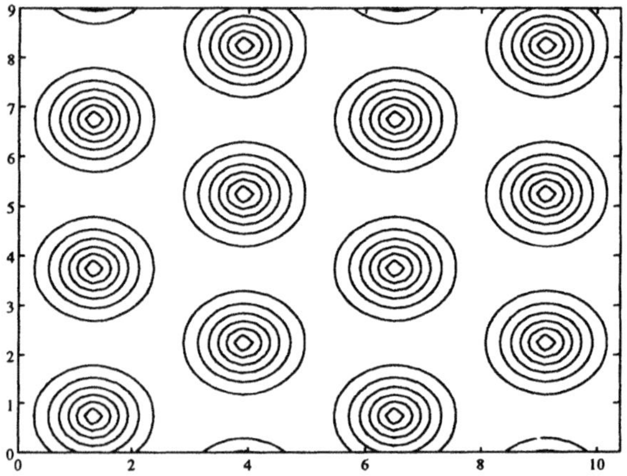

Fig. 5. **Vortex lattice**

Figure 5 shows the contour plot of $|\psi|$ for one such solution in a two-dimensional domain that models a thin-film superconductor as a normal magnetic field. The plot resembles vortices, and indeed superconducting currents are flowing along the contour lines in circular paths.

The first attempt to solve the Ginzburg-Landau equations numerically was by Doria, Gubernatis, and Rainer [10], who used the method of simulated annealing on a discrete version of the Helmholltz energy. Recently, our colleagues at Argonne applied more sophisticated optimization and linear algebra techniques to tackle three-dimensional versions of the Ginzburg-Landau equations that model layered superconductors [14]. Du et al. [8, 9] used finite element approaches in their work.

In our recent work [17, 19, 20] we proposed a new method, the sweeping algorithm, to solve the five-point stencil that results from discretizing the two-dimensional Ginzburg-Landau equations, thus leading to an alternative numerical method for solving the Ginzburg-Landau problem. The sweeping algorithm is of interest in itself, as it can be used to invert banded matrices and general five-point stencils arising from other partial differential equations. The method offers great potential in the exploitation of parallelism and the technique of domain decomposition.

In our numerical study of the Ginzburg-Landau model, the questions we have investigated include the effect of the aspect ratio of the rectangular region Ω on the free energy, the relation between the free energy and the average magnetic field, and the simulation of vortex pinning by impurities.

2.2. THE GAUGE TRANSFORM

The boundary conditions imposed on the problem are not the usual periodic type. In fact, the vector potential cannot be periodic if the external magnetic field is nonzero. The situation is further complicated by the fact that the Helmholtz energy functional, and hence also the Ginzburg-Landau equations, are invariant under a class of transformations called gauge transforms. More precisely, given ψ and $\mathbf{A}$, one can pick any real-valued function of position $\chi(x)$, called a gauge, and form a new pair

$$\overline{\psi} = \psi e^{i\chi(x)}, \quad \overline{\mathbf{A}} = \mathbf{A} + \nabla\chi(x). \tag{18}$$

One can easily verify that the new pair gives the same Helmholtz energy as the original pair. If the original pair is a solution of the Ginzburg-Landau equations, so is the new pair. Other physical quantities derivable from each pair, such as current and magnetic field, are also the same. The two pairs indeed are different representations of the same physical state. Periodicity in physical states demands only that the functions on opposite sides of the domain differ by a gauge. This leads to the so-called modified boundary conditions.

We confine our discussions to two-dimensional simulations. In this case, one can arrange to have strict periodicity in one direction and modified periodicity in the other. One can also make the simplification that χ is a linear function of the coordinate along the appropriate edge. For instance, for a rectangular domain, $[0, L_x] \times [0, L_y]$ that contains n vortices, the boundary conditions takes the form

$$\psi(0, y) = \psi(L_x, y), \quad \psi(x, 0) = \psi(x, L_y)e^{igx}, \tag{19}$$

$$A_1(0,y) = A_1(L_x,y), \quad A_1(x,0) = A_1(x,L_y) + g, \tag{20}$$

$$A_2(0,y) = A_2(L_x,y), \quad A_2(x,0) = A_2(x,L_y), \tag{21}$$

where $g = 2n\pi/L_x$, and A_1, A_2 are the two components of $\mathbf{A}$.

2.3. DOMAIN DECOMPOSITION TECHNIQUE

The value of g in the boundary conditions (20) and (21) depends on the number of vortices inside the domain, and is related to the total magnetic flux through the region. For a region that contains many vortices, g can be very large. This poses the following problem. Suppose that $\psi(x,0)$ is a smooth, slowly varying, but nontrivial function. Its counterpart on the opposite edge must be a wildly oscillating function, as a result of the phase factor e^{igx}. Likewise, if ψ is slowly varying on the right-hand edge, its value on the left-hand edge must oscillate wildly. It is impossible to find a gauge transform that gives slowly oscillating boundary values on both edges.

We illustrate this situation with the plots of the real and imaginary parts of ψ for a region that contains two of the vortices depicted in Fig. 5. The numerical simulation was performed using $\kappa = 5$. For the region shown, $L_x = 3\sqrt{3}$ and $L_y = 3$, and the number of grid points used is 24×24.

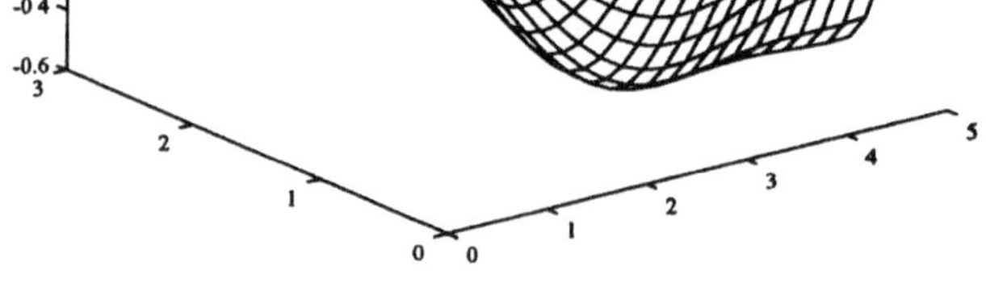

Fig. 6. **Real part of ψ**

As shown in the plots in Figs. 6 and 7, both the real and imaginary parts of ψ oscillate more and more as one approaches the right-hand boundary. For a larger rectangular region that involves more vortices, ψ will oscillate even more as we move to the right. For every horizontal distance equal to

$3\sqrt{3}$, the number of oscillations of ψ increases by 2, so that at the right-hand boundary, ψ has 20 maxima and 20 minima. It is no longer possible to approximate ψ adequately using only 24 grid points in the y direction.

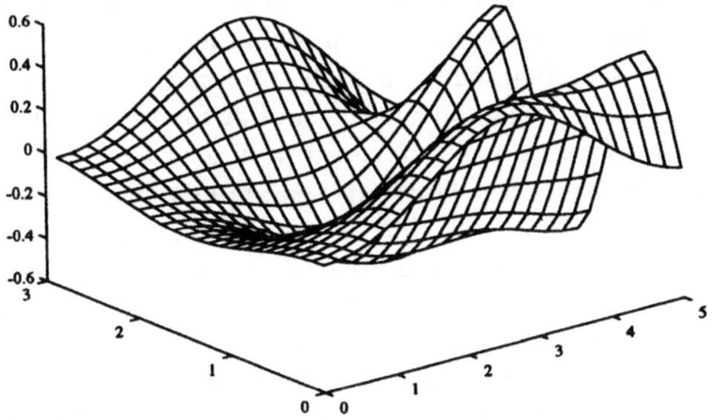

Fig. 7. **Imaginary part of ψ**

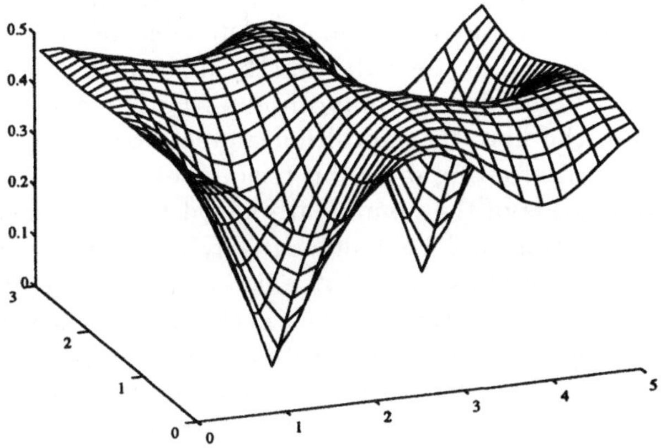

Fig. 8. **Absolute value of ψ**

Even though ψ oscillates wildly, the physical quantity we are interested in is represented by $|\psi|$, a plot of which is shown is Fig. 8. Unlike the real and imaginary parts of ψ, the oscillation of $|\psi|$ remains steady as one moves along the x direction. The plot shown in Fig. 8 is simply duplicated when a larger rectangular region is used.

The method we used to overcome the difficulty posed by the oscillation of ψ is to divide the domain into smaller subdomains. The Helmholtz energy is iteratively minimized over each subdomain while holding ψ and $\mathbf{A}$ fixed in the other subdomains (this process can be performed in parallel). The idea of partial minimization is simply a block Gauss-Seidel procedure. The important new idea is that we are not solving directly for ψ and $\mathbf{A}$ in each subdomain, but rather for an equivalent pair $\overline{\psi}$ and $\overline{\mathbf{A}}$ derived from a suitable gauge that can smooth out the oscillations in ψ on both edges of the subdomain (this approach is now possible because the subdomain is smaller). The gauge is associated with each subdomain and may vary from one subdomain to the next. The original ψ and $\mathbf{A}$ for the entire domain can be recovered by an inverse gauge transformation over each subdomain.

After each iterative cycle, the Ginzburg-Landau equations are satisfied in the interior of all the subdomains, but not on the common boundaries. At times the errors at these boundary points can be alarmingly huge, but one must remember that a better measure of the error is some suitable average error over the entire domain. The errors on the common boundaries decrease with the iterations. A usual technique in the method of domain decomposition to speed the improvement is to find an appropriate linking operator that can give a correction term to the solution found in a subdomain based on the solution value computed in the surrounding subdomains. Instead, we use a new approach involving two overlapping decompositions.

As an example, we can use, as the first decomposition, N vertical strips of uniform width cut off starting from the left edge of the domain. The second decomposition makes the cuts exactly midway between the edges of the subdomains in the first decomposition. Another example is to use N vertical strips in the first decomposition and N horizontal strips in the second (the familiar alternating direction technique). The two "half-strips" at the left and right ends of the domain are joined to form one subdomain.

The two decompositions are used alternatively in successive cycles of the partial minimization procedure. The rationale behind the proposed method is that the errors incurred in the previous cycle are now in the interior of the subdomains of the current decomposition, far away from the common boundaries, and so will have less effect in disturbing the solution values at these points. Numerical results show that the use of overlapping decompositions greatly improves the rate of convergence over the use of a fixed decomposition (without the help of a linking operator).

3. Conclusion

Domain decomposition is an excellent device to introduce and exploit parallelism in designing numerical algorithms for solving partial differential equations. It has also been realized that domain decomposition is sometimes called for because the problem has rich local structures; see, for example, the remarks by Chin and Hedstrom in [7]. Problems involving multiple scales belong to this category.

We have given in this article two new examples in which the technique of domain decomposition forces its way into the numerical treatment of the problems. Not only does domain decomposition provide an efficient means of computation, but it appears to be the only way to resolve the difficulties we encountered in the investigation.

Traditionally, domain decomposition methods are well suited for elliptic problems. Our first example involves a semilinear parabolic problem; the second example has application in the theory of superconductivity. In both examples, the main cause of the difficulties is not in the difference in length scales but in the singular and nonlinear nature of the underlying equations. Another characteristic of our examples is that different forms of the equations must be solved in different subdomains in order to avoid instability or singularity of the solutions.

Using sets of overlapping domain decompositions in alternative iterative cycles can effectively speed up convergence. Whether further improvements can be achieved by incorporating other traditional techniques such as conjugate gradient will the the subject of future investigations.

Although asymptotics does not seem to have played an explicit role in our discussion, it is instrumental in guiding us to choose the correct transformation needed in our first example and in the analytical work that Abrikosov did in [1]. Without the asymptotic results, no one would have ever guessed the lattice structure of vortices and the modified periodic form of the boundary conditions.

Future work will include three-dimensional domains, more general nonlinear terms, and more realistic refinements of the various models.

Acknowledgements

This work was supported by the Applied Mathematical Sciences subprogram of the Office of Energy Research, U. S. Department of Energy, under Contract W-31-109-Eng-38.

References

[1] Abrikosov, A. (1957) 'On the magnetic properties of superconductors of the second type', *Zh. Eksperim. i Teor. Fiz.*, **32**, 1442–1152. (English translation: *Soviet Phys. JETP*, **5** 1174–1182.)

[2] Bebernes, J., and Eberly, D. (1989) *Mathematical problems from combustion theory*, Springer-Verlag, New York.

[3] Bellout, H. (1992) 'A regularity result for a quasilinear equation and its consequences for blowing-up solutions', *J. Diff. Eqs.*, to appear.

[4] Berger, M., and Kohn, R. V. (1988) 'A rescaling algorithm for the numerical calculation of blowing up solutions', *Comm. Pure Appl. Math.*, 841–863.

[5] Bressan, A. (1990) 'On the asymptotic shape of blow-up', *Indiana Univ. Math. J.*, **39**, 947–960.

[6] Bressan, A. (1991) 'Blow-up asymptotics for the reactive-Euler gas model', *SIAM J. Math. Anal.*, **22**, 587–601.

[7] Chin, R. C. Y., and Hedstrom, G. W. (1991) 'Domain decomposition: an instrument of asymptotic-numerical methods', in H. G. Kaper and M. Garbey (eds.), *Asymptotic Analysis and the Numerical Solution of Partial Differential Equations*, Lecture Notes in Pure and Applied Mathematics, Vol. 130, Marcel Dekker, New York, 33–54.

[8] Du, Q., Gunzburger, M., and Peterson, J. (1992) 'Analysis and approximation of a periodic Ginzburg-Landau model for type-II superconductors', *SIAM Review,* to appear.

[9] Du, Q., Gunzburger, M., and Peterson, J. (1992) 'Modeling and analysis of a periodic Ginzburg-Landau model for type-II superconductors', preprint, 1992.

[10] Doria, M., Gubernatis, J., and Ranier, D. (1990) 'Solving the Ginzburg-Landau equations by simulated annealing', *Phys. Rev. B*, **41**, 6335–6340.

[11] Filippas, S., and Kohn, R. V. (1992) 'Refined asymptotics for the blowup of $u_t - \Delta u = u^p$ ', preprint.

[12] Friedman, A. B. (1991) 'Blow-up of solutions of nonlinear heat and wave equations', in H. G. Kaper and M. Garbey (eds.), *Asymptotic Analysis and the Numerical Solution of Partial Differential Equations*, Lecture Notes in Pure and Applied Mathematics, Vol. 130, Marcel Dekker, New York, 217–224.

[13] Friedman, A., and McLeod, B. (1985) 'Blow-up of positive solutions of semilinear heat equations', *Indiana Univ. Math. J.*, **34**, 425-447.

[14] Garner, J., Spanbauer, M., Benedek, R., Strandburg, K., Wright, S., and Plassmann, P. (1991) 'Critical fields of Josephson-coupled superconducting multilayers', Mathematics and Computer Science Division Preprint MCS-P281-1291, Argonne National Laboratory, Argonne, Illinois.

[15] Giga, Y., and Kohn, R. V. (1989) 'Nondegeneracy of blow-up for semilinear heat equations', *Comm. Pure Appl. Math*., **42**, 297–319.

[16] Herrero, M. A., and Velázquez, J. J. L. (1992) 'Blow-up profiles in one-dimension, semilinear parabolic problems', *Comm. Partial Diff. Eqs.*, to appear.

[17] Kaper, H. G. (1991) 'Mathematical models of superconductivity', Technical Memorandum MCS-TM-146, Mathematics and Computer Science Division, Argonne National Laboratory, Argonne, Illinois.

[18] Kaper, H. G., and Kwong, Man Kam (1992) 'The Ginzburg-Landau model for type II superconductors', preprint.

[19] Kwong, Man Kam (1992) 'Sweeping algorithms for five-point stencils and banded matrices', Technical Memorandum ANLmethodCS-TM-165, Mathematics and Computer Science Division, Argonne National Laboratory, Argonne, Illinois.

[20] Kwong, Man Kam (1991) *Sweeping algorithms for inverting the discrete Ginzburg-Landau operator*, (to appear in Applied Math. and Computation) Mathematics and Computer Science Division Preprint MCS-P307-0492, Argonne National Laboratory, Argonne, Illinois.

[21] Landman, M. J., Papanicolaou, G. C., Sulem, C., and Sulem, P. L. (1988) 'Rate of blowup for solutions of the nonlinear Schrödinger equation at critical dimension', *Physical Rev. A*, **38**, 3837–3843.

[22] Levine, H. A. (1989) 'Advances in quenching', *Proceedings of the International Conference on Reaction-Diffusion Equations and Their Equilibrium States*, Gregynog, Wales.

FACTORIZATION OF THE ADVECTION-DIFFUSION OPERATOR AND DOMAIN DECOMPOSITION METHOD

F. Nataf
CMAP
Ecole Polytechnique
91128 Palaiseau Cedex
France

and

F. Rogier
Division Calcul parallèle
ONERA
29 Av. de la division Leclerc
92322 Châtillon
France

ABSTRACT. We propose an iterative algorithm that involves, at each step, the solving of the advection-diffusion equation on each subdomain. Each subproblem is subject to boundary conditions issued from the approximate factorization of the advection-diffusion operator. The method is more efficient as the viscosity decreases. Numerical tests are shown.

1. Introduction

The solution to partial differential equations (PDEs) with complex geometries requires fast algorithms, suited to the new parallel computers. Domain decomposition methods offer an answer to this requirement. A problem set on a large domain is reduced to a series of subproblems set on small domains called subdomains. This series of subproblems leads, of course, to the solution of the initial problem set on the large domain. Data are exchanged between subdomains. Usually, these data are related to the boundary of the subdomains. The choice of the data to be exchanged is crucial for the efficiency of the method. It is not simply a matter of modifying the core of efficient codes. In fact, we have to design appropriate boundary conditions so as to have a fast convergence.

H. G. Kaper and M. Garbey (eds.),
Asymptotic and Numerical Methods for Partial Differential Equations with Critical Parameters 123–133.

We consider here the advection-diffusion equation. This equation, despite its simplicity, contains the main phenomena that occur in fluid mechanics in the equation of motion: a convective part and a diffusive part.

Recently, to solve the Helmholtz equation (and more generally the Maxwell system) by a domain decomposition method, B. Despres [3, 4] used the open boundary conditions of order 1 for the Helmholtz operator as transmission conditions. Moreover, in [1, 2, 5], the authors used the exact factorization of the operator to be inverted. This factorization involves nonlocal pseudo-differential operators. These nonlocal operators then must be implemented in the numerical method.

Our method is a generalization of these approaches in the case of the advection-diffusion equation. These factorizations involve only local operators. The transmission conditions are thus local and easy to implement in a numerical method.

2. One-Dimensional Case, Boundary Conditions: A Relevant Example

We consider the following advection-diffusion equation on a segment of line:

$$L(u) = a\frac{\partial u}{\partial x} - \nu\frac{\partial^2 u}{\partial x^2} = f \quad \forall x \in [0, L] \tag{1}$$

$$u(0) = u(L) = 0, \quad a = constant > 0. \tag{2}$$

We propose to analyze the convergence of a domain decomposition method with overlap at different boundary conditions. The general domain decomposition method with overlap consists of solving

$$L(u_1^{n+1}) = f, \forall x \in [0, l_2], u_1^{n+1}(0) = 0, B^+(u_1^{n+1})(l_2) = B^+(u_2^n)(l_2) \tag{3}$$

$$L(u_2^{n+1}) = f, \forall x \in [L_1, L], u_2^{n+1}(L) = 0, B^-(u_2^{n+1})(L_1) = B^-(u_1^n)(L_1), \tag{4}$$

where $0 < l_2 < L_1 < L$; B^+ and B^- are two operators determined by the choice of the method. To examine the convergence, we introduce the auxiliary unknowns $e_1^n = u_1^n - u, e_2^n = u_2^n - u$ (u is the exact solution of (1)). Since a is constant, we easily find the general solution of $L(e_1^{n+1}) = 0, e_1^{n+1}(0) = 0$, and $L(e_2^{n+1}) = 0, e_2^{n+1}(L) = 0$:

$$e_1^{n+1} = r_1^{n+1}(e^{ax/\nu} - 1), e_2^{n+1} = r_2^{n+1}(e^{a(x-L)/\nu} - 1).$$

The coefficients r_1^n, r_2^n are obtained recursively by application of the boundary conditions $B^+(e_1^{n+1}) = B^+(e_2^n)$, $B^-(e_2^{n+1}) = B^-(e_1^n)$.

We now define the convergence rate ρ as

$$\rho = \frac{r_1^{n+1}}{r_1^{n-1}} = \frac{r_2^{n+1}}{r_2^{n-1}}.$$

Table 1 gives the values of ρ for different choices of the operators B^+ and B^-, where $\phi(l_2) = \frac{e^{al_2/\nu}-1}{e^{a(l_2-L)/\nu}-1} \to +\infty$ as $l_2 \to L$.

Table 1. **Values for Different Choices of Operator**

Boundary Conditions	ρ
$B^+u = B^-u = u$	$\approx e^{-a\delta/\nu}$
$B^+u = B^-u = \dfrac{\partial u}{\partial x}$	1
$B^+u = \dfrac{\partial u}{\partial x}, B^-u = u$	$\phi(l_2)e^{-aL/\nu}$
$B^+u = \dfrac{\partial u}{\partial x}, B^-u = \dfrac{\partial u}{\partial x} - \dfrac{a}{\nu}u$	$e^{-aL/\nu}$

The first row of the table corresponds to the Dirichlet conditions. It is a well-known fact that the convergence rate of such methods depends on the thickness of the overlapping zone. The Neumann-Neumann conditions (second row) lead to a convergence rate of 1; the method does not converge. Physically more realistic conditions are given in the third row. In this case the position of l_2 is crucial to obtain a stable scheme. If l_2 is close to L, then the convergence rate is very high, and the domain decomposition method does not converge. The boundary conditions in the fourth row are completely optimal. As we will see in the following sections, they lead to an exact factorization of the operator $L(u)$ as $-\nu(\partial u/\partial x)(\partial u/\partial x - au/\nu)$.

3. Two-Dimensional Case

3.1. EXACT FACTORIZATION

In this section, we generalize the previous algorithm to the two-dimensional case and to the operator $\mathcal{L}$ arising from an implicit discretization in time of the advection-diffusion equation:

$$\mathcal{L} = \frac{1}{\Delta t} + a\frac{\partial}{\partial x} + b\frac{\partial}{\partial y} - \nu\left(\frac{\partial^2}{\partial x^2} + \frac{\partial^2}{\partial y^2}\right). \tag{5}$$

This operator has to be inverted at each time step. We need a factorization of the operator $\mathcal{L}$. By a factorization we mean writing (5) as the product of two first-order operators in the x-direction, one corresponding to a well-posed parabolic problem in the direction of positive x, the other to a problem in the direction of negative x. To find the explicit form of the factorization, we consider the constant coefficients operator. We take the Fourier transform in y of (5) (the dual variable is denoted by k):

$$\hat{\mathcal{L}} = \frac{1}{\Delta t} + a\,\partial x - i\,b\,k - \nu\frac{\partial^2}{\partial x^2} + \nu k^2,$$

which can be written as

$$\hat{\mathcal{L}} = -\nu(\partial x - \lambda^+)(\partial x - \lambda^-),$$

where

$$\begin{aligned}\lambda^+ &= \frac{a + \sqrt{a^2 + 4\nu/\Delta t - 4i\nu bk + 4k^2\nu^2}}{2\nu},\\ \lambda^- &= \frac{a - \sqrt{a^2 + 4\nu/\Delta t - 4i\nu bk + 4k^2\nu^2}}{2\nu}.\end{aligned} \tag{6}$$

Let Λ^+ and Λ^- be the operators defined by the inverse Fourier transforms of λ^+ and λ^-,

$$\Lambda^+(u) = \mathcal{F}^{-1}(\lambda^+(k)\ \hat{u}), \quad \Lambda^-(u) = \mathcal{F}^{-1}(\lambda^-(k)\ \hat{u}).$$

Then

$$\mathcal{L} = -\nu(\partial x - \Lambda^+)(\partial x - \Lambda^-). \tag{7}$$

The term λ^+ is positive and the term λ^- is nonpositive, whatever the sign of a. Thus, the operator $\partial_x - \Lambda^+$ (resp. $\partial_x - \Lambda^-$) corresponds to a well-posed parabolic problem in the direction of negative (resp. positive) x. The expressions of λ^+ and λ^- are not polynomial in k, so the operators Λ^+ and Λ^- are nonlocal.

We consider a domain $]0, L[\times \mathbf{R}$ covered by N vertical strips $[l_i, L_i] \times \mathbf{R}$ $(i = 1, \ldots, N)$ with a possible overlap of size δ. As in the one-dimensional case, this factorization suggests the following algorithm. Let u_i^n be an estimate of the solution in domain i. Then u_i^{n+1} is defined as

$$\begin{aligned}&\mathcal{L}(u_i^{n+1}) = f \text{ in domain } i\\ &(\partial x - \Lambda^+)(u_i^{n+1}) = (\partial x - \Lambda^+)(u_{i-1}^n) \quad \text{at } x = l_i\ (i \neq 1),\\ &(\partial x - \Lambda^-)(u_i^{n+1}) = (\partial x - \Lambda^-)(u_{i+1}^n) \quad \text{at } x = L_i\ (i \neq N).\end{aligned} \tag{8}$$

At $x = 0$ and $x = L$, we use the boundary conditions of the global problem. At each step, N independent subproblems have to be solved. This domain decomposition method is valuable for parallel computation.

In the case where the operators $\partial_x - \Lambda^+$ and $\partial_x - \Lambda^-$ are used as open boundary conditions at $x = 0$ and $x = L$, respectively, it can be proved that with or without overlap there is convergence in N steps. Nevertheless, the algorithm (8) cannot be used for practical purposes for two reasons. The first reason is the nonlocality of the boundary conditions in each subdomain. The second reason, which is much more severe, is that for variable coefficients a and b, the explicit form of the factorization is not known. Thus, we next consider approximations involving local operators.

To obtain approximate factorizations involving local operators, we take Taylor approximations of λ^+ and λ^- with respect to k. We are restricted to at most third-order approximations, since for higher-order approximations the sign of the approximations of λ^+ (resp. λ^-) is negative (resp. positive) for high Fourier numbers. These would yield boundary conditions that lead to ill-posed problems. We also do not consider the approximation of order three, since it yields a boundary condition of order three in y. The approximations we shall consider are

$$\lambda_0^+(k) = \frac{a+\sqrt{a^2+4\nu/\Delta t}}{2\nu}, \qquad \lambda_0^-(k) = \frac{a-\sqrt{a^2+4\nu/\Delta t}}{2\nu}, \tag{9}$$

$$\begin{aligned} \lambda_1^+(k) &= \frac{a+\sqrt{a^2+4\nu/\Delta t}}{2\nu} - ik\frac{b}{\sqrt{a^2+4\nu/\Delta t}}, \\ \lambda_1^-(k) &= \frac{a-\sqrt{a^2+4\nu/\Delta t}}{2\nu} + ik\frac{b}{\sqrt{a^2+4\nu/\Delta t}}, \end{aligned} \tag{10}$$

$$\begin{aligned} \lambda_2^+(k) &= \frac{a+\sqrt{a^2+4\nu/\Delta t}}{2\nu} \\ &\quad -ik\frac{b}{\sqrt{a^2+4\nu/\Delta t}} + \frac{\nu}{\sqrt{a^2+4\nu/\Delta t}}\left(1+\frac{b^2}{a^2+4\nu/\Delta t}\right)k^2, \end{aligned} \tag{11}$$

$$\begin{aligned} \lambda_2^-(k) &= \frac{a-\sqrt{a^2+4\nu/\Delta t}}{2\nu} \\ &\quad +ik\frac{b}{\sqrt{a^2+4\nu/\Delta t}} - \frac{\nu}{\sqrt{a^2+4\nu/\Delta t}}\left(1+\frac{b^2}{a^2+4\nu/\Delta t}\right)k^2. \end{aligned} \tag{12}$$

Since k occurs always in the combination νk, these approximations are also valid for small viscosity ν. Nevertheless, they are different from Taylor approximations of λ^+ and λ^- with respect to ν, because of the term $\nu/\Delta t$.

The approximate factorizations and the remainder terms are then (with obvious notation)

$$\begin{aligned} \mathcal{L} &= -\nu(\partial_x - \Lambda_0^+)(\partial_x - \Lambda_0^-) + b\partial_y - \nu\partial_{yy}, \\ \mathcal{L} &= -\nu(\partial_x - \Lambda_1^+)(\partial_x - \Lambda_1^-) - \nu\left(1+\frac{b^2}{a^2+4\nu/\Delta t}\right)\partial_{yy}, \\ \mathcal{L} &= -\nu(\partial_x - \Lambda_2^+)(\partial_x - \Lambda_2^-) - \frac{\nu^3}{a^2+4\nu/\Delta t}\left(1+\frac{b^2}{a^2+4\nu/\Delta t}\right)^2\partial_{yyyy}. \end{aligned}$$

We can now return to the operator with variable coefficients a and b. In this case, we simply propose as approximate factorizations

$$\mathcal{L} \simeq -\nu \left(\partial_x - \frac{a+\sqrt{a^2+4\nu/\Delta t}}{2\nu}\right)\left(\partial_x - \frac{a-\sqrt{a^2+4\nu/\Delta t}}{2\nu}\right), \tag{13}$$

$$\mathcal{L} \simeq -\nu \left(\partial_x - \frac{a+\sqrt{a^2+4\nu/\Delta t}}{2\nu} - \frac{b}{\sqrt{a^2+4\nu/\Delta t}}\partial_y\right)$$
$$\left(\partial_x - \frac{a-\sqrt{a^2+4\nu/\Delta t}}{2\nu} + \frac{b}{\sqrt{a^2+4\nu/\Delta t}}\partial_y\right), \tag{14}$$

$$\mathcal{L} \simeq -\nu \left(\partial_x - \frac{a+\sqrt{a^2+4\nu/\Delta t}}{2\nu} - \frac{b}{\sqrt{a^2+4\nu/\Delta t}}\partial_y \right.$$
$$\left. + \frac{\nu}{\sqrt{a^2+4\nu/\Delta t}}\left(1+\frac{b^2}{a^2+4\nu/\Delta t}\right)\partial_{yy}\right)$$
$$\left(\partial_x - \frac{a-\sqrt{a^2+4\nu/\Delta t}}{2\nu} + \frac{b}{\sqrt{a^2+4\nu/\Delta t}}\partial_y \right.$$
$$\left. - \frac{\nu}{\sqrt{a^2+4\nu/\Delta t}}\left(1+\frac{b^2}{a^2+4\nu/\Delta t}\right)\partial_{yy}\right). \tag{15}$$

The remainder terms are not the same as in the case where a and b are constants; they are of lower order with respect to ν.

Remark. In [6], in the case where $b = 0$ and $\Delta t = \infty$, the authors propose approximations of order 2, which lead to a remainder term still of order 3 with respect to the viscosity ν. For the sake of simplicity, these approximations will not be used here.

Algorithm (8) leads obviously to new algorithms, which depend on the approximate factorization used,

$$\begin{aligned} &\mathcal{L}(u_i^{n+1}) = f \text{ in domain } i \\ &(\partial x - \Lambda_j^+)(u_i^{n+1}) = (\partial x - \Lambda_j^+)(u_{i-1}^n) \quad \text{at } x = l_i \ (i \neq 1) \\ &(\partial x - \Lambda_j^-)(u_i^{n+1}) = (\partial x - \Lambda_j^-)(u_{i+1}^n) \quad \text{at } x = L_i \ (i \neq N), \end{aligned} \tag{16}$$

where $j = 0, 1$, or 2; $\Lambda_j^\pm$ are the operators arising from the approximate factorizations. At $x = 0$ and $x = L$, we use the boundary conditions of the global problem. Of course, contrary to the algorithm (8) based on the exact factorization, the convergence of (8) in a finite number of steps is no longer satisfied. In the next section, the convergence is studied in the case of constant coefficients with or without overlap and for variable coefficients in the case where there is no overlap.

3.3. CONVERGENCE RESULTS

3.3.1. With Overlap. When a and b are constant coefficients, we can carry out a Fourier analysis of the convergence of the method. For the sake of simplicity, we consider only the case $N = 2$ (two subdomains). The boundary conditions at $x = 0$ and $x = L$ are written in terms of the operators arising from the exact factorization (8), $(\partial x - \Lambda^+)(u) = g+$ at $x = 0$ and $(\partial x - \Lambda^-)(u) = g-$ at $x = L$. The error $u_2^n - u$ (resp. $u_1^n - u$) at the left (resp. right) boundary of domain 2 (resp. 1) at step n is denoted by e_2^n (resp. e_1^n). The convergence rate $\rho_j^i(k)$ (j denotes the order of the approximate factorization used in (8)) in the Fourier space is $\hat{e}_i^{n+2}(k) = \rho_j^i(k)\hat{e}_i^n(k),\ n \geq 1$. It turns out that $\rho_j^i(k)$ is independent of the subdomain i, so we omit the superscript. The convergence rate between the steps n and $n+2$ (not n and $n+1$) is considered, as only two steps are needed to achieve convergence when we use the exact factorization. A straightforward computation yields

$$\rho_j(k) = \frac{\lambda^-(k) - \lambda_j^-(k)}{\lambda^+(k) - \lambda_j^-(k)} \frac{\lambda^+(k) - \lambda_j^+(k)}{\lambda^-(k) - \lambda_j^+(k)} e^{(\lambda^-(k) - \lambda^+(k))\delta}. \tag{17}$$

We use the facts that $\lambda^-(k) + \lambda^+(k) = a/\nu$ and that the same identity holds for the approximations (i.e., $\lambda_j^-(k) + \lambda_j^+(k) = a/\nu$) to simplify (17),

$$\rho_j(k) = \left(\frac{\lambda^-(k) - \lambda_j^-(k)}{\lambda^+(k) - \lambda_j^-(k)}\right)^2 e^{(-a/\nu + 2\lambda^-(k))\delta}. \tag{18}$$

It can be proved that, for any k and δ, $|\rho_j(k)| < 1$. For small Fourier numbers, ρ_j tends to zero. For large Fourier numbers, ρ_j tends to 1 if $\delta = 0$ and to zero if $\delta \neq 0$ (see Fig. 1). Thus, overlap is important.

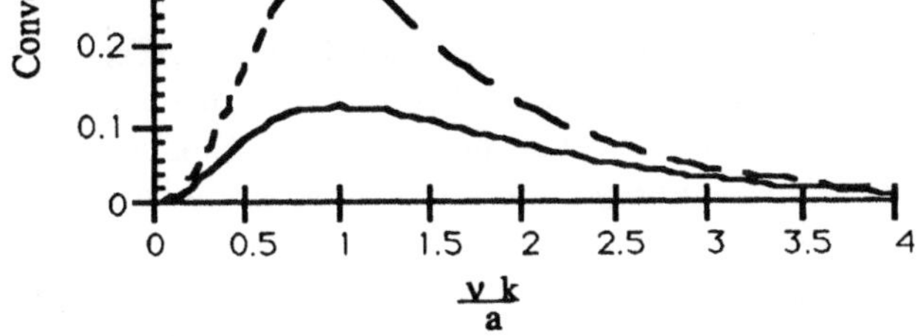

Fig. 1. **Convergence rate**

3.3.2. Without Overlap. We have some convergence results for variable a and b. We take the following boundary conditions for the global problem:

$$(\partial x - \Lambda_j^+)(u) = g_+ \quad \text{at } x = 0, \qquad (\partial x - \Lambda_j^-)(u) = g_- \quad \text{at } x = L. \tag{19}$$

We shall assume that the algorithm (16) is well defined, in the sense that for any $u_i^n \in H^2(]l_i, L_i[\times \mathbf{R})$ and any $i = 1, \ldots, N$, there exists a unique u_i^{n+1} satisfying (16). Using energy estimates, we prove the following theorems.

Theorem 1. *Let $a, b \in W^{1,\infty}$. If either (a) $\operatorname{div}(a,b) = 0$ or (b) $1/\Delta t > \frac{1}{2}\sup \operatorname{div}(a,b)$, then the algorithm (16) with $j = 0$ converges, in the sense that $\lim_{n\to\infty} \| u_i - u_i^n \|_{H^1} \longrightarrow 0$ for $i = 1, \ldots, N$.*

Theorem 2. *Let $a, b \in W^{3,\infty}$ satisfy the following condition: $A \geq a(x,y) \geq \alpha > 0$. Then, for all $\nu_0 > 0$, there exists Δt_0 such that for all ν, $0 < \nu < \nu_0$, for all Δt, $0 < \Delta t < \Delta t_0$, the algorithm (16) with $j = 2$ converges, in the sense that $\lim_{n\to\infty} \| u_i - u_i^n \|_{H^1} \longrightarrow 0$ for $i = 1, \ldots, N$.*

4. Numerical Results

To illustrate the validity of the method, we have performed a two-dimensional test problem. All the numerical tests have been implemented on an iPSC/2 with 32 processors. The problem that we have tested consists of finding the temperature of a fluid moving at the velocity $\mathbf{v} = (a, b)$. The crucial numerical difficulty is to capture with accuracy the boundary layer arising in the neighborhood of the line where the temperature is given. The computational domain is the square $[0,1] \times [0,1]$ (see Fig. 2); the viscosity ν is .001.

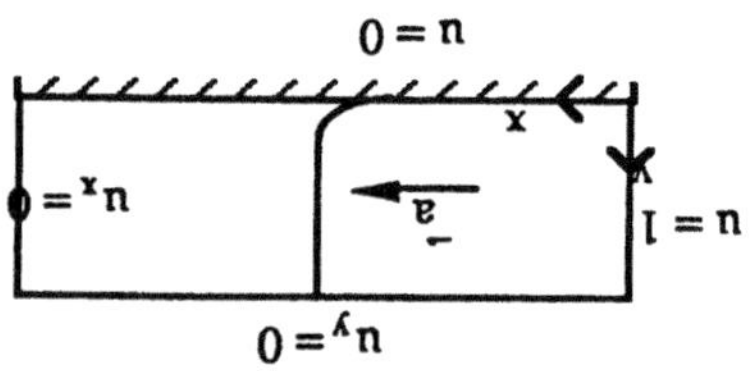

Fig. 2. **Computational domain**

Figure 3 shows the isovalues of the temperature at different time steps of the algorithm for a velocity $\mathbf{v} = (1, 0)$ and for an initial condition $u = 1$ in the middle of the domain. As the figure shows, the boundary layer is taken into account with accuracy. The curvature of the isovalues is due to the effect of the initial condition.

Figure 4 has been obtained by changing the order of the approximation. The logarithm of the residual between two iterations has been represented as a function of the number of iterations. It can be seen that, for the same residual, an increase in the order of the approximation by 1 results in a reduction of the number of iterations by a factor of 2.

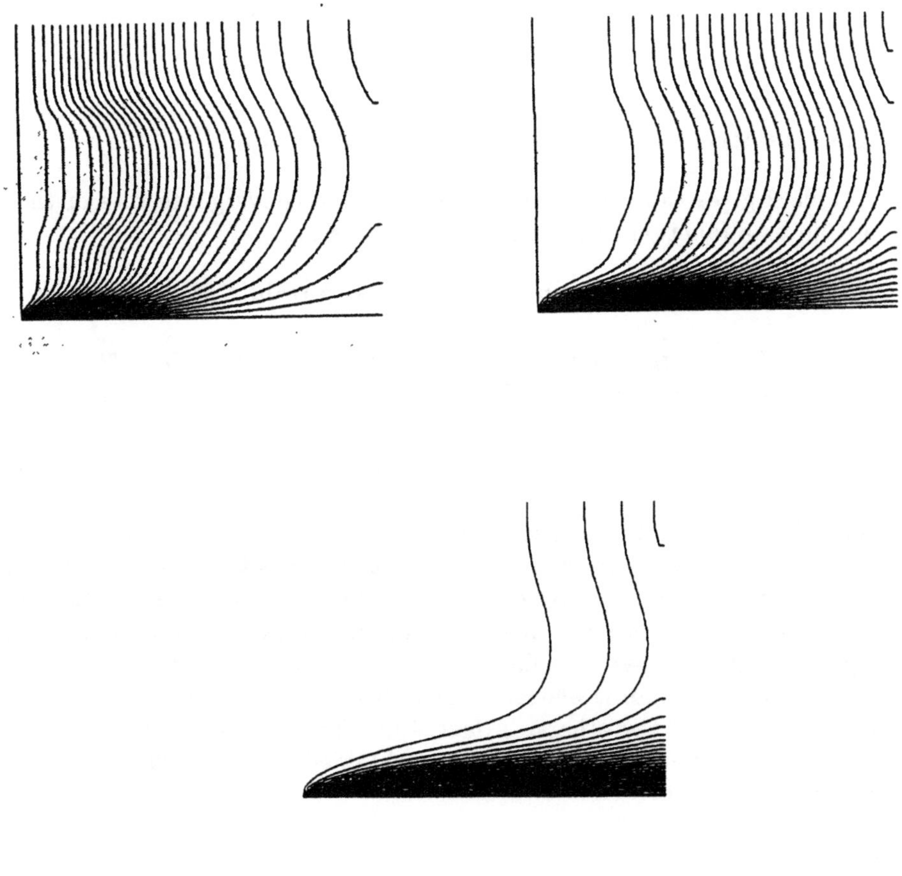

Fig. 3. **Isovalues of the temperature at time = .2, .4, and .8**

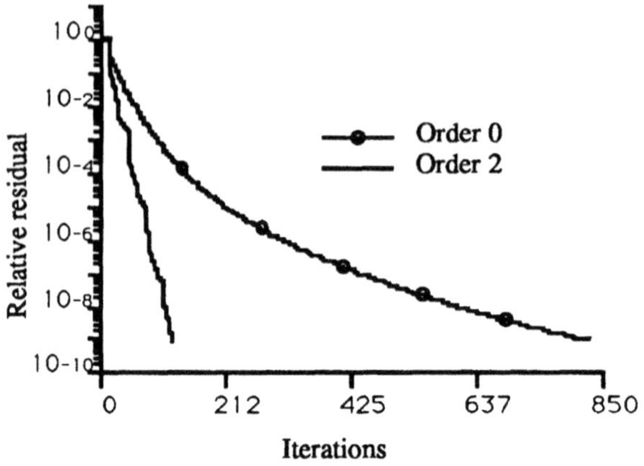

Fig. 4. **Relative residual vs. number of iterations**

Table 2 gives the computation time and the convergence rate of the algorithm as a function of the number of processors.

Table 2. Computation time (in sec) vs. number of processors

No. of processors	1	2	4	8	16	32
Computation time	170.0	51.0	26.7	11	6.0	6.1

The mesh size is 64×64. It can be seen that the computation time halved when the number of processors is doubled. But the computation time is the same when the number of processors is 16 and 32. This fact is due to the communication cost between the processors, which is too large in view of the computation time in each processor. Therefore, as the mesh size increases, the speedup approaches the number of processors. In other words, the convergence rate increases as a linear function of the number of subdomains, because the algorithm is efficient when each subdomain has exchanged information with all the others. In fact, this phenomenon is balanced by the gain (in terms of number of operations) obtained in the resolution and leads to optimal speedup.

References

[1] Chan, T., and Keyes, D. (1989) 'Interface preconditionings for domain-decomposed convection-diffusion operators', *Domain Decomposition Methods for Partial Differential Equations*, SIAM, Philadelphia, 245–262

[2] Chan, T., and Resasco, D. C. (1988) 'A framework for the analysis and construction of domain decomposition preconditioners', *Domain Decomposition Methods for Partial Differential Equations*, SIAM, Philadelphia, 217–230.

[3] Despres, B. (1991) 'Domain decomposition method and the Helmholtz problem', *Mathematical and Numerical Aspects of Wave Propagation Phenomena*, SIAM, Philadelphia, 44–52.
[4] Despres, B. (1990) 'Décomposition de domaine et problème de Helmohltz', *C.R. Acad. Sci., Paris*, **t. 311, Série I**, 313–316.
[5] Hagstrom, T., Tewarson, R. P., and Jazcilevich, A. (1988) 'Numerical experiments on a domain decomposition algorithm for nonlinear elliptic boundary value problems', *Appl. Math. Lett.*, **1**, no. 3, 299–302.
[6] Lohéac, J. P., Nataf, F., and Schatzman, M. (1992) 'Parabolic approximations of the convection-diffusion equation', *Math. Comp.*, to appear.

Part 3: Multiple-Scale Problems in Scientific Computing

Modulated Traveling Waves in Combustion

A. Bayliss, B. J. Matkowsky, H. Riecke

Asymptotics and Multiscale Simulation in a Numerical Combustion Laboratory

D. Scott Stewart, J. B. Bdzil

MODULATED TRAVELING WAVES IN COMBUSTION

Alvin Bayliss, Bernard J. Matkowsky, and Hermann Riecke
Department of Engineering Sciences and Applied Mathematics
Northwestern University
Evanston IL 60208

ABSTRACT. We describe the formation of spatial and temporal patterns in cylindrical premixed flames. In particular, we consider the formation of rotating cellular flames described by modulated traveling waves (MTWs) along the flame front. We find different types of MTWs, distinguished by their symmetries, in different parameter regimes of the Lewis number, *Le*, the ratio of thermal diffusivity to mass diffusivity of a deficient component of the mixture. The nature and evolution of the MTWs are described for each case. The adaptive pseudo-spectral method, which we developed for these and other computations, is motivated by the asymptotic notion of "stretching" a layer in the method of matched asymptotic expansions for treating singular perturbation problems.

1. Introduction

It has been long been known that premixed combustible gaseous mixtures do not necessarily burn in a uniform manner. In particular, under certain conditions flame fronts can exhibit a cellular structure, characterized by periodic arrays of crests along the flame front pointing in the direction of the combustion products. The pointed crests are connected by smooth troughs that are convex toward the fresh fuel mixture. The temperature is higher at the troughs, which therefore appear brighter, and lower at the crests, which therefore appear darker [33].

In many instances the transition from laminar to turbulent combustion occurs as a smooth flame breaks up into cells, which then undergo transitions leading to progressively more complex spatial and temporal behavior. Some of these transitions have been observed experimentally. For example, the cells can exhibit traveling wave (TW) behavior, for instance, a uniform translation along the flame front or a rotation about an axis, (see, e.g., [21, 22, 25, 33, 34]). Alternatively, the cells can exhibit nonperiodic dynamics (possibly quasi-periodic or chaotic) [21, 22, 25]. The symmetry along the front in which all cells are identical can also be broken so that different cells can have a different spatial structure [22]. Finally, the cellular flame can

H. G. Kaper and M. Garbey (eds.),
Asymptotic and Numerical Methods for Partial Differential Equations with Critical Parameters 137–162.

exhibit spatial chaos as well as temporal chaos, with the cells undergoing an erratic motion, in which cells are randomly created and annihilated [22, 33].

One scenario for the development of cellular flames is that of thermo-diffusive cellular instabilities which arise as certain parameters are varied [32, 37, 45]. In fact, many of the complex spatio-temporal patterns observed in flames can be qualitatively described as thermo-diffusive instabilities. Analytical methods have been successful in describing primary transitions from uniform combustion to cellular combustion (e.g., [17, 24, 32, 37, 39]). In some cases higher-order transitions can be found by analytical methods as well [17, 38, 39]. In most instances, however, secondary and higher-order transitions are difficult to find analytically and are more readily found by numerical solution of the systems of partial differential equations modeling combustion.

In this paper we consider cylindrical flames, in particular, adiabatic flames stabilized by a line source of fuel and the related problem of a flame established in the region between two coaxial cylinders, with the combustible mixture fed in through the inner cylinder and heat loss occurring through the outer cylinder. The computations build on and extend analytical results obtained for the problem of a flame stabilized by a line source of fuel. We first describe the analytical results.

We consider the diffusional thermal model for a flame stabilized by a line source of fuel of strength $2\pi\kappa$. In the diffusional thermal model the thermal expansion of the gas is assumed to be weak. As a result, the flow field is assumed to be unaffected by the chemical reaction and so can be specified beforehand from a calculation of the nonreacting gas flow problem. The model is described by a system of two nonlinear partial differential equations for the reduced temperature and the concentration of a deficient component of the fuel mixture [37]. This model has been analyzed for the limiting case of infinite activation energy [24, 37]. A stationary axisymmetric solution describing a circular flame front located at $r = \kappa$ has been found for all values of κ and the Lewis number, Le, the ratio of thermal diffusivity to mass diffusivity for a deficient component of the fuel mixture. This solution, called the basic solution, is subject to two different classes of instabilities depending on whether Le is greater than or less than 1.

There exists a critical value, $Le_{c1} > 1$, such that for $Le > Le_{c1} > 1$, the basic solution is unstable to small disturbances. The instability arises as two complex conjugate eigenvalues pass into the right half-plane, thus suggesting that small disturbances evolve to nonstationary cellular flames, such as axisymmetric pulsating flames, as well as traveling and standing waves along the flame front. Similar behavior has also been found in other geometries (e.g., [32, 35, 36, 40]). The regime Le sufficiently greater than 1, describing, for example, lean, heavy hydrocarbon/air mixtures, is often referred to as the pulsating regime.

Below a second critical value $Le_{c2} < 1$ there exists a κ_{c2} such that for $\kappa > \kappa_{c2}$ the basic solution is unstable to small perturbations. This instability occurs by a single real eigenvalue crossing into the right half-plane, and small perturbations evolve to stationary cellular flames [24, 37]. This behavior has also been found in other geometries [30, 32, 38, 45]. The regime Le sufficiently smaller than 1, describing mixtures where the deficient component is highly diffusive, for example lean hydrogen/air mixtures or rich, heavy hydrocarbon/air mixtures, is referred to as the cellular regime.

The analysis in [24, 37] provides an explanation for the onset of cellular flames as a diffusional thermal instability. For $Le > 1$ the analysis also explains the onset of nonstationary behavior. However, the analysis does not describe the subsequent evolution of these cellular flames into more complex spatial and temporal patterns and in particular the transition to chaotic behavior. Similarly for the cellular regime, $Le < 1$, the analysis does not describe many of the complex patterns observed in experiments (e.g., [22]). These complex patterns may arise as secondary or higher-order bifurcations and transitions from the primary states that were described analytically. These transitions can be found by numerically solving the mathematical model.

The analysis in [24, 37] did identify the roles of the two fundamental parameters in promoting diffusional thermal instabilities of cylindrical flames. For the cellular regime, $Le < 1$, the stationary circular flame front can be destabilized by either increasing κ or decreasing Le. Similarly in the pulsating regime $Le > 1$, increasing Le is destabilizing. Analytical methods have also identified heat loss as a destabilizing mechanism for planar flames in the pulsating regime, $Le > 1$ [16, 27]. In this paper we describe examples of complex spatial and temporal behavior for both the regimes $Le < 1$ and $Le > 1$. In both cases we concentrate on the development and properties of modulated traveling waves (MTWs) and their subsequent transitions to chaotic behavior. For the cellular regime we consider the behavior as Le is decreased for a fixed value of κ. For the pulsating regime we consider the role of heat loss in generating MTWs.

A variety of higher-order transitions in the cellular regime have been found numerically. The behavior of stationary cellular flames was considered in [12], where a numerical investigation of the effect of varying κ was described. It was found that if Le was held fixed and κ increased, modal transitions between stationary cellular flames of increasing mode number occurred. For the particular value of Le considered transitions from axisymmetry to a three-cell, to a four-cell, to a five-cell and to a six-cell stationary cellular flame were computed. All of the transitions, except for the first, are accompanied by a region of bistability where stationary cellular solutions of adjacent mode numbers coexisted, each with its own domain of attraction. We expect that unstable mixed-mode solution branches connect the computed (stable) cellular branches. Since we determine solutions by solving an initial-boundary value problem and marching forward in time until a steady state is achieved, we can generally compute only stable solutions.

In [6, 11] the effect of reducing the Lewis number was investigated. In [11] stationary cellular solutions were presented for a range of values of *Le*. It was shown that reducing *Le* led to a progressive deepening of the cells and more pronounced crests. Only stationary solutions were found for the parameter values used in the computation. In [6] we considered lower values of *Le*, and nonstationary 4-cell solutions were found as a secondary transition after the development of a branch of stationary 4-cell flames. These solutions described very slowly rotating cellular flames. The rotation occurred at a constant angular velocity and corresponded to a slowly traveling wave (TW) along the flame front. It was shown [9] that the transition from stationary to rotating behavior occurred as an infinite-period, symmetry-breaking bifurcation, in which the reflection symmetry of the cells was destroyed. We refer to this branch as the TW4 branch.

It was further shown in [9] that stable mixed-mode solutions exist after the transition to nonstationary behavior. The spatial behavior of these mixed-mode solutions was that of a 4-cell array in which at any fixed instant of time the cells differed in size and shape. These solutions were quasi-periodic in time. Two branches of mixed mode solutions were found in [9]. The first branch (MTW1) appeared to be stable over a small window in *Le* near the transition point from stationary to traveling cells. The second branch (MTW2) was found for values of *Le* smaller than the value of *Le* at which the TW4 branch loses stability. The branches MTW1 and MTW2 were characterized by the appearance of additional angular Fourier modes, of which modes 3 and 5 in the MTW1 solution and mode 3 in the MTW2 solution, appeared to be dominant.

In this paper the behavior of the mixed mode solution branches is studied further. These modes are identified as MTWs, in which after a certain time interval the solution recovers its structure along the front except for a phase shift. The structure of the MTWs is very different for the two mixed mode branches. Along the first branch (MTW1) the solution at any instant of time is composed both of nearly stationary, nearly symmetric cells and of nonstationary, asymmetric cells. Motion of the individual cell proceeds as a localized wave of asymmetry travels along the front. This behavior is directly analogous to the square dance mode observed in [22]. Along the second branch (MTW2) all cells undergo a basic rotation about the cylindrical axis, characteristic of the traveling wave branch (TW4); however, the cells undergo a periodic expansion and contraction during the motion. As *Le* is decreased further, this branch ends or loses stability to an attractor which seems to exhibit spatio-temporal chaos with the cells undergoing an erratic motion involving creation and annihilation of cells, characteristic of weakly turbulent flames.

Higher-order transitions have also been found for the pulsating regime $Le > 1$. In [2, 7, 8, 13] the behavior of pulsating adiabatic cylindrical flames stabilized by a line source of fuel was considered numerically. In particular, for fixed κ, as Le was increased, axisymmetric flames were shown to undergo transitions from stationary behavior to sinusoidal temporal behavior, clearly indicating a Hopf bifurcation. As Le was increased further, the dynamic behavior became more relaxational. Beyond a critical value of Le a period-doubling transition occurred. The behavior of pulsating cellular flames was also considered for Le fixed while κ increased. It was shown that as κ increased, transitions between traveling wave solutions and standing wave solutions of different mode number occurred with regions of bistability. This was analogous to the behavior found for stationary cellular flames in the cellular regime [12].

In [10] the behavior of nonadiabatic pulsating flames was considered. In this case the problem of a flame established in the region between two coaxial cylinders with feed through the inner cylinder and heat loss through the outer cylinder was computed. The solution behavior was determined as the heat transfer coefficient η through the outer wall was increased. Transitions from TW solutions in the case of no heat loss ($\eta = 0$) to MTW solutions occurred when η exceeded a critical value. Unlike the MTW solutions in the cellular regime, these transitions are due to the introduction of subharmonic or near-subharmonic modes in the solution as η increased. Strikingly different behavior was found depending on whether a true subharmonic mode could be excited (even number of cells) or whether only near-subharmonics (odd number of cells) could enter. In both cases a subsequent transition to chaos occurred when η became sufficiently large and the flame was near extinction. We illustrate some of the results from [10] here. We note that a transition to chaos via a period-doubling cascade has also been found for axisymmetric flames near extinction using a model that accounted for the thermal expansion of the gas [4].

Our computations are for the case of finite, though large, activation energy. In this case the chemical reaction terms are important only in a narrow region termed the reaction zone. Beyond the reaction zone the reaction terms are small, because the deficient component has been consumed, while ahead of the reaction zone the temperature is too low to sustain significant reaction. In the limit of infinite activation energy the reaction zone shrinks to a surface called the flame front. The analyses in [24, 37] were obtained in this limit. In the case of finite activation energy there is, strictly speaking, no flame front, although we will use this terminology to describe the solution.

The accuracy of the computed solution is very sensitive to the resolution of the reaction zone, the location of which is not known beforehand. As a result, adaptive procedures are needed to locate and resolve the reaction zone as the solution evolves in time. The problem of obtaining adequate resolution of the reaction zone is further complicated by the extremely deep cells that occur as Le is that. We employ an adaptive pseudo-spectral method which has been used successfully for a variety of other problems in combustion (see, e.g., [2, 3, 5,]). We give a description of the model in Section 2 and the numerical method in Section 3. In Section 4 we describe our results in detail.

2. Mathematical Model

We first consider the problem of a flame stabilized by a line source of fuel of strength $2\pi\kappa$. We assume that the reaction is limited by a single deficient component and is governed by one-step, irreversible Arrhenius kinetics. We denote dimensional quantities by $\tilde{}$. The unknowns are the temperature $\tilde{T}$ and the concentration $\tilde{C}$ of the deficient component. $\tilde{T}_u$ and $\tilde{T}_b$ are the temperatures of the unburned and burned fuel respectively and $\tilde{C}_u$ is the unburned value of $\tilde{C}$. Other dimensional quantities are the coefficient of thermal conductivity $\tilde{\lambda}$, the activation energy $\tilde{E}$, and the gas constant $\tilde{R}$. We introduce the nondimensional reduced temperature and nondimensional concentration by

$$\Theta = (\tilde{T} - \tilde{T}_u)/(\tilde{T}_b - \tilde{T}_u), \quad C = \tilde{C}/\tilde{C}_u.$$

The spatial and temporal variables are nondimensionalized by

$$t = \frac{\tilde{t}\tilde{U}^2}{\tilde{\lambda}}, \quad x_i = \frac{\tilde{x}_i\tilde{U}}{\tilde{\lambda}}.$$

Here $\tilde{U}$ is the planar adiabatic flame speed for the case of infinite activation energy in which the reaction term is replaced by a surface delta function. We employ polar coordinates (r, ψ), and the nondimensionalized flow velocity due to the fuel source is $\mathbf{V} = \frac{\kappa}{r}\hat{\mathbf{r}}$ where $\hat{\mathbf{r}}$ is the unit radial vector. The equations of the diffusional thermal model are [37]

$$\begin{aligned} \Theta_t &= \Delta\Theta - \frac{\kappa\Theta_r}{r} + C\Lambda \exp\left(\frac{N(1-\sigma)(\Theta-1)}{\sigma+(1-\sigma)\Theta}\right), \\ C_t &= \frac{\Delta C}{Le} - \frac{\kappa C_r}{r} - C\Lambda \exp\left(\frac{N(1-\sigma)(\Theta-1)}{\sigma+(1-\sigma)\Theta}\right). \end{aligned} \tag{1}$$

Here Δ is the Laplacian, $\sigma = \tilde{T}_u/\tilde{T}_b, N = \tilde{E}/(\tilde{R}\tilde{T}_b)$, Le is the Lewis number, and $\Lambda = M^2/(2Le)$, where $M = N(1-\sigma)$. We note that Λ, which is referred to as the flame speed eigenvalue, depends on the nondimensionalization. The value employed above arises from the use of the planar, adiabatic flame velocity in the infinite activation energy limit $M \to \infty$. A different nondimensionalization would change the spatial and temporal scales but would not alter the basic patterns exhibited by the solution. The boundary conditions are

$$\begin{aligned} &C \to 1, \quad \Theta \to 0 \text{ as } r \to 0, \\ &C \to 0, \quad \Theta \to 1 \text{ as } r \to \infty. \end{aligned} \tag{2}$$

In our computations these boundary conditions are imposed at points r_1 and r_2 far from the reaction zone where combustion occurs. The computed results were found to be insensitive to changes in r_1 and r_2.

The model for the case that the flame is established in the region between two coaxial cylinders is similar. The only difference is in the boundary conditions. We denote the inner and outer radii by δ and a, respectively. The combustible mixture is fed in through the inner cylinder with a purely radial velocity field given by

$$\hat{\mathbf{V}} = (\kappa/\mathbf{r})\,\hat{\mathbf{r}},$$

where κ is specified. At the inner cylinder $r = \delta$ we specify the concentration of the deficient component and the temperature to be that of the unburned mixture

$$C = 1, \quad \Theta = 0.$$

At the outer cylinder $r = a$ we assume complete combustion so that $C = 0$. We also allow for heat transfer through the outer wall so that

$$\partial\Theta/\partial r = -\eta(\Theta - \Theta_w), \tag{3}$$

where η is a nondimensionalized heat transfer coefficient and Θ_w is the nondimensionalized temperature exterior to the outer wall. The sequences of transitions that are presented for this model were obtained by increasing η for fixed values of the other parameters.

The solution to (1)–(2) has been studied analytically in the limit $M \to \infty$ [24, 37]. In this limit the reaction zone shrinks to a surface $r = \Psi(\psi)$, called the flame front. The following stationary, axisymmetric solution exists:

$$\begin{aligned} \Theta &= \begin{cases} (\frac{r}{\kappa})^{\kappa} + O(\frac{1}{M}), & r \le \kappa, \\ 1, & r \ge \kappa, \end{cases} \\ C &= 1 - \Theta + O(\frac{1}{M}), \\ \Psi &= \kappa, \end{aligned}$$

and is referred to as the basic solution.

This solution becomes unstable if $Le < Le_{c1} < 1$ and κ is sufficiently large, or if $Le > Le_{c2} > 1$. In the first case a single real eigenvalue crosses from the left half-plane into the right half plane and stationary cellular flames are predicted [37]. In the second case two complex conjugate eigenvalues cross from the left half-plane into the right half-plane, which suggests the existence of pulsating flames (traveling or standing waves along the flame front or axially symmetric pulsations) [24].

3. Numerical Method

We employ an adaptive Chebyshev pseudo-spectral method in r and the Fourier pseudo-spectral method in ψ. In order to enhance the resolution of the reaction zone, we adaptively transform r in order to minimize a functional monitoring the numerical error.

We first briefly discuss pseudo-spectral methods. Detailed descriptions of this class of numerical methods can be found in [18, 26]. We transform the domain for the variable r to the interval $(-1,1)$. Here we denote the transformed variable by r as well. The solution, u, is then expanded as a sum of basis functions

$$u \simeq u_J \equiv \sum_{j=0}^{J} a_j \phi_j(r). \tag{4}$$

In a Chebyshev method the basis functions ϕ_j are the Chebyshev polynomials, that is, $\phi_j(r) = T_j$ where T_j is the jth Chebyshev polynomial,

$$T_j(r) = \cos(j \cos^{-1}(r)).$$

The expansion coefficients a_j are obtained from collocation, that is the function u_J is forced to solve the equations at a set of $J+1$ collocation points r_j. We employ the Gauss-Lobatto points,

$$r_j = \cos(\frac{j\pi}{J}), \quad 0 \leq j \leq J,$$

as collocation points. Thus in a pseudo-spectral method the unknowns are the solution values at the collocation points. The expansion (4) is only used to compute spatial derivatives. The Fourier method is similar, except that trigonometric functions are used as basis functions.

The major advantage of pseudo-spectral methods over finite difference methods is enhanced accuracy for a fixed discretization size. In fact pseudo-spectral methods exhibit infinite-order accuracy when used to approximate infinitely differentiable functions; that is, the error can be shown to decrease faster than any inverse power of J. These methods are commonly used to approximate problems where a high degree of accuracy is required, for example, in the study of fluid dynamical instabilities (see, e.g., [18]). However, although these methods are highly accurate when used to approximate functions that exhibit relatively gradual spatial variation, they have difficulties in approximating functions exhibiting localized regions of rapid variation, such as the temperature rise across the narrow reaction zone. Severe spatial oscillations can occur in approximating rapidly varying functions which are not well resolved [18, 26]. These oscillations can affect the computed dynamics and in certain cases lead to the computation of spurious dynamics (for example, inadequately resolved computations can indicate chaos, whereas the exact solution is in fact periodic).

If the location of the reaction zone is known in advance, the oscillations can be eliminated by introducing a suitable change of coordinates so that in the new coordinate system the solution has a more gradual variation. However, typically the location of the reaction zone is not known in advance and is one of the objects of the computation. To realize the benefits (high accuracy) of the pseudo-spectral method in computing solutions that vary rapidly in localized regions of space, we developed an adaptive pseudo-spectral method. This method has proven to be effective in computing the rapidly varying solutions which occur in combustion problems. The method is described in detail in [2, 5]. The description we give here will be brief.

We introduce a family of coordinate transformations of the form

$$r = q(s, \boldsymbol{\alpha}), \tag{5}$$

where s is the new computational coordinate and $\boldsymbol{\alpha}$ represents a parameter vector that is typically of low dimension. We choose $\boldsymbol{\alpha}$ so that in the new coordinate system the solution exhibits a more gradual variation and thus is better approximated by a small number of basis functions. Since the behavior of the solution changes during the course of the computation, appropriate values of $\boldsymbol{\alpha}$ must be chosen adaptively so as to adapt to changes in the solution [2, 5]. The choice of the coordinate transformations, which we discuss below, is motivated by the asymptotic notion of "stretching" a layer in the method of matched asymptotic expansions [15, 28]. for treating singular perturbation problems.

To adaptively choose a coordinate transformation in which the spectral approximation is more accurate we employ error measures that are computed for each value of $\boldsymbol{\alpha}$ until a minimum is found. In this paper the error measure is the functional

$$I_2(g) = \left(\int_{-1}^{1} (L^2 g)^2 / w(s) ds \right)^{\frac{1}{2}}, \tag{6}$$

where

$$w(s) = \sqrt{1 - s^2}, \quad L = w(s)\frac{d}{ds}.$$

It can be shown that this functional gives an upper bound on the maximum norm of the error in approximating a function by its Chebyshev expansion [2].

A very important feature of the adaptive pseudo-spectral method is the particular family of mappings. The unknowns Θ and C vary gradually except near the reaction zone where rapid changes occur. We employ a family of mappings, introduced in [14], in which functions with properties similar to these are mapped into linear polynomials, which can be approximated using a relatively small value of J.

To describe this family of mappings, we let I denote the interval $-1 \leq r \leq 1$ and consider the family of functions

$$h(r, \alpha_1, \alpha_2) = s_0 + \tan^{-1}(\alpha_1(r - \alpha_2))/\lambda. \tag{7}$$

The parameters s_0 and λ are determined so that (7) maps the interval I univalently onto itself,

$$s_0 = \frac{\beta - 1}{\beta + 1}, \quad \beta = \tan^{-1}(\alpha_1(1 + \alpha_2))/\tan^{-1}(\alpha_1(1 - \alpha_2)),$$

$$\lambda = \tan^{-1}(\alpha_1(1 - \alpha_2))/(1 - s_0).$$

When α_1 is large (the analog of ϵ small in singular perturbation problems), (7) is nearly discontinuous with a region of rapid variation occurring near $r = \alpha_2$, i.e., there is an internal layer at $r = \alpha_2$. We observe that the function tanh(z) can be used in place of the function $\tan^{-1}(z)$.

The inverse of (7),

$$r = q(s, \alpha) = \alpha_2 + \tan((s - s_0)\lambda)/\alpha_1, \tag{8}$$

describes a two-parameter family of mappings of I that are one to one and onto, with the property that $h(q(s, \alpha_1, \alpha_2), \alpha_1, \alpha_2)$ is a linear function. If α_1 and α_2 are properly chosen, the temperature and concentration profiles will be sufficiently similar to (7) that the composite function can be represented by a small number of Chebyshev polynomials. The parameter α_1 is related to the rapid rate of change of the function, while α_2 is related to the location of the layer, namely, the region of rapid variation. In our method these parameters are determined adaptively by minimizing the functional (6).

In the computation of cellular flames, the location of the reaction zone depends on ψ. In this case the optimal values of α_1 and α_2 will also depend on ψ. A two-dimensional adaptive pseudo-spectral method that allows for this dependence has been developed and applied to problems in gasless condensed phase combustion in [3]. The resulting coordinate system is non-separable, thereby requiring additional computation as described in [3]. The problem considered here is posed in a coordinate system in which the Laplacian is separable, and it is more efficient to use one-dimensional coordinate transformations that maintain separability (i.e., α_1 and α_2 independent of

ψ), even at the expense of not using values of α_1 and α_2 which are optimal for any particular value of ψ. We therefore compute the functional (6) for all angular points and minimize the average, rather than minimizing for each value of ψ. We are nevertheless able to obtain a high degree of resolution of the wrinkled reaction zone, because of the effectiveness of the family of mappings (8) in concentrating resolution in a fixed region. As a result, after the mapping is applied Θ and C are no longer rapidly varying in any angular direction, although the parameters α_1 and α_2 are generally not optimal for any particular direction.

4. Results

We first describe results obtained in the cellular regime, $Le < 1$. The following parameters were held fixed, $N = 20, \sigma = .615, \kappa = 14.8$. The time steps were typically $O(10^{-3})$. In all cases 101 Chebyshev collocation points were sufficient to accurately compute solutions in the radial direction. We generally used 128 collocation points in the angular direction, although in some cases we employed 256. There was virtually no effect in increasing the number of collocation points in either independent variable. In the radial direction the computational domain was taken as $r_1 \leq r \leq r_2$ with $r_1 = 1, r_2 = 41$. We found virtually no effect in changing these values. The computations presented here were obtained at the NCSA and the NERSC.

An overview of the different solution branches is given in Fig. 1 where we plot the maximum norm of the difference between the computed cellular solution and the (unstable) axisymmetric solution for the different solution branches that we have found. We generally concentrate on determining the nature of the transitions that occur and on the properties of solutions along different solution branches rather than on determining the precise numerical values of the control parameter (in this case Le) at which the transitions occur. We note that we solve the initial value problems so that in general we compute only stable solutions. However, unstable solutions can sometimes be computed by imposing certain symmetries in the angular direction. For example, unstable axisymmetric solutions can be computed by allowing no variation in ψ. Unstable 4-mode solutions can be computed by restricting the computation to the interval $0 \leq \psi \leq 2\pi/4$ and imposing periodicity. Unstable reflection symmetric solutions can be computed by restricting to the interval $0 \leq \psi \leq 2\pi/8$ and imposing reflection symmetry.

We first comment on the branches illustrated in Fig. 1. We see that the axisymmetric solution loses stability to a branch of stable 4-cell solutions (S4) for a value of Le slightly larger than 0.52. The solutions along this branch are reflection symmetric, and the branch is bistable with a branch of stationary symmetric 5-cell solutions (S5). Our results indicate that the branch S5 does not emanate stably from the axisymmetric solution branch, although there may be unstable connecting branches between S5 and either S4 or the axisymmetric branch.

A branch of traveling wave solutions (TW4) emanates from S4 at a value of *Le* near 0.44. As shown in [9], transition is accompanied by a breaking of the reflection symmetry, sometimes referred to as parity breaking. Beyond the transition point the branch S4, although unstable, can be computed by imposing reflection symmetry. For values of *Le* near 0.4 we can no longer compute solutions on this branch even with the imposition of reflection symmetry. As *Le* is decreased, the first harmonic (Fourier mode 8) becomes stronger, while the strength of the fundamental decreases. At about $Le = 0.39$ the branch S4 ends on the branch S8 so that there is a continuous transition from the branch S4 to the branch S8. Alternatively, as *Le* increases, the branch S4 bifurcates from the branch S8. The branch S8 is unstable when the restriction of reflection symmetry is removed over the whole range of *Le* indicated in the figure. We do not follow the branch S8 for larger values of *Le*.

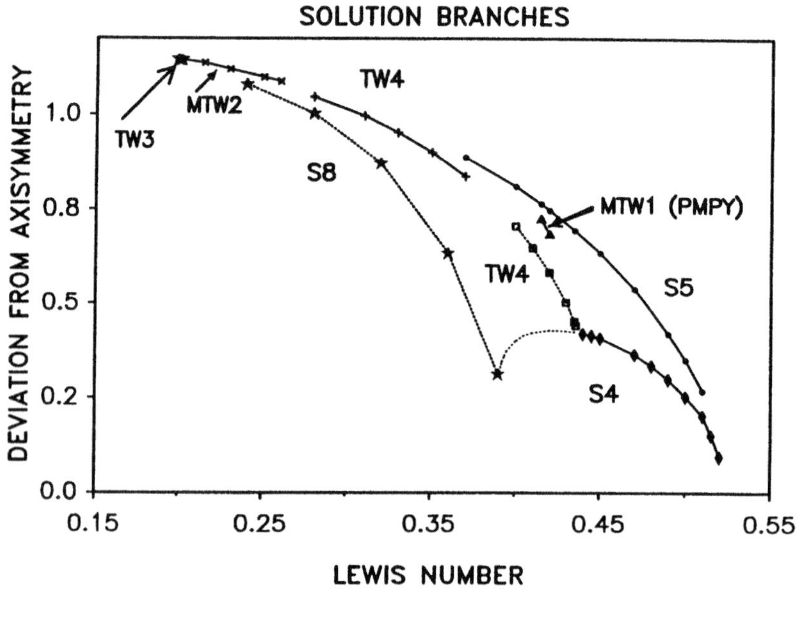

Fig. 1. **Solution branches as *Le* is varied**

Our results indicate that at some distance beyond the transition point the TW4 branch is unstable. While it is stable to disturbances of period $2\pi/4$, it is unstable to disturbances of other mode numbers until Le is decreased to around 0.37 when it becomes stable. Over most of the region of instability for the branch TW4 arbitrary initial data evolves to the branch S5. There is, however, a small interval in Le where initial data evolves to a mixed mode branch MTW1. The solution along this branch is quasi-periodic. It is the first of the MTW branches that we describe.

The branch TW4 becomes stable for Le near 0.37 and can be computed for values of Le down to 0.28. Beyond this value of Le we no longer find stable 4-cell TW solutions. Rather stability is transferred to the branch MTW2. This is a mixed mode solution with modes 3 and 4 (as well as mode 0) predominating. It is the second of the MTW branches that we describe, and its evolution leads to spatio-temporal chaos. Finally, we note that there is a branch of 3-cell traveling wave solutions (TW3) that is stable over a very narrow interval around $Le = 0.2$. This branch is indicated in Fig. 1, although the maximum norm of the difference between these solutions and the axisymmetric solution is indistinguishable from the corresponding values for the MTW2 branch. It may well be the case that the branch TW3 extends to higher values of Le although it is unstable and that the MTW2 branch results from interaction between unstable continuations of the branches TW3 and TW4.

We now describe the behavior of the branch MTW1 in more detail. We consider the case $Le = 0.42$. In Fig. 2 we present a contour plot of Θ for a fixed value of t.

The contours have been chosen to emphasize the values of Θ near 0.98 where the cellular structure of the solution is most pronounced. It can be seen from the figure that the solution is composed of 4 cells. None of the cells are exactly alike; however, two of the cells are nearly reflection symmetric, characteristic of the behavior along the branch S4, while two of the cells are not symmetric, characteristic of the behavior along the branch TW4. (This is brought out more clearly in [9], where the cellular structure is illustrated in color).

It would be expected that the nearly symmetric cells are nearly stationary while the asymmetric cells undergo motion. The dynamics of this solution can be seen in Fig. 3, where we have plotted $\Theta(r_*, \psi, t)$ as a function of ψ for different values of t. Here r_* is a fixed point in the reaction zone. From this figure we see that as the forward crest (temperature minimum) of a given cell (cell 1) is jerked clockwise (to the left in Fig. 3), the cell expands, becomes asymmetric, and moves clockwise. This, in turn, pulls the cell behind it (cell 2) so that it too expands, becomes asymmetric, and moves clockwise. At the same time cell 1 pushes the cell ahead of it, compressing it and causing it to become nearly symmetric and nearly stationary. This process then repeats with cell 2, 3, etc., in turn, so that individual cells undergo a motion in sequence, which leads to a localized wave of asymmetry propagating counter clockwise through the cellular array.

This behavior is analogous to the mode observed experimentally in [22], where it was termed a square dance mode. We employ the descriptive terminology "pushmi - pullyu" (PMPY) [31] to describe this solution. Similar patterns have been observed in directional solidification [23], in directional viscous fingering [42] and in Taylor vortex flow [46]. It can be seen from Fig. 3 that the behavior can also be characterized as that of an MTW, since after a certain time interval the pattern reemerges with a phase shift. Alternatively, the solution would be periodic in a moving coordinate system.

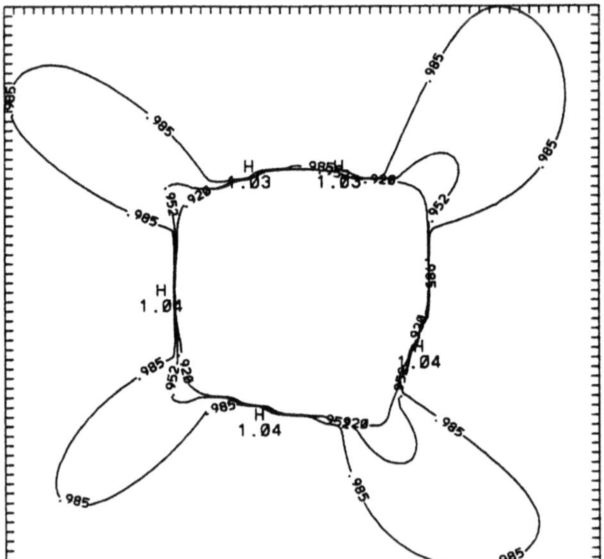

Fig. 2. **Contour plot for Θ for solution on MTW1 branch,** $Le = 0.42$

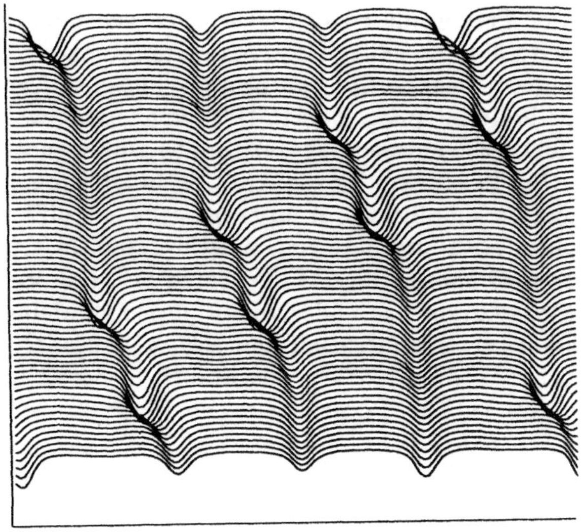

Fig. 3. $\Theta(r_*, \psi, t)$ for solution on MTW1 branch, Le = 0.42. Time t (angle ψ) increases along vertical (horizontal) axis.

We next describe the behavior along the MTW2 branch. This behavior is also that of an MTW, although very different from that of the MTW1 branch. In Fig. 4 we plot $\Theta(r_*, \psi, t)$ as a function of ψ for different values of t where r_* is a point in the reaction zone. The solution is composed of 4 distinct cells that undergo a counterclockwise motion (to the right in the figure). To understand this motion, we note that the general effect of decreasing Le is to increase the size of the cells, where by the size of the cells we refer to the elevated high-temperature region. In fact, decreasing Le increases the diffusivity of the reactant relative to the diffusivity of heat. As a result, reactant diffuses into a region of higher temperature where a higher degree of burning has occurred, thus increasing the size of the cells. For the values of Le considered, the radial location of the reaction zone does not change significantly; and denoting r_* as a generic point in the reaction zone, we find that the cells must fit into a circumference $2\pi r_*$. As Le is decreased, progressively larger cells are forced to fit in the given circumference. At some point a more stable configuration is that illustrated in Fig. 4 where the cells are of different size. The following dynamical behavior is observed. There is an overall rotation of the cellular array. During the evolution, one cell expands compressing its neighbor to the left. After another time interval the neighbor to the right expands, compressing the original cell. In this way there is a modulation in cell size (and also in rotational velocity) superimposed on the TW behavior leading to MTW behavior.

For the value of *Le* considered in Fig. 4, each cell undergoes the same modulation but with a phase shift. Thus, in an appropriately chosen moving coordinate system, with each cell appropriately phase shifted, the solution would be $2\pi/4$ periodic in space. The PMPY mode exhibits the same symmetry. As *Le* is further decreased, a transition to an alternating modulated traveling wave (AMTW) occurs. In the AMTW, in addition to phase differences, the modulation amplitudes of adjacent cells are different, while the modulation amplitudes of alternate cells are the same. In a moving coordinate system, chosen as above, the AMTW would be $2\pi/2$ periodic in space. That is, the MTW solution has undergone a period doubling. Presumably this transition occurs by a bifurcation and reflects a branch different from the MTW2 branch. We do not trace this branch in detail but indicate an example of a solution on this branch in Fig. 5 where we consider the case $Le = 0.18$.

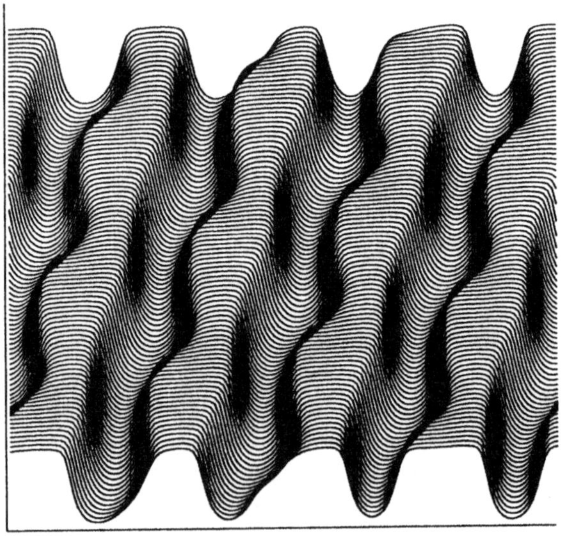

Fig. 4. $\Theta(r_*, \psi, t)$ **for solution on MTW2 branch,** $Le = 0.23$**. Time** t **(angle** ψ**) increases along vertical (horizontal) axis**

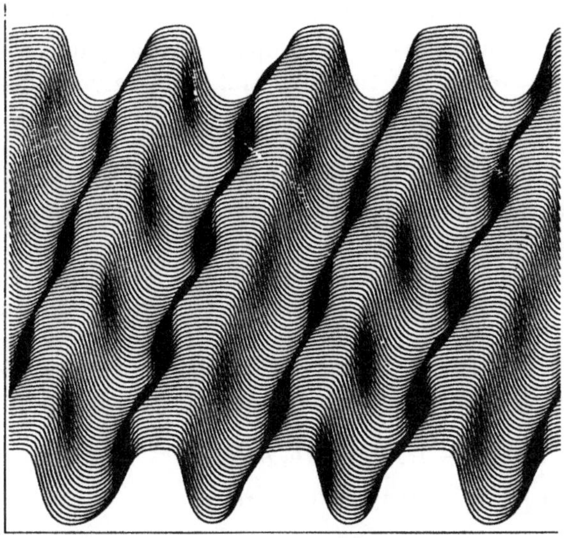

Fig. 5. $\Theta(r_*, \psi, t)$ **for MTW solution** , Le = 0.18 **Time** t **(angle** ψ**) increases along vertical (horizontal) axis.**

Fig. 6. $\Theta(r_*, \psi, t)$ **for solution apparently exhibiting spatio-temporal**

Upon decreasing Le yet further, the MTWs appear to become unstable, and a transition occurs to a complex spatial pattern which seems to exhibit erratic movement of the cells and seemingly random creation and annihilation of cells. This is illustrated in Fig. 6 for the case $Le = 0.165$.

There is a connection between the $S4$, $S8$, and $TW4$ branches as well as between the $TW4$ and $MTW1$ branches which follows from considerations we now describe. The merging of the $S4$ mode with the $S8$ mode suggests considering the resonant interaction of a mode with amplitude A and wave number q and a mode with amplitude B and wavenumber $2q$ corresponding to $S4$ and $S8$, respectively. In the vicinity of a codimension-2 point in parameter space at which both these modes destabilize the axisymmetric state simultaneously, the dynamics of the system restricted to $2\pi/4$ periodic solutions can be described by two complex amplitude equations

$$\begin{aligned} \dot{A} &= \mu_1 A + c_1 A^* B + a_1 A|A|^2 + b_1 A|B|^2, \\ \dot{B} &= \mu_2 B + c_2 A^2 + b_2 B|A|^2 + a_2 B|B|^2, \end{aligned} \tag{9}$$

where * denotes complex conjugate, and where $\dot{}$ denotes differentiation with respect to a slow time. The parameters μ_1 and μ_2 are the control parameters. These equations, which have been studied extensively (e.g., [1, 20, 41]), exhibit a parity breaking transition from a stationary q mode branch to a TW branch before the q mode branch merges with the $2q$ mode branch, just as in this paper. Though the present simulations are not close to such a codimension-2 point, calculations in directional solidification [43] and in Taylor vortex flow [44] showed that the regime of existence of the TW branch can extend far from this special point in parameter space. Thus the amplitude equations for the $q : 2q$ mode interaction appear to describe transitions from $S4$ to $TW4$ and to $S8$, respectively. The amplitude equations also describe MTW solutions, but these are different from the MTW solutions described in this paper. This is due to the fact that the amplitude equations are restricted to a description of the behavior of four identical cells, while in our MTW solutions MTW1 and MTW2 as well as in our AMTW solution, the four cells do not behave identically.

To understand the connection between the TW and the $MTW1$ branches, it is useful to take a different approach. In the vicinity of a parity breaking instability the dynamics can be described by two coupled real equations for the amplitude A of the antisymmetric part of the solution, which measures the extent to which the reflection symmetry is broken, and the local phase ϕ of the solution, which gives the instantaneous position of each cell [19]. In the context of the present computation, the symmetric solution $S4$ corresponds to $A = 0$. In an extended system, A and ϕ can depend slowly on space (i.e., on the angular coordinate), as well as time, so that they satisfy

$$\dot{\phi} = A, \tag{10}$$

$$\dot{A} = (\lambda + \lambda_1\phi' + \lambda_2\phi'^2)A + d_2A'' - A^3 + gAA' + (b_2 + b_{21}\phi')\phi'',$$

where $\dot{}$ and $'$ denote differentiation with respect to the slow time and space variables, respectively, and *lambda* is the control parameter. An investigation of these equations for an infinite system shows that at onset the *TW* branch is in general unstable with respect to sideband instabilities. The *TW* branch can, however, become stable further away from onset, that is, for somewhat larger amplitudes. This is consistent with the present computations, where we find that the TW4 solutions change from unstable to stable as *Le* is decreased along the TW4 branch.

The coupled phase-amplitude equations (10) can also be shown to possess stable *localized* solutions, in which the reflection symmetry is broken in only part of the system. The pattern then exhibits a number of asymmetric traveling cells which are embedded in symmetric stationary cells [44]. This solution is identified with the *MTW*1 branch. A characteristic feature of this solution is that the individual cells travel in a direction (to the left in Fig. 3) which is opposite to that of the envelope of the antisymmetric part of the solution.

We now consider the behavior of nonadiabatic pulsating cellular flames with $Le > 1$. We describe two sets of computations with the basic solution, that is, with the heat transfer coefficient $\eta = 0$, corresponding to a 7-cell TW and an 8-cell TW, respectively. For both sets of computations the following parameters were held fixed, $N = 20, \sigma = 0.5$, $Le = 4.4$, $\Theta_w = 0.4$. We obtained accurate radial resolution with 61 collocation points. The number of angular collocation points was either 256 or 512. In the first set of computations $\kappa = 20$, $\delta = 27.3$, $a = 40$. For the second set $\kappa = 24$, $\delta = 34.25$, $a = 47$. For both sets of computations we consider the effect of increasing η until extinction occurs. The predominant effect of increasing η is to promote the development of subharmonic modes along the front. This leads to different behavior for the two cases considered.

We first consider the case where the basic solution is an 8-cell TW. In this case the introduction of the subharmonic breaks the symmetry corresponding to the $2\pi/8$ periodicity. The introduction of mode 4 leads to a symmetry based on $2\pi/4$ periodicity. There is now an 8-cell array in which cells alternate in a regular fashion. Correspondingly the period (in time) of the solution at any fixed point in space is doubled as two cells have to pass a fixed point before the solution can repeat a given value. Thus there is a space/time period doubling. Each cell remains unaltered as it rotates, and the resulting solution is still a TW. We illustrate this behavior in Fig. 7 where we plot $\Theta(r_*, \psi, t)$ as a function of ψ, for r_* in the reaction zone and for two different values of t. Increasing η yet further leads to additional transitions in which subharmonic modes enter. Mode 2 enters for a value of η between 15 and 18.5 , leading to a second space/time period doubling.

Mode 1 enters for a value of η between 18.5 and 18.8, leading to a third space/time period doubling. This sequence of transitions cannot continue beyond this point, as there are no additional subharmonics that can enter. We find that shortly after the third period doubling the solution exhibits apparently chaotic temporal dependence [10]. The spatial structure is still that of an 8-cell rotating solution where each cell is of a different size (as is also the case after the third period doubling); however, each cell now undergoes a low-level, apparently chaotic, oscillation in size and maximum temperature. We have been unable to find any state between the periodic state after the third period doubling and the apparently chaotic state.

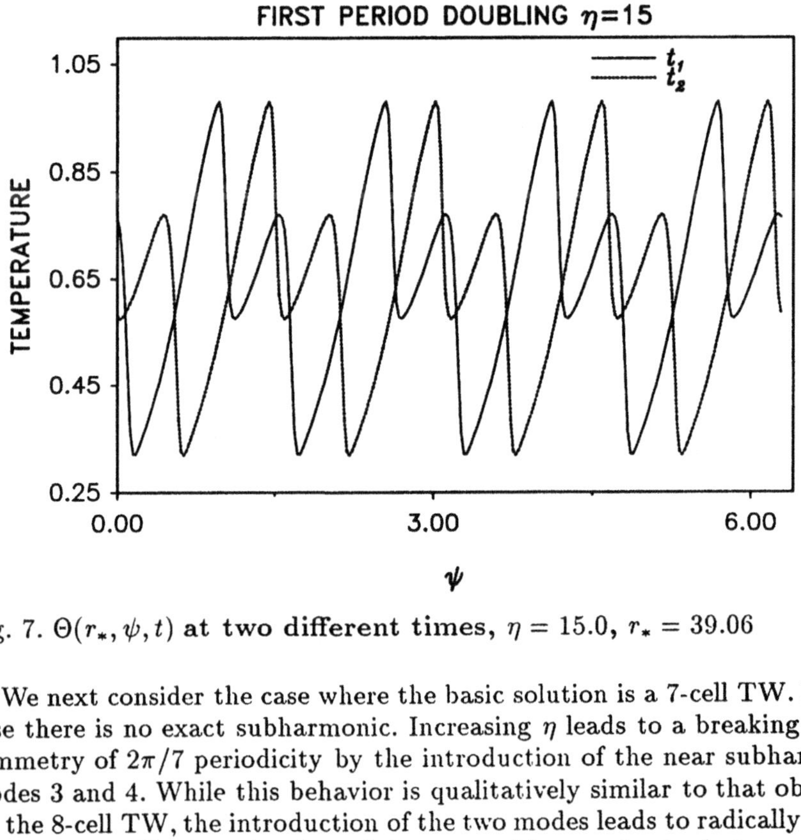

Fig. 7. $\Theta(r_*, \psi, t)$ **at two different times,** $\eta = 15.0$, $r_* = 39.06$

We next consider the case where the basic solution is a 7-cell TW. In this case there is no exact subharmonic. Increasing η leads to a breaking of the symmetry of $2\pi/7$ periodicity by the introduction of the near subharmonic modes 3 and 4. While this behavior is qualitatively similar to that observed for the 8-cell TW, the introduction of the two modes leads to radically different behavior. The introduction of the near subharmonics leads to a roughly alternating behavior; however, the pattern cannot be regular because of the odd number of cells. Thus at any fixed instant of time the spatial pattern looks roughly period doubled in ψ (i.e., there is a rough alternating behavior in the size of the cells as ψ increases from 0 to 2π).

This alternating behavior is superimposed on a TW so that the resulting structure rotates about the cylindrical axis. Over short time scales the temperature (or concentration) at a fixed point reflects this rough period doubling in ψ. In particular, as the alternating array of cells rotates past the fixed point, the time evolution of the temperature also exhibits a rough period doubled type of behavior. However, the spatial period doubling is not maintained over long time scales, as an alternating pattern is not compatible with an odd number of cells. Over long time scales each cell undergoes an oscillation in both cell size and maximum temperature. This cell oscillation is such that there is a constant phase difference between each cell and its neighbors. Near the transition point, the period of the oscillation is long compared to the rotation time of the cellular flame. Because of the rotational symmetry of the system, the two frequencies are incommensurate with one another. The result is a quasi-periodic MTW.

As heat loss is increased, further transitions occur in the cell oscillation. Specifically there are two period doublings of the cell oscillation. The resulting solution still exhibits two frequency quasi-periodic dynamics. As heat loss is further increased, there is a transition to apparently chaotic behavior. We have computed Lyapunov exponents for this solution using a code based on embedding techniques [29]. For all choices of the input parameters (e.g., embedding dimension and time delay), the largest Lyapunov exponent was found to be clearly positive. As in the case of the 8-cell TW, we are unable to find any other solution prior to the onset of chaos.

The dynamic response, for example the temporal dependence of the temperature at a fixed point in space, is extremely complex. The spatial structure along the front exhibits considerably more order. At any instant of time there are seven cells along the front. As time increases, these cells rotate about the cylindrical axis $r = 0$ and also change shape as they rotate. An additional effect of increasing heat loss is an increase in the amplitude of the higher Fourier modes. This results in a sharpening of the cells so that very rapid changes in temperature occur over a narrow interval in ψ. For large heat losses the effect is almost that of a series of fronts rotating in the angular direction. The degree of sharpening changes as each cell oscillates. When the peak temperature of the cell is high, the cells are extremely sharp, while when the peak temperature is reduced the cellular structure relaxes or becomes less sharp. In this sense the behavior of the MTW is analogous to the sharpening and relaxation of the reaction zone that occurs in axisymmetric or planar pulsating flames where there is no structure along the flame front [4].

We illustrate the dynamics in Fig. 8 for the case $\eta = 14.9$, which is as close to the transition point from TW to MTW that we have computed, by plotting the maximum temperature of each individual cell, obtained at a fixed radial location in the reaction zone in the manner described in [10]. We note that the cell oscillation is nearly sinusoidal. Increasing η leads to a progressive sharpening of the cell pulsation. We illustrate this in Fig. 9, where we plot the same quantities for $\eta = 30.0$. We note that for this value of η no period doubling of the cell oscillation has occurred, but rather, two localized pulses per period occur, because of the strengthening of even harmonics. Increasing η yet further leads to two period doublings in the oscillation of each individual cell. There is some indication that these period doublings are connected with the further development of modes 2 and 1 in the solution although this is not clear as nonlinear interactions can generate all modes once modes 3, 4, and 7 are present. In Fig. 10 we illustrate the cell modulation for $\eta = 50.0$ after the second period doubling. Increasing η further leads to apparently chaotic behavior. This is illustrated in Fig. 11 for $\eta = 54.0$. The chaotic behavior is primarily temporal. The spatial structure consists of 7 cells rotating about the cylindrical axis undergoing an erratic pulsation in temperature and size.

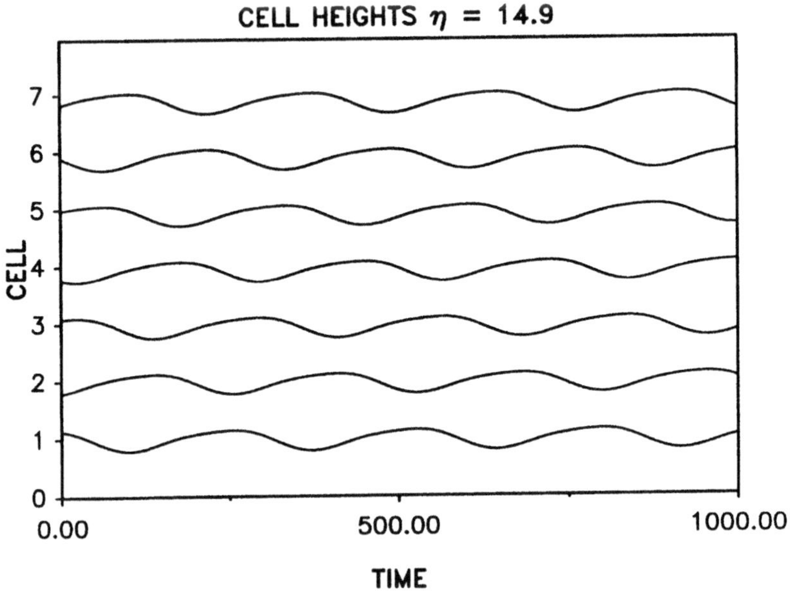

Fig. 8. **Cell temperatures for all 7 cells,** $\eta = 14.9$, $r_* = 32.714$

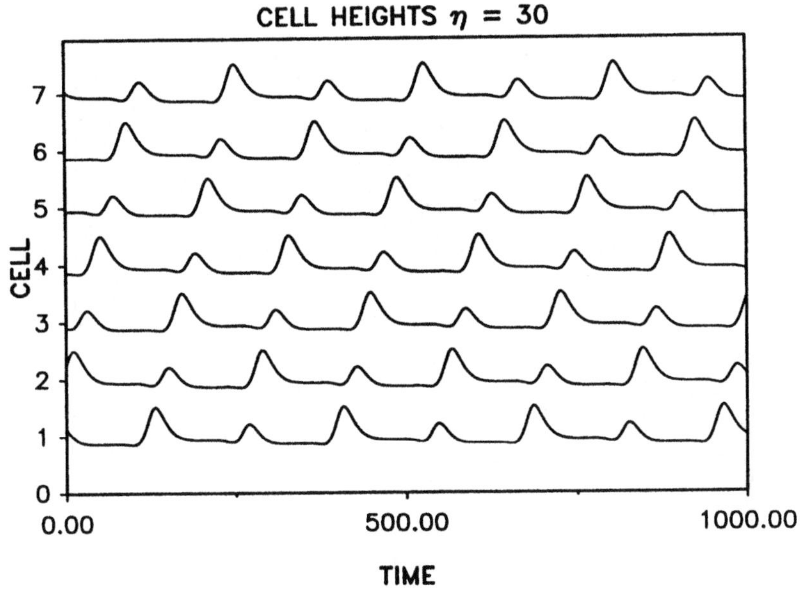

Fig. 9. Cell temperatures for all 7 cells, $\eta = 30.0$, $r_* = 32.714$

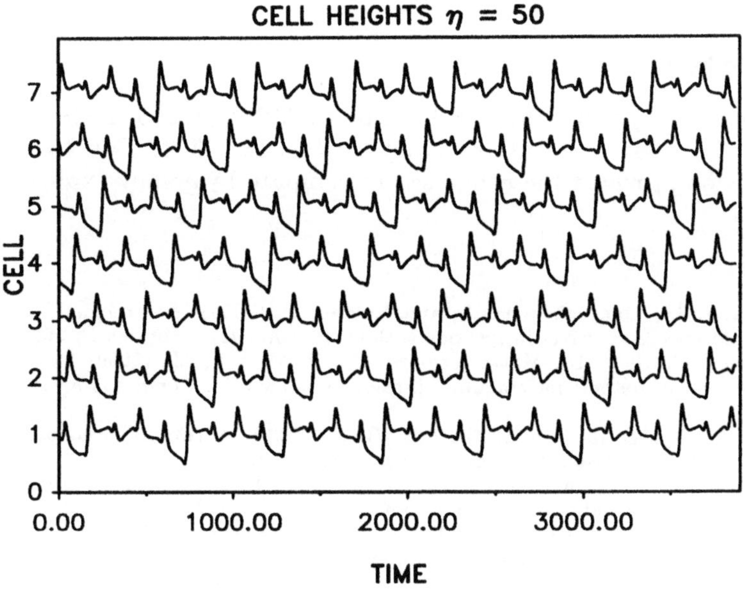

Fig. 10. Cell temperatures for all 7 cells, $\eta = 50.0$, $r_* = 32.714$

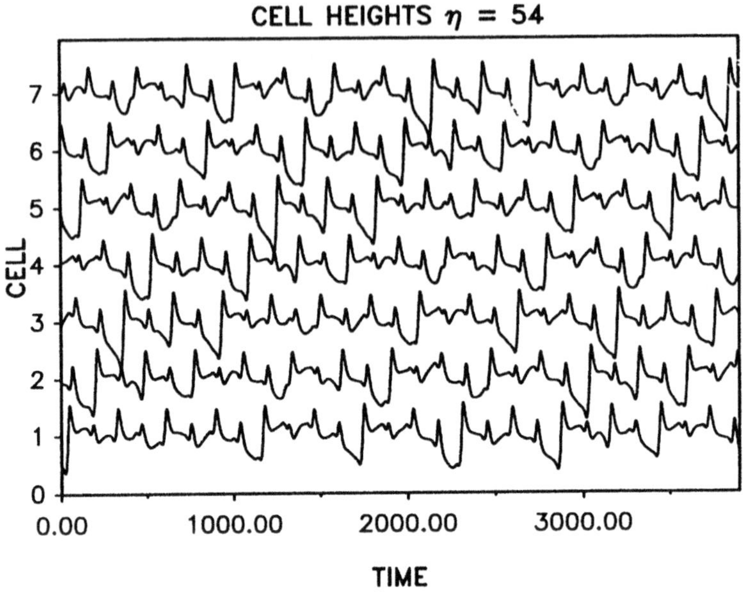

Fig. 11. **Cell temperatures for all 7 cells,** $\eta = 54.0$**,** $r_* = 32.714$

Acknowledgements

This work was supported by N.S.F. grants DMS 89-22716, DMS 90-20289 and MSS 91-02981 and D.O.E. grant DEFG02ER25027. We thank Eric Kostelich who provided the code used to compute Lyapunov exponents.

References

[1] Armbruster, D., Guckenheimer, J., and Holmes, P. (1988) 'Heteroclinic cycles and modulated travelling waves in systems with O(2) symmetry', *Physica D,* **29**, 257–282.

[2] Bayliss, A., Gottlieb, D., Matkowsky, B. J., and Minkoff, M. (1989) 'An adaptive pseudo-spectral method for reaction diffusion problems', *J. Comput. Phys.*, **81**, 421–443.

[3] Bayliss, A., Kuske, R., and Matkowsky, B. J. (1990) 'A two-dimensional adaptive pseudo-spectral method', *J. Comput. Phys.*, **91**, 174–196.

[4] Bayliss, A., Leaf, G. K., and Matkowsky, B. J. (1992) 'Pulsating and chaotic dynamics near the extinction limit', *Comb. Sci. and Tech.*, **84**, 253–278.

[5] Bayliss, A., and Matkowsky, B. J. (1987) 'Fronts, relaxation oscillations, and period doubling in solid fuel combustion', *J. Comput. Phys.*, **71**, 147–168.

[6] Bayliss, A., and Matkowsky, B. J. (1990) 'Spinning cellular flames', *Appl. Math. Letters*, **3**, 75–79.

[7] Bayliss, A., and Matkowsky, B. J. (1991) 'Bifurcation pattern formation and chaos in combustion', in Fife, Linan, and Williams (eds.), *Dynamical Issues in Combustion Theory*, Springer-Verlag, Berlin, 1–36.

[8] Bayliss, A., and Matkowsky, B. J. (1991) 'Bifurcation pattern formation and transition to chaos in combustion', in Seydel, Schneider, Kupper, and Troger (eds.), *Bifurcation and Chaos: Analysis, Algorithms, Applications,* Birkhauser, Basel, 36–51.

[9] Bayliss, A., and Matkowsky, B. J. (1992) 'Nonlinear dynamics of cellular flames', *SIAM J. Appl. Math.,* **52**, 396–415.

[10] Bayliss, A., and Matkowsky, B. J. (1992) 'From traveling waves to chaos in combustion', preprint.

[11] Bayliss, A., Matkowsky, B. J., and Minkoff, M. (1987) 'Adaptive pseudo-spectral computation of cellular flames stabilized by a point source', *Appl. Math. Lett.,* **1**, 19–24.

[12] Bayliss, A., Matkowsky, B. J., and Minkoff, M. (1989) 'Cascading cellular flames', *SIAM J. Appl. Math.,* **49**, 1421–1432.

[13] Bayliss, A., Matkowsky, B. J., and Minkoff, M. (1989) 'Numerical computation of bifurcation phenomena and pattern formation in combustion', in Dervieux and Larrouturou (eds.), *Numerical Combustion, Lecture Notes in Physics,* Vol. 351, Springer-Verlag, Heidelberg, 187–198.

[14] Bayliss, A., and Turkel, E. (1992) 'Mappings and accuracy for Chebyshev pseudospectral approximations', *J. Comput. Phys.,* to appear.

[15] Bender, C. M., and Orszag, S. A. (1978) *Advanced mathematical methods for scientists and engineers,* McGraw-Hill, New York.

[16] Buckmaster, J. (1983) 'Stability of the porous plug burner flame', *SIAM J. Appl. Math.,* **43**, 1335–1349.

[17] Buckmaster, J. (1984) 'Polyhedral flames – an exercise in bimodal bifurcation analysis', *SIAM J. Appl. Math.,* **44**, 40–55.

[18] Canuto, C., and Hussaini, M. Y., A. Quarteroni, and Zang, T. A. (1987) *Spectral methods in fluid dynamics,* Springer-Verlag, New York.

[19] Coullet, P., Goldstein, R. E., and Gunaratne, G. H. (1989) 'Parity-breaking transitions of modulated patterns in hydrodynamic a breaking of the symmetry of $2\pi/7$ periodicity by systems', *Phys. Rev. Lett.,* **63**, 1954–1957.

[20] Dangelmayr, G. (1986) 'Steady-state mode interactions in the presence of O(2)-symmetry', *Dyn. Stab. Sys.,* **1**, 159–185.

[21] el-Hamdi, M., Gorman, M., Mapp, J. W., and Blackshear, J. I., Jr. (1987) 'Stability boundaries of periodic modes of propagation in burner-stabilized methane-air flames', *Comb. Sci. and Tech.,* **55**, 33–40.

[22] el-Hamdi, M., Gorman, M., and Robbins, K. A. (1992) 'A picture book of dynamical modes of flat, laminar premixed flames', preprint.

[23] Flesselles, J. M., Simon, A. J., and Libchaber, A. (1991) 'Dynamics of one-dimensional interfaces: An experimentalist's view', *Adv. Phys.,* **40**, 1.

[24] Garbey, M., Kaper, H. G., Leaf, G. K., and Matkowsky, B. J. (1989) 'Linear stability analysis of cylindrical flames', *Quarterly of Applied Math.,* **47**, 691–704.

[25] Gorman, M., el-Hamdi, M., and Robbins, K. A. (1992) 'Spatiotemporal chaotic dynamics of premixed flames', in *First Experimental Conference on Chaos,* World Scientific Publishers, 403–407.

[26] Gottlieb, D., and Orszag, S. A. (1977) *Numerical analysis of spectral methods: Theory and applications, CBMS - NSF Conference Series in Applied Mathematics,* SIAM, Philadelphia.

[27] Joulin, G., and Clavin, P. (1979) 'Linear stability analysis of non-adiabatic flames', *Comb. and Flame,* **35**, 139–153.

[28] Kevorkian, J., and Cole, J. D. (1981) *Perturbation Methods in Applied Mathematics,* Springer-Verlag, New York.

[29] Kostelich, E. J., and Yorke, J. A. (1990) 'Noise reduction: Finding the simplest dynamical system consistent with the data', *Physica D,* **41**, 183–196.

[30] Kuske, R., and Matkowsky, B. J. (1992) 'Two dimensional cellular burner stabilized flames', *Quart. Appl. Math.,* to appear.

[31] Lofting, H. (1922) *The Voyages of Doctor Dolittle,* J. B. Lippincott, Philadelphia.

[32] Margolis, S. B., and Matkowsky, B. J. (1983) 'Nonlinear stability and bifurcations in the transition from laminar to turbulent flame propagation', *Comb. Sci. and Tech.*, **34**, 45–77.
[33] Markstein, G. H., ed. (1967) *Nonsteady Flame Propagation*, Pergamon Press, Elmsford, New York.
[34] Markstein, G. H., and Schwartz, D. (1956) 'Cellular slot burner flames', in D. K. Fleming (ed.), *Proceedings of the Gas Dynamics Symposium on Aerothermochemistry*, Northwestern University, 83–94.
[35] Matkowsky, B. J., and Olagunju, D. O. (1982) 'Traveling waves along the front of a pulsating flame', *SIAM J. Appl. Math.*, **42**, 486–501.
[36] Matkowsky, B. J., and Olagunju, D. O. (1990) 'Spinning waves in gaseous combustion', *SIAM J. Appl. Math.*, **42**, 1138–1156.
[37] Matkowsky, B. J., Putnick, L. J., and Sivashinsky, G. I. (1980) 'A nonlinear theory of cellular flames', *SIAM J. Appl. Math.*, **38**, 489–504.
[38] Olagunju, D. O., and Matkowsky, B. J. (1990) 'Burner-stabilized cellular flames', *Quart. J. Appl. Math.*, **48**, 645–664.
[39] Olagunju, D. O., and Matkowsky, B. J. (1991) 'Polyhedral flames', *SIAM J. Appl. Math.*, **51**, 73–89.
[40] Olagunju, D. O., and Matkowsky, B. J. (1992) 'Coupled complex Ginzburg-Landau type equations in gaseous combustion', *Stab. and Appl. Analysis of Continuous Media*, to appear.
[41] Proctor, M. R. E., and Jones, C. A. (1988) 'The interaction of two spatially resonant patterns in thermal convection. Part 1. Exact 1:2 resonance', *J. Fluid Mech.*, **188**, 301–335.
[42] Rabaud, M., Michalland, S., and Couder, Y. (1990) 'Dynamical regimes of directional viscous fingering: Spatiotemporal chaos and wave propagation', *Phys. Rev. Lett.*, **64**, 184–187.
[43] Rappel, W. J., and Riecke, H. (1992) 'Parity-breaking in directional solidification: Numerics versus amplitude equations', *Phys. Rev. A*, **45**, 846–859.
[44] Riecke, H., and Paap, H. g. (1992) 'Parity breaking and Hopf bifurcation in axisymmetric Taylor vortex flow', *Phys. Rev. A*, **45**, 8605–8610.
[45] Sivashinsky, G. I. (1983) 'Instabilities, pattern formation, and turbulence in flames', *Ann. Rev. Fluid Mech.*, **15**, 179–199.
[46] Wiener, R. J., and McAlister, D. F. (1992) 'Parity-breaking and solitary waves in axisymmetric Taylor vortex flow', preprint.

ASYMPTOTICS AND MULTISCALE SIMULATION IN A NUMERICAL COMBUSTION LABORATORY

D. Scott Stewart
Theoretical and Applied Mechanics
National Center for Supercomputing Applications
University of Illinois
Urbana IL 61801

and

John B. Bdzil
Los Alamos National Laboratory
Los Alamos NM 87545

ABSTRACT. The combustion problems that arise in complex engineering applications contain many important yet disparate-length scales. The device size is typically many orders of magnitude larger than the heat-release zone size. To get an accurate rendering of the device, one needs to accurately model the physics that occurs on all these scales. We discuss three problems related to this issue. First, by direct numerical simulation we show how an asymptotic submodel derived from the full model (reactive Euler equations) provides both a simple and useful description of a thin, stable, detonation heat-release zone. Next, we describe a detailed numerical study of a highly unstable detonation reaction zone. From the results for this problem, we find that the solutions are very sensitive to the type of numerical algorithm that is used to get the solution. Finally, we discuss the computational costs associated with obtaining highly resolved solutions to multiscale problems.

1. Introduction

In this article we describe some of our recent efforts to construct a numerical laboratory for computation of combustion problems. The numerical combustion laboratory envisioned is an integrated collection of simulation codes that use modern numerical methods on large-memory parallel/vector supercomputers, combined with postprocessing and visualization in the networked environment. In particular, our aim has been to determine the quality, accuracy, and robustness of certain asymptotic approximations for multidimensional detonation wave propagation. But the interdisciplinary issues involving fluid mechanics, asymptotics, and computational science raised by such

H. G. Kaper and M. Garbey (eds.),
Asymptotic and Numerical Methods for Partial Differential Equations with Critical Parameters 163–187.

an exercise are generic to the broader class of computational-asymptotics problems found in multidimensional combustion theory.

We start the discussion with an illustration of the validation of a multidimensional asymptotic theory, that is, detonation shock dynamics for curved, near-Chapman-Jouguet detonation, by direct numerical simulation. This is an example of where the same mathematical model defines both the asymptotic theory and the computations. Any ambiguities found in the comparison can arise only from questions about the validity of the asymptotics and the accuracy and resolution of the computations. The importance of such a comparison arises from the simplicity of the asymptotic theory and its direct application to engineering problems. Also, if it is correct, the asymptotic theory has implications for the explanation of physical mechanisms and the extension of the underlying physical theory. Verification of the asymptotic theory by direct simulation is not a matter of choice; rather, it is necessary since the result is imbedded in a much more complex system that is generally not suited to direct analytical study.

We next turn to some difficult issues in multiscale computations that are almost always found in combustion problems and in other problems in nonlinear science. We present a workshop problem recently used in a comparison of numerical methods for reactive gasdynamics. The workshop problem exhibits multiscale dynamics [17]. We briefly discuss the issues raised by the lack of accuracy in combustion hydrocodes and the implications for calculating long-time scale dynamics.

Based on some recent calculations, we project the computational requirements for difference methods solutions to flows with multiple scales (the example is from reactive gasdynamics) and do some simple arithmetic for CPU and memory requirements. We argue that direct simulation strategies based solely on brute force are likely to fail; computational strategies must take full advantage of the analytic (hence asymptotic) properties of multiscale problems in order to make rapid progress.

It is easy, especially for theoreticians, to underestimate the impact of advanced computations in engineering and science. Calculations and physical simulations can now be done that were inconceivable even a few years ago, with greater spatial and temporal resolution and much larger domains in both space and time. At the same time, it is nearly impossible to study phenomena as important and complex as combustion and detonation flow in multidimensions without the use of a priori analytical estimates to structure the calculations. Asymptotics will continue to play an important role in determining these estimates. The choice of problems addressed by theorists using asymptotics will be strongly driven by new phenomena uncovered by direct simulation of "more realistic" physical situations. In the context of combustion phenomena, the most useful asymptotic results are destined to be derived from a symbiotic relationship with direct simulation.

2. Validation of Multidimensional Asymptotics by Direct Simulation: Weakly Curved, Near-CJ Detonation

In [16], we presented a summary of our analysis of multidimensional detonation wave propagation. We will repeat only the essential ingredients here; the reader is referred to the previous article and its references.

A detonation is a traveling wave comprising a shock wave that ignites an ambient, premixed combustible, which in turn is supported by the combustion zone that travels behind the shock. The region of combustion that supports the shock has a finite length. This is the standard mathematical model of detonation called the Zeldovich-von Neumann-Doering (ZND) model. It is based on the Euler equations of gasdynamics modified to include heat-releasing chemical reactions. The Euler equations are the governing equations used for the calculations discussed here:

$$\frac{D\rho}{Dt} + \rho(\nabla \cdot \mathbf{u}) = 0, \tag{1}$$

$$\rho\frac{D\mathbf{u}}{Dt} + \nabla P = 0, \tag{2}$$

$$\frac{De}{Dt} + P\frac{D(1/\rho)}{Dt} = 0, \tag{3}$$

$$\frac{D\lambda}{Dt} = r, \tag{4}$$

where we restrict our attention to a model with a single, exothermic reaction. The shock relations take the form (only the jump relation for conservation of mass is given)

$$[\rho(u_n - D_n)] = 0, \dots , \tag{5}$$

where ρ is the density, u_n is the particle velocity normal to the shock, and D_n is the normal shock velocity.

The constitutive relations defining the internal energy function e in terms of the pressure, density, and reaction progress variable, P, ρ, λ, and the reaction rate r, to the same, are the standard simple forms:

$$e = \frac{1}{\gamma - 1}\frac{P}{\rho} - Q\lambda, \;\; R_g T = P/\rho, \;\; r = k(1-\lambda)^{\nu} e^{-E/(R_g T)}. \tag{6}$$

The parameters that appear are γ, the polytropic exponent; Q, the heat of reaction; E, the activation energy; k, the rate constant; and ν, the depletion coefficient. The reaction progress variable, λ, is zero at the shock and one at complete reaction.

This simple model admits a continuum of one-dimensional, steady traveling wave solutions parameterized by D_n. The admissible values of D_n are bounded from below by D_{CJ}, the Chapman-Jouguet (CJ) velocity. The CJ detonation has the special property that its state is sonic at the end of the supporting reaction zone. Thus, a supersonic wave head comprising the shock and the reaction zone can become causally disconnected from a trailing flow.

Multidimensional detonation in real explosives seldom exhibits this simple solution; the detonation shock is nearly always cellular. The number of cells observed can vary in both space and time. In explosive mixtures with state-sensitive reaction rates, like gaseous hydrogen and oxygen, many cells are observed on the detonation shock. For condensed phase explosives, usually only one cell is seen: a single, broadly curved, (often symmetric) shock front.

We have developed an asymptotic theory for a single-cell, curved front, in the combined limit of near-CJ detonation and weak shock curvature, as measured on a characteristic reaction-zone length scale. We call the theory "detonation shock dynamics" (DSD), because of its similarity to Whitman's theory of "geometrical shock dynamics." We have recorded this work in a series of papers whose references are listed in [16]. A recent extension of this work by Klein and Stewart [13] also gives an extensive set of references.

The main features of the asymptotic analysis are as follows. The detonation shock is assumed locally to be a suitably smooth surface. An intrinsic coordinate representation is used for both the shock surface and the following reaction zone, with a unit vector, $\hat{n}$, in the shock normal direction and with the normal shock velocity, D_n, in that direction. At any point on the shock, the total shock curvature is denoted by κ. In the asymptotic description, the detonation is near-CJ, and an expansion of the following form holds:

$$D_n = D_{CJ} + D'_n, \text{ where, } (D'_n/D_{CJ}) << 1. \tag{7}$$

The curvature, κ, is taken to be small when measured on the scale of a characteristic reaction-zone length, ℓ_{rz}. Hence the limit of weak curvature is defined by

$$\kappa \ell_{rz} << 1. \tag{8}$$

Repeating from [16], the asymptotic analysis requires that the flow be quasi-steady. In particular, this requirement means that the corrections to a locally planar and steady flow, resulting from curvature (κ) and unsteadiness ($\partial/\partial t$), are comparable.

The analysis (see [20, 13]) derives a weakly curved, slow-time theory and leads ultimately to a discussion of ODEs in the direction of the shock normal (this is the principal flow direction). For the equation of state and reaction rate considered here, these equations reduce to a single ODE for u_n, the relative particle velocity, in terms of λ the reaction progress variable. The solution of this reduced equation can be discussed in a u_n^2, λ phase plane. The total curvature κ appears explicitly in the formulation. From a simple, numerical point of view, the ODE can be solved by shooting in the u_n^2, λ plane. Starting from the shock, one integrates towards the equilibrium point which is near complete reaction. This equilibrium point is a saddle point in the u_n^2, λ phase plane. For each fixed κ, D_n is varied until the solution passes through the saddle point. By varying κ, the function $D_n(\kappa)$ is obtained.

The mathematical problem of solving for the structure through the reaction zone can also be posed as a singular perturbation problem with $\kappa \ell_{rz}$ taken as the small parameter for expansion. This is the most powerful representation. To ensure a uniform solution through the detonation reaction zone, one finds that there must be a relationship between D_n and $\kappa \ell_{rz}$. The asymptotic solution of this problem, for $\kappa << 1$, is found to have two distinct layers, where the asymptotic expansions for u_n in terms of κ have different forms: hence, the singular character of the perturbation problem. The largest layer in the u_n^2, λ plane starts near the shock and ends near the equilibrium point. Klein and Stewart call this the "main reaction layer." This layer represents the regular perturbation of the 1-D, planar solution. The second layer is called the "transonic layer." It lies close to the equilibrium point where perturbations resulting from the curvature and those resulting from the reaction balance. The function $D_n(\kappa)$ is derived by matching the main reaction layer and the transonic layer. The first calculation of this type appears in [20].

The preceding 1-D analysis in the shock normal direction yields the dependence of the shock velocity on the local shock curvature. This time-independent expression serves as a constitutive equation; alone it does not describe the shock's time evolution. The time dependence enters the problem through the shock surface kinematics (a purely geometric relation between D_n, time, and the shock surface shape). Combining $D_n(\kappa)$ and the surface kinematics with an explicit coordinate representation (Cartesian, intrinsic, polar, etc.) yields an evolution equation for the shock. When D_n depends on κ, this equation is parabolic and can be thought of as a nonlinear heat equation. The simplest manifestation of this equation is a Burgers' equation for the shock slope relative to a fixed orientation. This was first discussed in [5].

To complete the specification of the surface evolution problem, one must give initial and boundary data for the shock. No special restrictions exist on the form of the initial data. However, many unanswered questions remain about the nature of the boundary conditions. An analysis of the flow in the vicinity of the shock and explosive-edge intersection is needed to get the proper boundary conditions for DSD theory. To date, such an analysis has yet be done. In its place, we have formulated boundary conditions that neglect the reaction zone. These boundary conditions are based on the classical interaction problem for an inert oblique shock and a material boundary. The type of boundary condition applied depends on the nature of the confining material adjacent to the explosive and on the flow character (supersonic or subsonic) as seen by an observer riding along the shock/edge intersection locus. A discussion of the issues surrounding the boundary conditions can be found in [3], [4], and [18].

We have distilled the following rule from this matching problem. If a vacuum is adjacent to the explosive interface, we bound the angle between the shock normal and the edge tangent, σ, from below by the sonic flow value, σ_c. When $\sigma > \sigma_c$, (a supersonic flow) we apply no boundary condition. When a perfectly rigid material confines the explosive, the critical angle, σ_c, is set to zero ($\sigma_c = 0$).

2.1. VALIDATION OF ASYMPTOTIC THEORY AND NUMERICAL EXPERIMENT

The validation of the asymptotic theory of detonation shock dynamics can be accomplished by direct numerical simulation of the full multidimensional equations. This is an unambiguous test in the sense that both theory (asymptotics) and experiment (direct numerical simulation) are carried out on precisely the same underlying mathematical model. This is in contrast to comparisons of asymptotics with physical experiment or of numerical simulation with physical experiment. In the cases when the numerics are of sufficiently high quality, there is the very real chance of extending the asymptotic (i.e., analytical) theory by using the simulation to capture the loss of asymptotic validity of the approximations. This process leads to insights on how to extend the theory to include additional physical mechanisms not contained within the original asymptotic description. New forms of asymptotic expansions and information about the boundary asymptotics could be obtained in this way.

Here we present a result obtained by direct 2-D numerical simulation of the Euler equations (1)–(6), for a plane Chapman-Jouguet detonation, propagating down a channel of fixed width, which suddenly expands or contracts. To solve this problem, we modified a version of a Los Alamos hydrocode CAVEAT [1], [15]. CAVEAT uses a conservative, finite-volume numerical technique, with cell-centered variables. The scheme is based on a second-order Godunov method, which we modified to include an exact Riemann solver. An Eulerian mesh (with 400×200 cells) and van Leer advection limiting are used. The chemical rate used for these comparisons is state independent and hence has the activation energy $E = 0$. The poly-

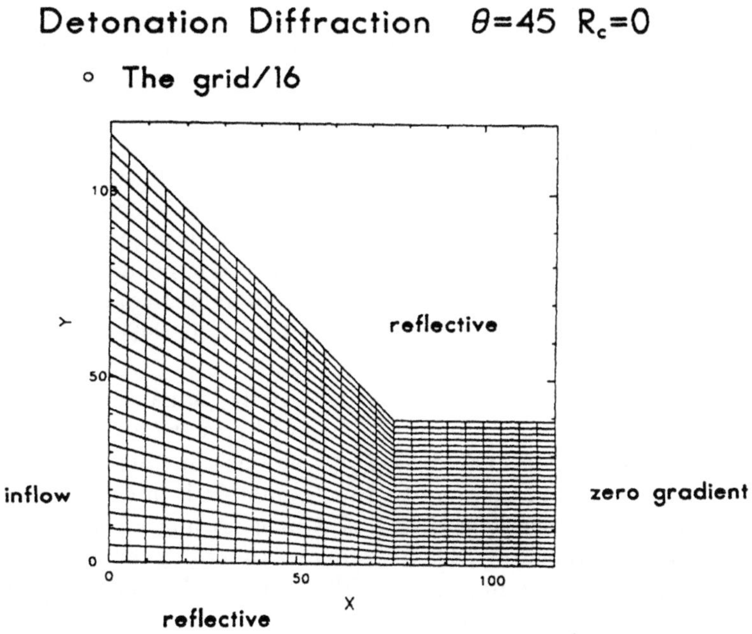

Fig. 1. **Typical grid for expanding wedge calculation for CAVEAT.** (The detonation runs from the right towards the left.)

tropic exponent is $\gamma = 3$, the heat release is $Q = D_{CJ}^2/16 = 4\ \mathrm{mm}^2/\mu\mathrm{s}^2$, and the upstream states are $P_0 = 0$ and $\rho_0 = 2$ g/ cm^3. The depletion factor is $\nu = 1/2$; consequently the reaction-zone length is finite.

In the simulations, the detonation runs from the right towards the left or from the bottom towards the top. In the right-left configuration, the boundary conditions at the top and bottom wall of the wedge are reflective; the right boundary is an outflow; the left (ambient state) boundary is taken to be a constant state (inflow). A typical grid is shown schematically in Fig. 1, scaled down by a factor of 16.

The leading form of the $D_n - \kappa$ relation for this material can be derived from the theory and has the form

$$D_n = D_{CJ} - \alpha\kappa, \tag{9}$$

where $\alpha = 8.35 D_{CJ} x_{1/2}$ and is computed from the formulas found in [20] using the parameters cited above. The constant $x_{1/2}$ is the distance from the shock ($\lambda = 0$) to the point $\lambda = 1/2$ in the steady-state, one-dimensional, unsupported ($D_n = D_{CJ}$), ZND reaction zone. We use it as our distance scale (see Section 3.1). A front-tracking code written by Bdzil and Fickett [2] was used to solve the DSD problem. This code can solve the shock propagation problem efficiently for a wide variety of explosive geometries. (Less than a minute of CRAY Y-MP time was needed to solve the problems we consider here.) The code uses an intrinsic coordinate representation of the shock, with the arc length along the shock ξ and time t as the independent variables. The dependent variable is ϕ, the inclination of the shock, defined as the angle between the outward normal in the direction of propagation of the shock and the vertical (axis) direction. The angle ϕ is positive for clockwise rotation of the normal from the vertical. The boundary conditions imposed are that the shock be normal to the top and bottom reflective boundaries ($\sigma_c = 0$).

We first consider the diffraction of an initially planar detonation (with shock parallel to the y axis) around a sharp 45° corner (see Fig. 1). We use $x_{1/2}$ as our distance scale and $(x_{1/2}/D_{CJ})$ as our time scale. The velocity scale is D_{CJ}. Expressed in these units, the detonation is initially located near the bottom at approximately $y = 20$ and is propagating upwards at $D_{CJ} = 1$. When the detonation enters the diverging section of the channel, the shock curvature at the edge becomes infinite. The numerical simulation shows the shock pressure, P_{shock}, dropping to P_{CJ} at this instant (i.e., $D_n = 0.7D_{CJ}$). To capture this behavior and to prevent (9) from developing a singularity, we modify (9) to $D_n = 1 - 8.35\kappa/(1 + 28\kappa)$ for this example only. This limiting factor, $(1 + 28\kappa)^{-1}$, gives $D_n = 0.7$ when $\kappa \to \infty$.

Figures 2a and 2b show a comparison of the pressure contours and the DSD shock location at the initial time and at a later time, when the detonation has propagated into the expanded section approximately 22 1-D reaction-zone lengths ($\simeq 90x_{1/2}$). In addition, the shock location calculated from a Huygens construction is shown for comparison. (For the Huygens construction, the normal velocity, D_n, is always assumed to be $D_{CJ} = 1$, and there is no correction for curvature.) Both the qualitative and the quantitative comparison of the direct numerical simulation and the asymptotic theory are good and provide some confidence in the usefulness of the asymptotic theory in such cases.

Next we consider a converging channel with a turning angle of $\pi/8$ radians. The corner is rounded ($R_c = 40$), and (9) is used without the limiting factor. Figure 3 shows a comparison of the pressure contours and the DSD shock location after the detonation has entered into a converging channel. In this case the Huygens shock is not influenced by the convergence and remains flat. Again, both the qualitative and the quantitative comparisons of the direct numerical simulation and the asymptotic theory are good and provide some confidence in the usefulness of the asymptotic theory in such cases. This result is particularly interesting since for this model [12] it can be shown that, unlike the expanding case which always maintains a sonic

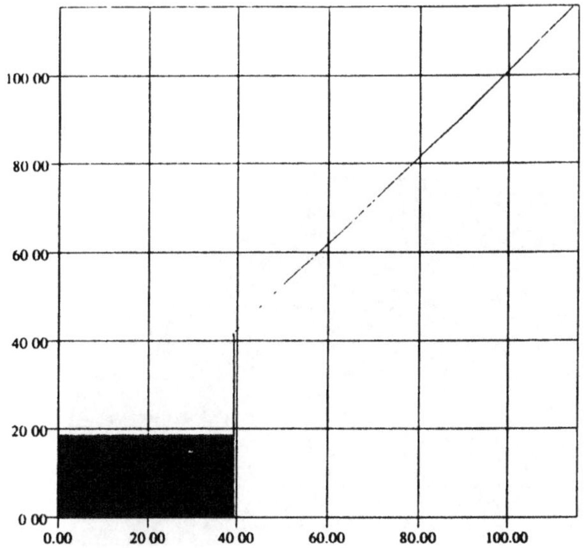

(2a)

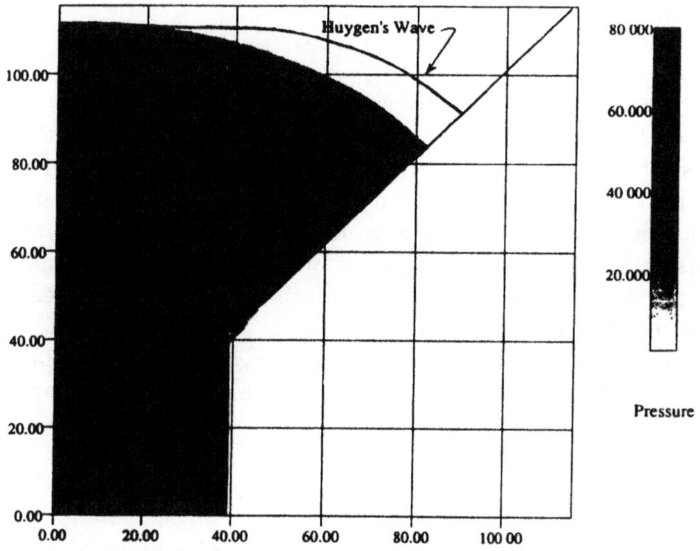

(2b)

Fig. 2. **Expanding wedge calculation.** (The detonation runs from the bottom towards the top.)

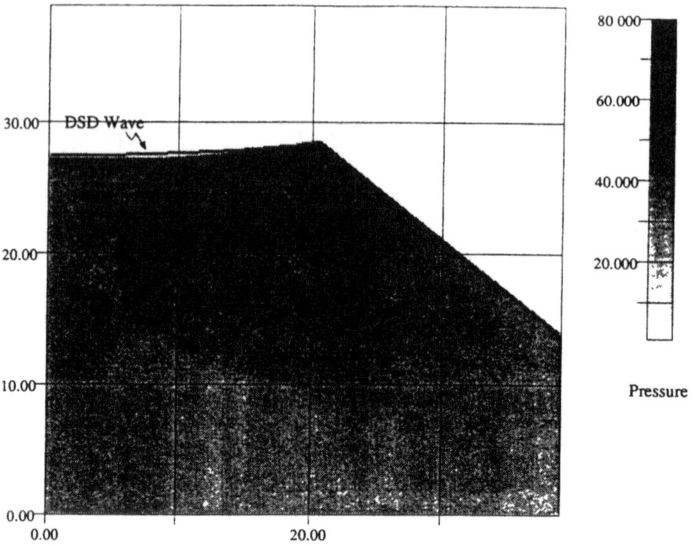

Fig. 3. **Converging wedge calculation.** (The detonation runs from the bottom towards the top.)

locus behind the main shock, a fully subsonic layer forms behind the shock in the region where the shock curvature is negative. Hence, while the $D_n - \kappa$ relation holds for a finite time, there is the possibility of secular growth of a higher-order perturbation which can destroy the leading-order approximation. However, for the calculation shown, the appearance of such a secularity is not obvious in its effect on the motion of the shock. In this case, direct simulation suggests that the $D_n - \kappa$ relation may have an extended validity and usefulness beyond its original derivation.

It is noteworthy that the "Mach" wave is curved in this calculation and that a prominent reflected shock is not seen. These effects are due principally to finite-length reaction zone effects.

3. Multiple Important Scales in Physical Simulation

Next we consider the issue of complexity within the physical simulation (still focused on the model of reactive gasdynamics). The aim is to argue that direct simulation of a rather large class of important problems requires that a new class of numerical algorithms be developed that use analytic information about the solution.

Here we briefly describe recent computations on an unstable 2-D detonation problem. This problem was the subject of a recent workshop on numerical combustion [17]. It was computed by a number of different researchers with different codes that use different algorithms. Although simply stated, this problem has extremely complex dynamics. In addition, the physics of the problem contains a number of important scales associated with the shock thickness, shock dynamics (cell width), and reaction zone related thicknesses (induction length and heat-releasing region). Currently, this problem is difficult to compute reliably, especially at long time. It is better seen as a computational challenge rather than a standard test problem. The overdriven 2-D detonations, recently computed by Bourlioux and Majda, have less complex dynamics and may serve as better test problems [6].

The workshop problem makes the point that complex multidimensional combustion problems with sensitive chemistry are difficult to calculate accurately by using conventional direct numerical simulation methods. These are the types of problems that are likely to be encountered in engineering. Even when the domains are limited to a small number of reaction-zone thicknesses and to a rather small number of characteristic passage times through the reaction zone, the calculations are just barely feasible. The workshop test problem was for the physically important case of Chapman-Jouguet detonation, with large temperature sensitivity in the reaction rate. The attempts at solving the problem with different state-of-the-art codes produced a wide variation in results among the participants. A main objective of the workshop was to illustrate the variation in results obtained by the different methods. Our experience with CAVEAT and CMHOG shows that there can be profound qualitative and quantitative differences in the solutions obtained with different codes.

The governing equations for the workshop problem are those given above. A modern reference for this model is found in [14]. Erpenbeck's model was chosen because it is the most-studied model: stability and asymptotic results are available for it [14, 6]. Erpenbeck's model serves as a basis for a "numerical test suite" for detonation simulation codes. The term "test suite" refers to the series of analytical and/or numerical solutions that a simulation code must successfully compute before it is deemed accurate enough to be used for further engineering/scientific purposes.

3.1. DESCRIPTION OF THE 2-D WORKSHOP PROBLEM: 2-D-STRONG TRANSIENT

The natural-length scale is based on the one-dimensional, steady ZND solution, which is reviewed below. The ZND solution can be found in two steps. The first step is algebraic and derives the Rankine-Hugoniot relations for a partially reacted gas with $0 < \lambda < 1$. The pressure, density, and relative velocity are expressed in terms of λ, and the steady detonation velocity D. Then one goes to the reaction equation in the steady wave coordinate, replaces P, ρ, and $u - D$ in terms of λ, and solves the resulting first-order, separable, differential equation for λ as a function of distance from the shock, x. For a stationary upstream velocity, the pressure and density are given by P_0 and ρ_0, respectively. The ZND profile can be calculated by the following relations in sequence:

$$m \equiv -\rho_0 D \tag{10}$$

$$t_1 \equiv \left(\frac{m^2}{\rho_0} + P_0\right)\frac{1}{\gamma + 1}, \tag{11}$$

$$t_2 \equiv \left(P_0\frac{\gamma}{\gamma + 1} + \frac{m^2}{\rho_0(\gamma + 1)}\right)^2 \tag{12}$$

$$t_3 \equiv \frac{2m^2Q(\gamma - 1)}{(\gamma + 1)}\lambda \tag{13}$$

$$P = t_1 + (t_2 - t_3)^{1/2} \tag{14}$$

$$\rho = \left[\frac{1}{\rho_0} - \frac{P - P_0}{m^2}\right]^{-1} \tag{15}$$

$$u - D = \frac{m}{\rho}. \tag{16}$$

The distribution of the mass fraction, λ, is then determined from the integral

$$x = \int_0^{\lambda} \frac{u - D}{k(1 - \lambda)e^{-E/(R_gT)}}d\lambda. \tag{17}$$

The dimensional pressure and density scales are chosen to be the quiescent upstream values P_0 and ρ_0, respectively. The length scale is chosen to be the $\lambda = 1/2$-reaction length as defined by the steady, one-dimensional ZND structure. The velocity scale is chosen to be $(P_0/\rho_0)^{1/2}$. The time scale is the length scale divided by the velocity scale.

With these choices, the dimensionless equations have the same form as the dimensional equations with the following conventions and interpretations: (1) The scaling requires that the upstream pressure and density are both set equal to one. (2) In the reaction rate, T in the Arrhenius term $e^{E/(R_gT)}$ is replaced to obtain $e^{E/(P/\rho)}$. The quantity E represents the activation energy scaled relative to the upstream temperature and is the dimensionless activation energy found in all of the previous stability analyses of Erpenbeck's model. (3) With these scalings, Q is the scaled heat release normalized with respect to the upstream temperature and is the same quantity referenced in the stability studies. These scales are the same ones used by Erpenbeck.

If we use the choice of scales mentioned above, the rate constant, k (now representing a dimensionless group) has a precise value. In particular, we calculate the dimensionless rate constant k by using formula (17) and set $\lambda = 1/2$ and $x = 1$ to obtain the following expression for k:

$$k = \int_0^{1/2} \frac{u - D}{(1 - \lambda)e^{-E/(P/\rho)}} d\lambda. \tag{18}$$

In summary, the upstream state has $P_0 = 1, \rho_0 = 1$, and the rate constant is calculated from (18). The equations for the workshop problem then take the same form as their dimensional counterparts.

The workshop problem is referenced to certain steady ZND solutions and is specified in terms of the parameter set $f \equiv (D/D_{CJ})^2, \gamma$ and Q and E. The Rankine-Hugoniot relations define the steady detonation velocity D. Likewise D_{CJ} is uniquely defined.

3.1.1. Computational Domain and Boundary Conditions. The domain is rectangular with left, right, top, and bottom boundaries. The left boundary is an (upstream) inlet boundary. There the fluid is to be injected towards the right at velocity $u = D$ with $\lambda = 0$ and $P_0 = 1$, $\rho_0 = 1$. The x direction is the main flow direction, from left to right. The right boundary is an outlet boundary with the x gradients set equal to zero. The top and bottom boundaries are reflecting walls. Hence the y velocity component (i.e., velocity component normal to the wall) is set equal to zero.

3.1.2. The Steady State. The steady state assumes that $P_0 = \rho_0 = 1$ and is defined by $\gamma = 1.2, Q = 50, E = 50$, and $f = 1$ (a Chapman-Jouguet detonation). Thus fluid is injected at the left boundary at $u = D_{CJ}$.

3.1.3. The Initial Conditions. The initial conditions are those for an unsupported ZND detonation facing towards the left and stationary in the inlet flow. The flow is disturbed immediately ahead of the shock. If the flow were not disturbed, the steady, plane ZND wave would remain stationary in the computational domain.

The initial conditions are given in a box of length 12 in the x direction and length 5 in the y direction: an aspect ratio of 12:5. A steady planar ZND detonation is placed at the mid-plane of the channel (a distance of 6 to the right of the inlet at $x = 0$). Initially the plane ZND shock runs vertically in the y direction with the shock at $x = 6$, and the reaction zone occupies the region $x > 6$. It is necessary to numerically integrate (17) to obtain the initial ZND solution at the grid points. The inlet stream velocity must be added to the ZND solution.

The initial disturbance is a rectangular block of high density ($\rho = \rho_{pert.} = 1.05$) placed immediately ahead of the shock. The high-density region convects into the detonation shock to create an initial disturbance. The perturbed density block has a length of 0.4 in the x direction and 4.0 in the y direction. The block is centered in the computational domain in the y direction, and the right edge of the block is placed at the location of the unperturbed shock.

The basic grid resolution for a uniform, cell-centered mesh over the entire domain was 240 cells in the x direction and 100 cells in the y direction. For this resolution there are 20 points per one unit of scale distance (half a reaction-zone length). If we number the cell index (i, j), with (1,1) being the lower left corner cell of the computational domain, then the shock will be located at the interface between cells with i=120 and cells with i = 121. The corner cells of the block of high-density material in the unreacted zone ahead of the shock are (113,11), (120,11), (113,90), (120,90), respectively.

3.2. SUMMARY OF WORKSHOP PROBLEM RESULTS

The collective experience of the workshop group can be summarized as follows. For the case of a high-density (low-temperature) block, the onset of a transverse wave instability is delayed. The detonation wave attaches to an unstable mode corresponding to a decaying, traveling wave that is produced by the drop in the pressure and the temperature as the density perturbation enters the reaction zone. The shock decouples from the reaction front, leaving in its wake an increasing quantity of unburned fuel. It was difficult and expensive to calculate the solution at long times. In most cases the detonation failed, traveling out of the computational domain through the outflow boundary.

Although the group consensus was that the detonation failed, it is by no means certain that this is the true solution. Different codes produced dramatically different results. In our own case, we observed different results with two codes we developed, CAVEAT and CMHOG. CAVEAT is described above. The inert version of CMHOG is a code recently developed and written by Jim Stone at NCSA [21] and modified for combustion by Stewart and Stone. CMHOG can be run in 1-D, 2-D, or 3-D and uses the PPM algorithm of Colella and Woodward [8] and the splitting scheme for combustion suggested by Colella, Majda, and Roytburd [7]. CMHOG is both spatially and temporally split and thus differs substantially from CAVEAT, which is based on a particle-in-cell method in its formulation. CAVEAT is formally second-order accurate, while CMHOG on a uniform grid is thought to be fourth-order accurate.

The CMHOG calculation showed a response to the high-density perturbation that was different from that of CAVEAT. Initially the shock advanced a short distance, and a weak transverse instability developed. The transverse instability grew in strength but then died out after a short time. The detonation shock then receded from the inflow boundary and left the computational domain before $t = 10$. The CAVEAT calculation showed the development of a large-amplitude transverse wave that initially advanced rapidly towards the inflow boundary. Eventually, the mean position of the shock stabilized, which kept the unstable reaction zone in the computational domain through time $t = 10$. The dramatic differences in the solutions obtained by these two codes, which had both passed many tests and which were being used with reasonable (numerical) parameters, are disconcerting at best. The sharp differences in the qualitative behavior obtained with these codes illustrate the extreme sensitivity of computed results to the numerical algorithms used and the consequent loss in accuracy. Advanced algorithms must reduce accuracy errors wherever possible, if realistic engineering problems are to be solved reliably. This last point is widely understood in the applied mathematics and numerical analysis communities, but it is not widely understood in the engineering and aerospace communities. These groups are the main users of such codes and the consumers of their results.

Another disturbing observation is the apparent limitation of these calculations to small physical domains. In the workshop problem, the domain is 12 half-reaction zones long and 5 wide. For a gas-phase detonation with a typical ZND reaction-zone length of 0.1 mm, this translates to a channel that is 1.2 mm long and 0.5 mm wide. A combustor for an engine may be on the order of a meter in its characteristic dimension. From the reaction-zone thickness to the combustor size there are typically two to four orders of magnitude differences in size. The implications for direct simulation are daunting.

Figure 4 shows an example of the workshop problem for the case when a low density block, $\rho_{pert.} = 0.95$ (high temperature), was used as the initial disturbance. The contour plots from CMHOG for the 240×100 zone calculation are shown. First the detonation wave undulated; then a large transverse wave started up the channel. However, the detonation eventually seemed to fail; it was flushed out of the channel at large times. The multiscale structures that are observed are quite complex and include shocks, induction zones (regions of unburnt fuel caught between the shock and the energy release zone), and the energy release zone. Generally speaking, extremely high frequency acoustics are generated by shock-shock and shock-wall collisions, while lower-frequency acoustics are generated near the end of the reaction zone. In addition, Fig. 4 shows unburnt pockets of fuel that are "cut off" from the main combustion zone by the interacting transverse shocks. These pockets react only at late times and large distance behind the lead shock.

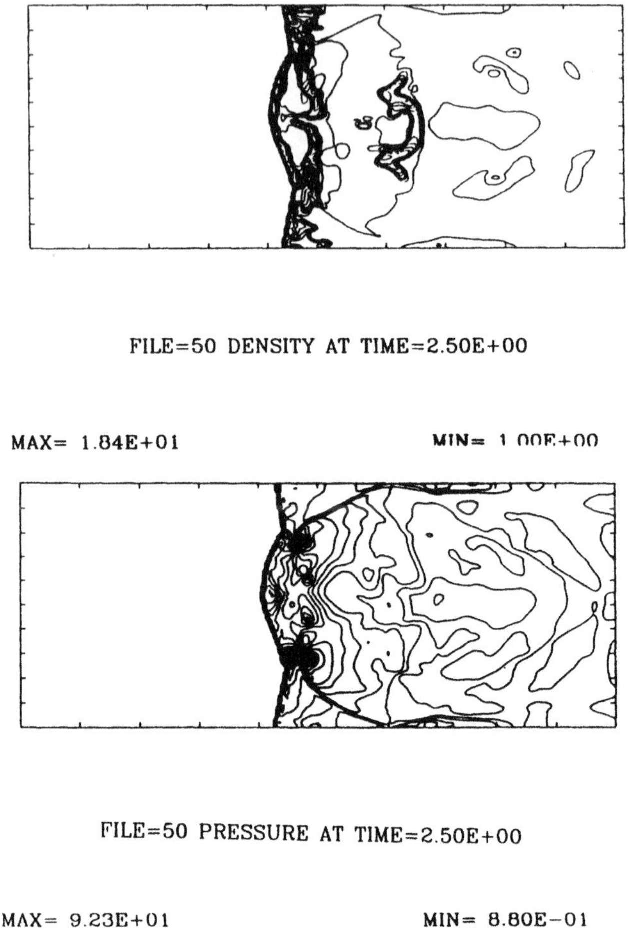

Fig. 4. The CMHOG density and pressure fields at t = 2.5 for the workshop problem. Note the pocket of cold (high density) explosive approximately 3 units behind the lead shock.

For these rate-sensitive calculations, a probable explanation for the loss of accuracy is the errors generated by the shock waves and other discontinuities. Generally, discontinuities generate first-order errors when they propagate across a fixed grid. Even though such an error may be small, when the error is produced at the main shock, it is immediately introduced into the reaction zone, where it undergoes a tremendous nonlinear amplification as a result of the chemistry. For large activation energy asymptotics, it is well known that algebraically small perturbations are often converted into order one perturbations through Arrhenius kinetics.

The consequences of these errors is the quick loss of phase accuracy; that is, the positions of interfaces whose speeds are sensitive to the chemistry (like the heat-releasing zone that moves behind the shock) are inaccurate. In calculations like the workshop problem, all known sources of error in the hydrocode must first be understood and then minimized. For example, it is commonly thought that it is sufficient to use approximate Riemann solvers in a method like PPM. Yet even the small errors produced by a very good approximate Riemann solver such as Colella's "two-shock" approximation can lead to large phase errors.

Figure 5 shows a comparison of two simulations, calculated to the same time, with the only difference being that the first calculation used the approximate "two-shock" Riemann solver, while the second calculation used an exact Riemann solver. The flows look similar, but on closer inspection show differences. Figure 5c shows a contour plot of the results found by taking the difference of the pressure fields $(1 - P_{2shock}/P_{exact})$ for the two simulations at a fixed time. Part of the explanation of the large (order one) errors is the difference in the speed of the Mach stems imbedded in the flow. This is a loss of accuracy as a result of a difference in the phase velocities of the imbedded waves. A similar calculation, using a different approximate Riemann solver of Dukowicz [9] that assumes the strong shock limit holds for *all* shocks, is not even in qualitative agreement with the exact solver solution.

Clearly, if phase accuracy is easily lost, the results of long-time simulations are questionable at best. Can a statistical measure be calculated accurately? Should we abandon the traditional measures of accuracy? What are the correct norms for an accurate solution? Should we use norms for the numerical solutions that reflect approximations to the dynamical system of the underlying PDEs? These are open questions.

Implementation of the computational rule of thumb that all physically important high-frequency features must be resolved, argues for the use of a uniform grid. But this simple-minded prescription, while sensible, carries an extreme cost. (We argue this in the next section). Rapid progress towards simulation and prediction of the exact behavior of the underlying PDE's is likely to be made only by limiting the computational domain to one that can be resolved numerically; this limitation restricts the physical dimensions and times for the calculations. To determine what calculations can be resolved with fixed computational resources requires a priori approximations of the solutions to the PDEs, hence asymptotic analysis.

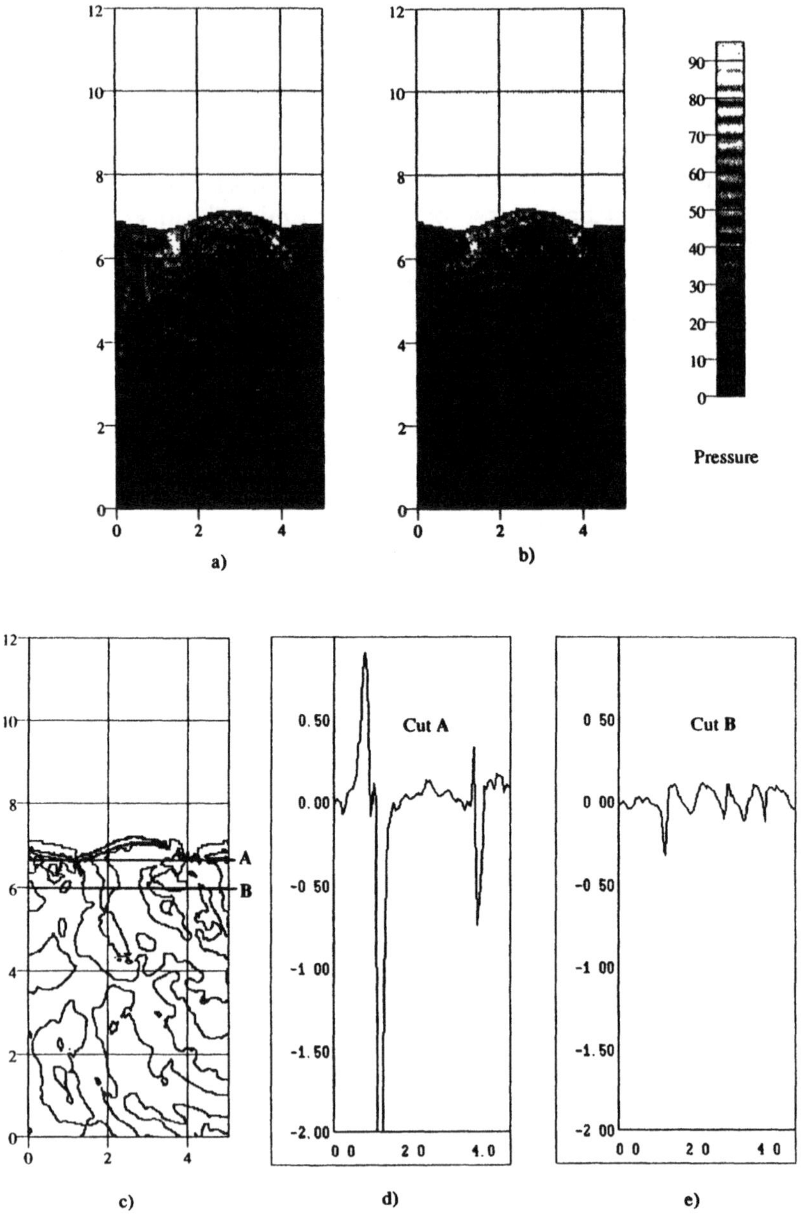

Fig. 5. **Comparison of the pressures for the workshop problem computed with CMHOG and using an exact (Fig. 5a) and the approximate (Fig. 5b) "two-shock" Riemann solver.** A contour plot of $1 - P_{2shock}/P_{exact}$ is displayed in Fig. 5c. Two horizontal slices through the data of Fig. 5c, with slice-A near the shock and slice-B well into the reaction zone, are shown in Fig. 5d. and Fig. 5e.

4. The Economics of Multiscale Calculations

Here we discuss the CPU and memory arithmetic for multidimensional, multiscale combustion calculations. We do this to illustrate that direct simulation of such problems will require advances in both algorithms and computer hardware. Multigrid methods or other adaptive gridding strategies that utilize asymptotic solutions derived from the PDEs could play a major role in the new developments. They have the potential for providing orders of magnitude reduction in the direct simulation efforts. These methods would be analytically based and lead naturally to improved physical understanding. This is an important point: If all the advances in direct simulation come strictly from hardware improvements, our understanding of the underlying science and engineering would not be advanced. Even now, large computer simulations produce unimaginably large datasets. To interpret these results, simple submodels need to be simulated that highlight a single physical mechanism. This reduction is almost always obtained through asymptotic arguments. Our belief is that asymptotics applied to complex multiscale simulations will invariably play a central role as the computing technology continues to develop.

To understand the massive size of a multiscale, direct simulation, consider the following scenario for a simulation in a physical domain. A typical multiscale engineering problem is represented in Figure 6. The largest scale (the device scale) is called the macro scale, L. The next, imbedded scale is the micro scale, ℓ, and the smallest scale is the nanoscale, η. The total number of spatial cells in the calculation, which depends on the problem's dimensionality, call N_{cells}. Likewise, call the total number of cycles required to achieve the final simulated time N_{cycles}. Let (CPU/cell cycle) be the average CPU time spent per cell per cycle in the calculation. The total CPU time is calculated by the formula

$$T_{CPU} = N_{cells} \left(\frac{CPU}{cell\,cycle} \right) N_{cycles}, \tag{19}$$

and the number of words in memory in a two-level scheme is approximately $16N_{cells}$. The total CPU time for a single simulation is both hardware and software dependent. For a reasonably well written code, (CPU/cell cycle) is not strongly dependent on N_{cells}. The quantities N_{cells} and N_{cycles} are thus mostly algorithm dependent, and (CPU/cell cycle) is mostly hardware dependent.

The statement of the multiscale character of the problem is represented by the scalings

$$\ell = \epsilon L, \quad \text{and,} \quad \eta = \delta \ell. \tag{20}$$

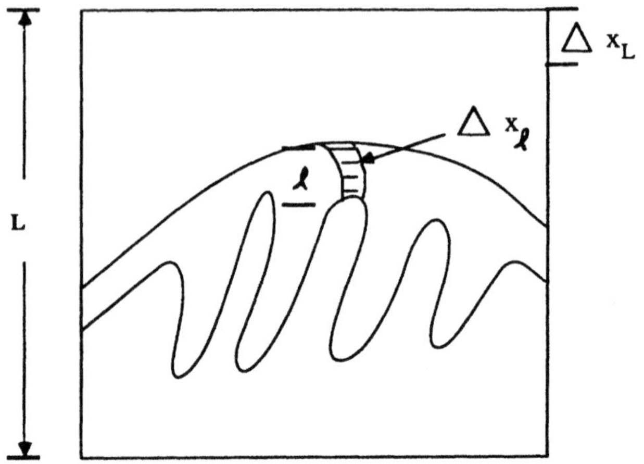

Fig. 6. **Schematic diagram showing the representative dimensions for the simulation of an engineering device.** Two scales L and ℓ are shown.

A three-scale problem (two imbedded scales) is represented by the limit

$$\epsilon << 1, \delta << 1. \tag{21}$$

A two-scale problem is obtained by choosing $\delta = 1$, or $\ell \equiv \eta$.

For an accurate calculation, the smallest scale must be resolved. Hence, we write

$$\eta = N_{Acc}\Delta x, \tag{22}$$

where Δx is the smallest spatial interval. Explicit schemes usually have a stability requirement, which we take here to be the Courant condition. Suppose the local Courant number is fixed and is represented by $c = \Delta x / \Delta t$. Then the choice of Δx determines Δt in these estimates.

The problem's physical time determines the CPU and memory required, with

$$T_{final} = \alpha L / D, \tag{23}$$

where D is the steady, traveling-wave speed in the system (in our case a detonation wave, or a reactive front), and L is the macroscopic device dimension. The parameter α measures the complexity of the problem, for example, whether a single interaction of fronts or the long-time dynamics of fronts is desired. When $\alpha = 1$, the formula $T_{final} = L/D$ represents the time that it takes a wave (e.g., a detonation) to sweep across the system. If the wave is a standing wave, centered in the computational domain, and we consider $\alpha \to \infty$, this is the limit of long-time dynamics.

For a simulation to a fixed time on a uniform grid, the following formula estimates the total CPU time required for a *single* simulation, of a *single* physical realization:

$$T_{CPU} = \left(\frac{CPU}{cell\,cycle}\right)\left(\frac{N_{Acc}}{\epsilon\delta}\right)^{d+1}(\alpha c/D). \tag{24}$$

Note that d is the physical dimension of the simulation, $d = 1, 2,$ or 3. For the purpose of our estimates we can think of $c \sim D \sim O(1)$.

While the estimate for T_{CPU} is linear in $\left(\frac{CPU}{cell\,cycle}\right)$, c, D, and α, it varies to the power $(d+1)$, in $(N_{Acc}/\epsilon\delta)$. These last dependencies make resolved, uniform-grid calculations for simulations on the scale of the device size nearly impossible with the current technology. To see this even more clearly, consider the following. Suppose we take a recent average from a CRAY Y-MP run of CMHOG for which $\left(\frac{CPU}{cell\,cycle}\right) = 3 \times 10^{-4}\frac{sec}{cell\,cycle}$. Table 1 shows the resulting estimates for one and two imbedded scales and for a single wave interaction ($\alpha = 1$), several interactions ($\alpha = 10$), and many interactions ($\alpha = 100$).

Table 1. Estimates of T_{CPU}

	One Interaction ($\alpha = 1$)	
T_{CPU}	$d = 2$	$d = 3$
one scale $\epsilon = 0.1$	100× 100 0.3×10^3 sec or 0.1 hr	100×100×100 30×10^3 sec or $\sim$ 10 hr
two scales $\epsilon = 0.1, \delta = 0.1$	1000×1000 30×10^3 sec or $\sim$ 10 hr	1000×1000×1000 30×10^6 sec or $\sim 10^4$ hr
	Several interactions ($\alpha = 10$)	
T_{CPU}	$d = 2$	$d = 3$
one scale $\epsilon = 0.1$	100×100 1 hr	100×100×100 100 hr
two scales $\epsilon = 0.1, \delta = 0.1$	1000×1000 100 hr	1000×1000×1000 10^5 hr
	Many interactions ($\alpha = 100$)	
T_{CPU}	$d = 2$	$d = 3$
one scale $\epsilon = 0.1$	100×100 10 hr	100×100×100 1000 hr
two scales $\epsilon = 0.1, \delta = 0.1$	1000×1000 1000 hr	1000×1000×1000 10^6 hr

For the purpose of the estimates, we take $N_{Acc} = 10, \epsilon = 0.1$, and $\delta = 1$ and 0.1. Then for a uniform grid with three resolved scales ($\epsilon = 0.1, \delta = 0.1$), one interaction ($\alpha = 1$) in three dimensions ($d = 3$) requires $1000 \times 1000 \times 1000 = 10^9$ computational cells. This problem would require approximately 10^4 CPU hours to compute (i.e., one year). A many-interactions calculation ($\alpha = 100$) would require 10^6 CPU hours (i.e., 100 years). Even a two-dimensional calculation with $N_{Acc} = 100$, $\alpha = 100$ and with one resolved scale ($\epsilon = 1, \delta = 1$), requires 10^4 cells and approximately 10 hours of CPU time.

In order to be useful for the purpose of carrying out a parametric study, the simulation of each case should take no longer than about one day's work (i.e., 8 to 10 hours). Under the best of circumstances, 100-hour simulations are not suited to parametric studies. Hardware advances may soon bring two or three orders of magnitude improvement, but not likely six. Clearly, improved algorithms combined with advances in hardware are a necessity for rapid progress in multiscale calculations. A simple estimate of the improvements that are likely with multigrid adaptive methods is given below.

4.1. ESTIMATE OF IMPROVEMENT FROM A MULTIGRID METHOD

We will restrict our attention to one imbedded scale and two or three dimensions. The macro-length scale is still taken to be L, and the one subscale is $\ell = \epsilon L$. We take the coarse-grid scale to be $\Delta x_L \equiv \ell$, and the fine scale to be Δx_ℓ. Assuming the simplest asymptotic relation between the scales (a linear relation) yields

$$\Delta x_\ell = \epsilon \Delta x_L. \tag{25}$$

The accuracy requirement is that $\ell = N_{Acc} \Delta x_\ell$. For consistency we have to take

$$\epsilon^{-1} \approx N_{Acc}. \tag{26}$$

Consider that in a 2-D calculation, the area of the domain occupied by the small-scale features can be approximated by an arc length β_x times the thickness ℓ, where β_x is measured in units of ℓ. Then it is easy to show that the estimate for the number of cells is simply the sum of the number of coarse-grid cells and the fine-grid cells and is represented by

$$N_{cells} = \left[\beta_x (N_{Acc})^3 + (N_{Acc})^2\right]. \tag{27}$$

The number of cycles goes like

$$N_{cycles} = (\alpha c / D)(N_{Acc})^2, \tag{28}$$

and the CPU estimate follows from (19).

Having sketched these new estimates, we now compare them to the same ones obtained for a uniform grid (using $\epsilon = 1/N_{Acc}$ and retaining only the leading order in N_{Acc}). The label UG refers to uniform grid, while MG refers to the multigrid. A multigrid strategy may offer the reduction of one order, measured in powers of N_{Acc}. Whether this saving is realized, however, is highly dependent on the geometrical structure of the multiscale feature that is imbedded in the larger scale. This is measured by the arc length of these features, β_x and β_y; these could be large numbers in the exact solutions to the PDE's. Multigrid methods may or may not be adequate to the ultimate task, but offer a potential advantage.

4.2. TWO-DIMENSIONAL

$$UG: \quad T_{CPU} = \left(\frac{CPU}{\textit{cell cycle}}\right)(N_{Acc})^6(\alpha c/D) \tag{29}$$

$$MG: \ T_{CPU} \approx \left(\frac{CPU}{\textit{cell cycle}}\right)(N_{Acc})^5\beta_x(\alpha c/D) \tag{30}$$

4.3. THREE-DIMENSIONAL

$$UG: \quad T_{CPU} = \left(\frac{CPU}{\textit{cell cycle}}\right)(N_{Acc})^8(\alpha c/D) \tag{31}$$

$$MG: \ T_{CPU} \approx \left(\frac{CPU}{\textit{cell cycle}}\right)(N_{Acc})^7\beta_x\beta_y(\alpha c/D) \tag{32}$$

5. Closing

Here we list some of the elements of a numerical laboratory environment.

- A basic Fortran or C code that solves a class of initial-boundary value problems in a generic domain with a state of the art numerical method. In a high-performance computing environment, such a code should take advantage of the distributed-computing environment.

In our case, CMHOG solves the reactive Euler equations in 1, 2, and 3 dimensions in line, rectangle, and rectangular parallelpiped geometries. It uses a fixed Eulerian mesh and can treat inflow, outflow, and periodic boundary conditions. CMHOG can run on a number of different platforms, including Cray, Convex, and Connection Machine supercomputers. Thought needs to be put into the code structure to allow enough flexibility to evolve with the new machine architectures. Groups that support a numerical laboratory should be willing to restructure their code when necessary.

- Code verification against a "test suite" of analytical and numerical standard test problems.

For example, CMHOG has been tested on a series 1-D and 2-D test problems that have either analytic solutions or well-documented numerical solutions in the literature. Comparison against all reasonable test cases is essential in order to use the results of the code as a high-quality approximation to the exact solution of the underlying PDEs. One needs to be able to distinguish between physical effects and numerical artifacts. For this reason, the linear stability and DSD results for detonation play an important role in code verification.

– Standardized graphics, visualization, and animation output.

CMHOG generates a standard graphics output file that can be postprocessed: contour plots, surface plots, time histories, 2-D color animations, 3-D isosurfaces, and volume-rendered animations. Standardization of the graphics environment and modularization of the code into single function subroutines are important to take advantage of the modern computing infrastructure, particularly when there are many members within the group.

– A sufficiently large mass-storage environment for storage of the data. In our case, we have access to the mass storage facilities of NCSA and LANL.

None of the basic elements of the numerical laboratory mentioned above is conceptually new or is newly identified by us. These elements have been in place at most of the national laboratories in the United States and in Europe, and more recently within the various academic supercomputer centers like NCSA. However, the combination of enhanced computational power, advanced algorithms, easy-to-use advanced graphics, visualization, and animation software, and large quantities of mass storage creates a new scientific tool to examine the complex internal flows in combustion. The synergism of these elements has provided us a powerful new tool, the numerical combustion laboratory.

Acknowledgements

D. S. Stewart was supported by the U.S. Air Force Armament Laboratory, Eglin AFB, Florida, and the National Center for Supercomputer Applications. J. B. Bdzil is supported by the U.S. Department of Energy. D. S. Stewart acknowledges NCSA's hospitality and support during his sabbatical, with special thanks to M. Norman, J. Stone, and the NCSA staff.

References

[1] Addessio, F. L, Baumgardner, J., Dukowicz, J. K., Johnson, N. L., Kashiwa, B. A., Rauenzahn, R. M., and Zemach, C. (1990) 'CAVEAT: A computer code for fluid dynamics with large distortions and internal slip', Los Alamos Report, LA-10613.

[2] Bdzil, J. B., and Fickett, W. (1992) 'DSD technology - A detonation reactive-Huygens code', Los Alamos Report, LA-12235-MS.

[3] Bdzil, J. B., Fickett, W., and Stewart, D. S. (1990) 'Detonation shock dynamics: A new approach to modeling multi-dimensional detonation waves', in *Proceedings of the*

9th (International) Symposium on Detonation, Portland, Oregon, August 28–Sept. 1, 1989, 730–742.

[4] Bdzil, J. B., and Stewart, D. S. (1989) 'Modeling of two-dimensional detonation with detonation shock dynamics', *Physics of Fluids A,* **1,** no. 7, 1261–1267.

[5] Bdzil, J. B., and Stewart, D. S. (1986) 'Two dimensional, time-dependent detonation: The interaction of edge rarefactions with finite length reaction zones', *J. Fluid Mechanics,* **171,** 1-026.

[6] Bourlioux, A., and Majda, A. J. (1992) 'Theoretical and numerical structure of unstable detonations', *Progress in Energy and Combustion Science,* to appear.

[7] Colella, P., Majda, A. J., and Roytburd, V. (1986) 'Theoretical and numerical structure for reacting shock waves', *SIAM J. Sci. Stat. Comput.,* **7,** no. 4, 1059–1080.

[8] Colella, P., and Woodward, P. R. (1984) 'The piecewise parabolic method for gasdynamical simulations', *J. Comput. Phys.,* **54,** 174–201.

[9] Dukowicz, J. K. (1985) 'A general, non-iterative Riemann solver for Godunov's method', *J. Comput. Phys.,* **61,** 119–137.

[10] Erpenbeck, J. J. (1964) 'Stability of idealized one-reaction detonation', *Physics of Fluids,* **7,** no. 7, 684–696.

[11] Fickett, W., and Davis, W. C. (1979) *Detonation*, University of California Press.

[12] Klein, R. (1991) 'On the dynamics of weakly curved detonations', in P. C. Fife, A. Linan, and F. Williams (eds.), *Dynamical Issues in Combustion Theory, IMA Volumes in Mathematics and Its Applications,* Vol. 35, Springer-Verlag, New York, 127–166.

[13] Klein, R., and Stewart, D. S. (1992) 'The relation between curvature, rate state dependence and detonation velocity', preprint.

[14] Lee, H. I., and Stewart, D. S. (1990) 'Calculation of linear detonation instability, one-dimensional instability of plane detonation', *J. Fluid Mechanics,* **216,** 103–132.

[15] Stewart, D. S. (1991) 'Running NCSA/DETONATION/CAVEAT', Los Alamos National Laboratory, preprint.

[16] Stewart, D. S. (1991) 'Surface evolution equations from detonation theory', in H. G. Kaper and M. Garbey (eds.), *Asymptotic Analysis and the Numerical Solution of Partial Differential Equations, Lecture Notes in Pure and Applied Mathematics,* Vol. 130, Marcel Dekker, Amsterdam, 175–183.

[17] Stewart, D. S. (1991) 'Workshop problems for the comparison of numerical methods for detonation simulations', Minisymposium on Comparison of Numerical Methods for Detonation Simulations, Fourth International Conference on Numerical Combustion, sponsored by SIAM/INRIA, St. Petersburg Beach, Florida.

[18] Stewart, D. S., and Bdzil, J. B. (1990) 'Examples of detonation shock dynamics for detonation wave spread applications', in *Proceedings of the 9th (International) Symposium on Detonation,* Portland, Oregon, August 28–September 1, 1989, 773–783.

[19] Stewart, D. S., and Bdzil, J. B. (1988) 'A lecture on detonation shock dynamics', *Mathematical Modelling in Combustion Science, Lecture Notes in Physics,* Vol. 249, Springer-Verlag, New York, 17–30.

[20] Stewart, D. S., and Bdzil, J. B. (1988) 'The shock dynamics of stable multi-dimensional detonation', *Combustion and Flame,* **72,** 311–323.

[21] Stone, J. M., and Norman, M. L. (1992) 'The three-dimensional interaction of a supernova remnant with an interstellar cloud', *Astrophysical J. Lett.*, **390**, L17–L19.

Part 4: Applied and Asymptotic Analysis

VISCOELASTIC FLUID FLOW: CRITICAL PARAMETERS AND ASYMPTOTICS

L. Pamela Cook, D. O. Olagunju, and G. F. Schleiniger
Department of Mathematical Sciences
University of Delaware
Newark DE 19716

ABSTRACT. This article discusses critical parameters and the use of asymptotic analysis in viscoelasticity. The discussion centers on two specific problems: the first examines the entrance layer structure that accounts for memory effects of a fluid with "small elasticity" as it moves through the inlet into a channel; the second examines secondary flows and stability of the flow of a viscoelastic fluid in a cone-plate apparatus of small angle.

1. Introduction

Viscoelastic fluids exhibit behavior that lies midway between that of a viscous fluid and an elastic solid. These fluids have a fading memory, so that the flow history is important, not just the present state. Polymeric materials exhibit viscoelastic behavior. Since these materials are in widespread use today, there is a significant interest in their properties, especially in their efficient processing, and therefore in their flow simulations.

In addition to the Reynolds number, Re, which is the ratio of inertial to viscous forces, viscoelastic flows are characterized by the Weissenberg number, We, which is the ratio of the relaxation time scale of the fluid to that of the particular flow. Alternatively, and especially in time-dependent flows, this is known as the Deborah number, De. Instead one could use the Mach number, M, the ratio of the characteristic flow speed to the propagation speed of shear waves in the fluid, and the elasticity number to characterize the flow. There are clearly several interesting limits. As We $\rightarrow$ 0, the fluid becomes Newtonian and hence loses its memory. This feature is of especial note in what follows for the discussion of boundary conditions at a flow inlet. As Re $\rightarrow$ 0, the flow becomes inertialess, and the linearized momentum equation takes over. The real practical interest is in the range We $= O(1)$, but studying the limiting situations clarifies the role of each of the parameters.

H. G. Kaper and M. Garbey (eds.),
Asymptotic and Numerical Methods for Partial Differential Equations with Critical Parameters 191–205.

Several models have been proposed to describe viscoelastic fluids and the resultant constitutive relations derived. For models of differential type, the flow-governing equations form a coupled system of nonlinear partial differential equations. For steady flows (with no Newtonian solvent), this system is composite (consisting of both elliptic and hyperbolic components at a given point) and may be of mixed type (that is, components changing from elliptic to hyperbolic within the flow region) and may degenerate at the boundaries [2],[3]. Serious questions arise as to the correct boundary conditions to pose, the existence and uniqueness of solutions of the resultant problems, and the stability of solutions. Also, an analysis of the behavior of a variety of flows governed by each model must be carried out and the results from the various models contrasted. Asymptotic and perturbation techniques can clarify features of these complicated nonlinear systems and can be used in numerical schemes.

In this review we focus on two particular problems. First, we consider the boundary conditions and resultant flow at an inlet, and second, we consider secondary flows and stability of a cone-plate viscometer. The first problem is of crucial importance in numerical simulations, the second in experimental determination of the properties of a given polymer.

2. Governing Equations

We consider viscoelastic flows that are isothermal, incompressible, and governed by the Maxwell or the Oldroyd-B models. These are basic constitutive models that show some but not all characteristics of viscoelastic flows. Our results generalize, at least for the inlet flow, to other models of Maxwell type as well as to some other models, for example, the Tanner model and the Bird-DeAguiar model.

The nondimensional equations governing the flow of such a fluid are

$$\text{conservation of mass} \quad \nabla \cdot \mathbf{q} = 0\,, \tag{1}$$

$$\text{conservation of momentum} \quad \mathrm{Re}\,\frac{D\mathbf{q}}{Dt} + \nabla p = \operatorname{div}\mathbf{T}\,, \tag{2}$$

and the

$$\text{constitutive relation} \quad \boldsymbol{\tau} + \mathrm{We}\,\boldsymbol{\tau}_{(1)} = 2\beta\,\mathbf{D}. \tag{3}$$

The extra stress $\mathbf{T}$ and the Maxwell stress $\boldsymbol{\tau}$ are related by

$$\mathbf{T} = 2(1-\beta)\mathbf{D} + \boldsymbol{\tau}\,, \tag{4}$$

and p is the isotropic pressure, $\mathbf{q}$ is velocity, $(\nabla\mathbf{q})_{ij} = \partial q_i/\partial x_j$, $\mathbf{D}$ is the symmetric part of $\nabla\mathbf{q}$, D/Dt is the material derivative, and $(\quad)_{(1)}$ denotes the upper convected derivative

$$\boldsymbol{\tau}_{(1)} = \frac{D\boldsymbol{\tau}}{Dt} - \nabla\mathbf{q}\,\boldsymbol{\tau} - \boldsymbol{\tau}\,\nabla\mathbf{q}^t.$$

Here, $\beta = \eta_p/\eta$; η, the total viscosity, is given by $\eta = \eta_s + \eta_p$, where η_s is the solvent viscosity and η_p is the polymer viscosity. If there is no solvent, we have a polymer melt $\beta = 1$, and we have the upper-convected Maxwell (UCM) model. If there is no polymer, we have a Newtonian fluid, $\beta = 0$. In general, $\mathrm{Re} = U\ell/\nu, \mathrm{We} = \lambda U/\ell$, where U, ℓ, ν, and λ are characteristic speed and length of the flow, kinematic viscosity, and relaxation time of the fluid, respectively.

Note that, in 3-D flow, the equations (1)–(4) give ten equations for the ten unknowns q_i, T_{ij}, $i, j = 1, 2, 3$, and p. The system of coupled equations is quasilinear. In 2-D flow, the equations (1)–(4) give six equations in six unknowns.

Consider (1)–(4) with $\beta = 1$ for a simple 2-D steady shearing motion of the form $\mathbf{q} = (u, v)$, where

$$u = u(y), \quad v = 0, \quad \text{and} \quad \tau = \begin{pmatrix} \sigma & \tau \\ \tau & \gamma \end{pmatrix}.$$

Then the system reduces to

$$p_x = \sigma_x + \tau_y$$

$$p_y = \tau_x + \gamma_y$$

$$\sigma + \mathrm{We}\, u\, \sigma_x - 2\mathrm{We}\, u_y\, \tau = 0$$

$$\tau + \mathrm{We}\, u\, \tau_x - \mathrm{We}\, u_y\, \gamma = u_y$$

$$\gamma + \mathrm{We}\, u\, \gamma_x = 0.$$

In particular, we have a Poiseuille-type flow,

$$u = \mathrm{Re}\,(1 - y^2)/2, \quad v = 0,$$

with

$$\tau = -\mathrm{Re}\, y\,, \quad p = -\mathrm{Re}\, x\,,$$

$$\sigma = 2\mathrm{We}\, \mathrm{Re}^2 y^2\,, \quad \gamma = 0\,.$$

Note that $\dot{\gamma} = 2\mathbf{D}$ has only one nonzero entry, which we denote by $\dot{\gamma}$ (shear rate), so the flow is viscometric. There are several features to note of this flow. First, it does not show shear thinning, which is a behavior typical of viscoelastic fluids. That is, if the apparent viscosity is defined as

$$\eta(\dot{\gamma}) \equiv \frac{\tau(\dot{\gamma})}{\dot{\gamma}},$$

then typically for a viscoelastic fluid η decreases as $\dot{\gamma}$ increases. Other, more complicated models do in fact show this effect. On the other hand, if we denote the first normal stress difference by

$$N_1 = \sigma - \gamma,$$

we find

$$N_1 = 2\mathrm{We}\,\mathrm{Re}^2 y^2.$$

Note that if We $\neq 0$, N_1 is not zero as it is in Newtonian (We $= 0$) flow. It is this normal stress difference that is responsible for rod climbing, or the Weissenberg effect. Typically for a viscoelastic fluid in 3-D simple shear, the second normal stress difference ($\zeta =$ the third normal stress component)

$$N_2 = \gamma - \zeta$$

is nonzero, and in fact $N_1 \gg N_2 \neq 0$. The properties of the two normal stress differences and of the viscosity, in a viscometric flow, are thought to characterize a particular fluid. These flows present a situation where each particle has a constant history, so that measured properties are indeed fluid characteristics. Viscometers are designed to measure these properties and hence to imitate viscometric flows as closely as possible.

It is important to note one other property of the equations (1)–(3). In steady flow these equations form a composite coupled nonlinear system of partial differential equations that can change type within the flow [2],[3]. In particular, one can identify the Mach number, $\mathrm{M} = \sqrt{\mathrm{ReWe}} = U/\sqrt{\nu/\lambda} =$ speed of disturbances relative to the speed of shear waves in the fluid. For $\mathrm{M} < 1$ the first-order system has one repeated real eigenvalue, two purely imaginary eigenvalues, and two complex eigenvalues. For $\mathrm{M} > 1$ the system has four real eigenvalues (one repeated) and two purely imaginary eigenvalues [3]. Furthermore, the system degenerates on no-slip boundaries [2].

In what follows we first discuss inlet boundary conditions (Section 3) and then secondary flows in a cone-plate viscometer (Section 4).

3. The Inlet Layer

An important question for viscoelastic flow models is what constitutes a well-posed boundary value problem. Here we are specifically interested in what boundary conditions to pose at the inlet of a 2-D truncated channel, a problem that arises in numerical simulations of channel flow. The question was answered in part by Renardy [7], who showed that for small perturbations of slow channel flow, subject to certain compatibility conditions, a well-posed problem is to prescribe the velocity components u, v as well as the normal stresses σ, γ at the inlet boundary, no-slip conditions on the side walls, and compatible conditions at the outflow boundary. This was proved for Maxwell-like models using an iterative method. Here, using a singular perturbation technique as the Weissenberg number We $\to 0$, we are able to consider other models, other boundary conditions, and to understand the evolution of the flow through the inlet.

For the UCM model, consider first the problem as posed by Renardy,

$$\begin{aligned} \nabla \cdot \mathbf{q} &= 0 \\ \mathrm{Re}\, \mathbf{q} \cdot \nabla \mathbf{q} &= \nabla p + \mathrm{div} \boldsymbol{\tau} \qquad 0 < x < L, -1 < y < 1 \\ \boldsymbol{\tau} + \mathrm{We}\, \boldsymbol{\tau}_{(1)} &= \dot{\gamma} \end{aligned}$$

with the boundary conditions on $-1 \leq y \leq 1$

$$u(0, y; \mathrm{We}) = U_0(y) + \cdots \tag{5}$$
$$v(0, y; \mathrm{We}) = V_0(y) + \cdots \tag{6}$$
$$\sigma(0, y; \mathrm{We}) = \Sigma_0(y) + \cdots \tag{7}$$
$$\gamma(0, y; \mathrm{We}) = \Gamma_0(y) + \cdots \tag{8}$$
$$u(L, y; \mathrm{We}) = U_e(y) + \cdots \tag{9}$$
$$v(L, y; \mathrm{We}) = V_e(y) + \cdots \tag{10}$$

and

$$u(x, \pm 1; \mathrm{We}) = v(x, \pm 1; \mathrm{We}) = 0 \quad \text{on} \quad 0 \leq x \leq 1 \tag{11}$$

subject to conservation of mass,

$$\int_{-1}^{1} U_0(y) dy = \int_{-1}^{1} U_e(y) dy .$$

We also require, as did Renardy, that the prescribed conditions be continuous at the corners in order to avoid singularities, for example, to this order $U_0(\pm 1) = V_0(\pm 1) = V_0'(\pm 1) = U_e(\pm 1) = V_e(\pm 1) = V_e'(\pm 1) = 0$, and that $U_0(y) \geq 0$, so that indeed $x = 0$ is an inlet boundary. Compatibility conditions on $\Sigma_0(\pm 1), \Sigma_0'(\pm 1), \Gamma_0(\pm 1)$ and $\Gamma_0'(\pm 1)$ are derived from (3).

As We $\to 0$, a regular expansion of the solution,

$$\mathbf{f}(x, y; \mathrm{We}) = \mathbf{f}_0(x, y) + \mathrm{We}\, \mathbf{f}_1(x, y) + \cdots , \tag{12}$$

clearly gives (on substitution into (1)–(3)) that the lowest-order flow is governed by the Navier-Stokes equations. For that system only two variables can be prescribed at the boundaries (e.g., u, v given all around). Clearly the regular expansion is nonuniform near $x = 0$, since it cannot satisfy the boundary conditions there. Hence, an inlet layer is necessary to account for this.

3.1. THE INLET LAYER

$$(X = x/\mathrm{We}, y) \quad \text{fixed as} \quad \mathrm{We} \to 0$$

The failure of the regular expansion at the inlet is due to the fact that we have failed to account for the large derivatives there. We now stretch the x variable $X = x/\mathrm{We}$ so that $\partial/\partial x = (1/\mathrm{We})\partial/\partial X$. The expansion in the inlet then has the form

$$\mathbf{f}(x, y; \mathrm{We}) = \tilde{\mathbf{f}}_0(X, y) + \mathrm{We}\,\tilde{\mathbf{f}}_1(X, y) + \cdots .$$

Substituting into the governing equations, we find that to lowest order the governing equations are

$$\tilde{U}_{0,X} = 0 \tag{13}$$
$$(\tilde{\sigma}_0 - \tilde{p}_0)_{,X} = 0 \tag{14}$$
$$\mathrm{Re}\,\tilde{U}_0\,\tilde{V}_{0,X} = \tilde{\tau}_{0,X} \tag{15}$$
$$\tilde{\sigma}_0 + \tilde{U}_0\,\tilde{\sigma}_{0,X} = 2\tilde{U}_{1,X} \tag{16}$$
$$0 = \tilde{V}_{0,X} \tag{17}$$
$$\tilde{\gamma}_0 + \tilde{U}_0\,\tilde{\gamma}_{0,X} = 2\tilde{V}_{0,y}. \tag{18}$$

From (13), (14), (15), and (17),

$$(\tilde{U}_0, \tilde{\sigma}_0 - \tilde{p}_0, \tilde{V}_0, \tilde{\tau}_0) = (U_0(y), (\tilde{\sigma}_0 - \tilde{p}_0)(y), V_0(y), \tilde{\tau}_0(y)).$$

That is, $\tilde{U}_0, \tilde{\sigma}_0 - \tilde{p}_0, \tilde{V}_0$, and $\tilde{\tau}_0$ carry through the layer with no variation in X. To determine $\tilde{\sigma}_0$, we must go to higher order to determine $\tilde{U}_{1,X}$ but, as expected from continuity,

$$\tilde{U}_{1,X} = -\tilde{V}_{0,y}\,.$$

Solving (13)–(18) with the prescribed boundary conditions (5)–(11) gives in the inlet layer

$$\tilde{U} = U_0(y) + \cdots \tag{19}$$
$$\tilde{V} = V_0(y) + \cdots \tag{20}$$
$$\tilde{\sigma} = -2V_0'(1 - e^{-X/U_0}) + \Sigma_0\, e^{-X/U_0} + \cdots \tag{21}$$
$$\tilde{\gamma} = 2V_0'(1 - e^{-X/U_0}) + \Gamma_0\, e^{-X/U_0} + \cdots \tag{22}$$
$$\tilde{\tau} = \tilde{\tau}_0(y) + \cdots \tag{23}$$
$$\tilde{p} = -2V_0'(1 - e^{-X/U_0}) + \Sigma_0\, e^{-X/U_0} + P_0(y) + \cdots, \tag{24}$$

where $\tilde{\tau}_0, P_0$ are as yet unknown and must be determined from matching. Clearly, the normal stresses (as well as the pressure) relax exponentially across the layer.

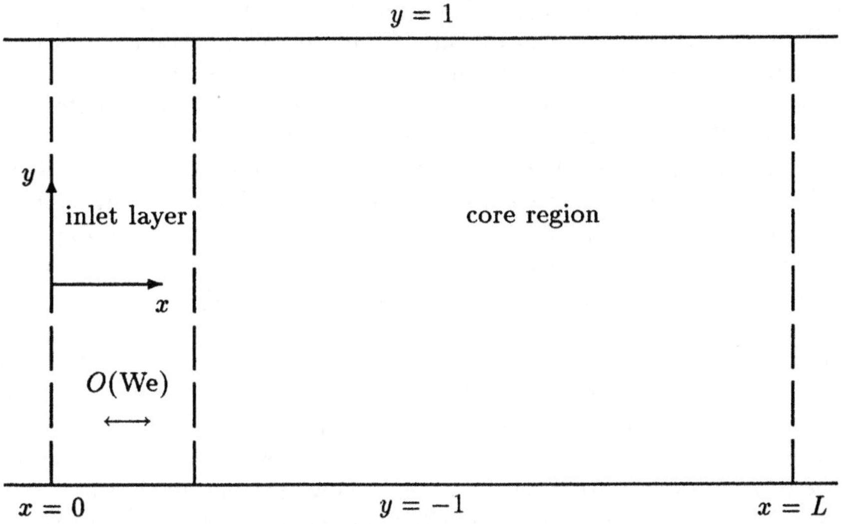

Fig. 1. **Channel flow: inlet layer and core region**

3.2. THE CORE REGION

(x, y) fixed as We $\to 0$

In the core region the regular expansion (12) holds, so that to lowest order we have

$$u_{0,x} + v_{0,y} = 0 \tag{25}$$

$$\mathrm{Re}\,\{u_0 u_{0,x} + v_0 u_{0,y}\} + p_{0,x} = \Delta u_0 \tag{26}$$

$$\mathrm{Re}\,\{u_0 v_{0,x} + v_0 v_{0,y}\} + p_{0,y} = \Delta v_0\,. \tag{27}$$

Here we have substituted the explicit expressions obtained for the stress from the constitutive equation. A well-posed problem for the Navier-Stokes equations (25)–(27) is to give u_0, v_0 on all boundaries. In particular, by matching we expect

$$u_0(0, y) = U_0(y)\,, \quad v_0(0, y) = V_0(y)\,.$$

Note that the explicit relationship for the normal stresses are

$$\sigma_0 = 2u_{0,x}\,, \quad \gamma_0 = 2v_{0,y}\,.$$

Since from continuity $u_{0,x} = -v_{0,y}$ it is clear from (21), (22) that the normal stresses relax in the boundary layer to exactly their Newtonian values as required for the core region. Once the Newtonian problem is solved, the values of $\tau_0(0, y)$ and $p_0(0, y)$ will, by matching, determine $\tilde{\tau}_0, P_0$.

The matching process can be carried out to higher orders [1]. Note that to first order the compatibility conditions at the corners ensure that no side layers are needed. Note also that this expansion is valid only for Re fixed as We $\rightarrow$ 0. That is, the inner layer terms e^{-X/U_0} have as exponent, for perturbation of Poiseuille flow, $-X/U_0 = -2x/[\mathrm{Re}\,\mathrm{We}\,(1 - y^2)]$. Thus, if $\mathrm{ReWe} = O(1)$, these terms are no longer exponentially small out of the inlet layer. In fact, since $\sqrt{\mathrm{ReWe}} = \mathrm{M}$ (the Mach number), as we expect this analysis holds for subcritical flow, $\mathrm{M} < 1$, but fails once the flow becomes critical (that is, once they change type). Further analysis must be carried out for this case.

These results are valid for all Maxwell fluids as well as some other fluids, for example, Tanner and Bird-DeAguiar. Perhaps more interesting is that we can use this analysis to consider other boundary conditions. For example, a physically more realistic problem would be to prescribe traction conditions at the inlet. However, from our inlet analysis we see clearly that $u, v, \sigma - p, \tau$ are carried directly from any prescribed values across the inlet to the core. On the core we can prescribe only two conditions, so $u, v, \sigma - p, \tau$ cannot all be prescribed at the inlet. One other inlet prescription that does result in a well-posed problem is $\sigma, \gamma, \tau, \sigma - p$, although this does not seem physical.

4. Secondary Flows in a Cone-Plate Viscometer

The geometry of the cone-plate viscometer is shown in Fig. 2. The cone axis is normal to the plate with the test fluid between the cone and the plate. We consider the cone rotating about its axis at angular speed, Ω, and a stationary plate. The cone angle α, measured from the plate, is assumed small.

The advantage of this arrangement is that the flow is almost viscometric and that the flow is realizable. Unfortunately, the flow is not exactly viscometric, and the effect of secondary flows must be understood, as must the onset of instability in the flow. The latter is observed to occur earlier for viscoelastic flows than for Newtonian flows.

The boundary value problem for this flow consists of the dimensionless equations (1)–(4) together with the no-slip boundary conditions which, in spherical coordinates (r, θ, φ) with θ the polar angle, are

$$\mathbf{q} = 0 \quad \text{for} \quad r = 0\,,$$

$$v_r = v_\theta = 0\,, \quad v_\varphi = r \sin\theta \quad \text{on} \quad \theta = \frac{\pi}{2} - \alpha\,,$$

$$\mathbf{q} = 0 \quad \text{on} \quad \theta = \frac{\pi}{2}\,.$$

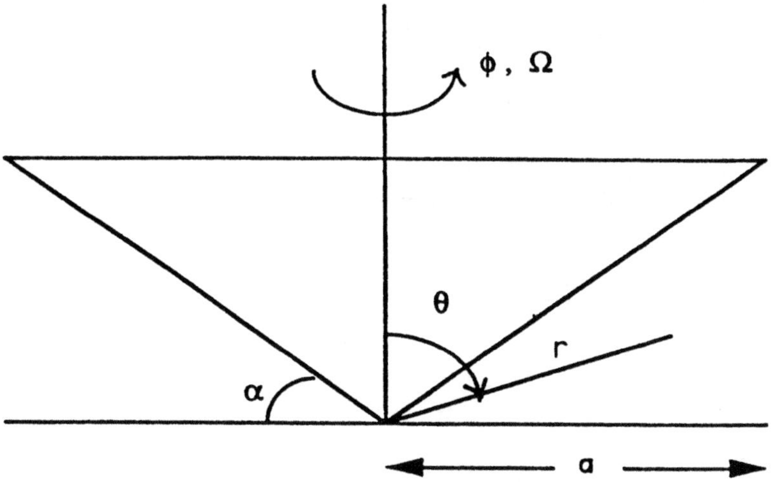

Fig. 2. **Cone and plate arrangement**

Note that we will ignore the free surface in this analysis—although clearly its effect should be analyzed in the future. The nondimensionalization here is by some typical length a (e.g., the plate radius), velocity $a\Omega$, and time $1/\Omega$. Pressure and stresses are scaled by $\eta\Omega$, so that

$$\mathrm{Re} = \frac{a^2\Omega}{\nu}, \quad \mathrm{De} = \lambda\Omega .$$

For simplicity, in this note we consider the case $\beta = 1$ (the UCM model) although general β are discussed in [4]. The governing equations are then (1)–(4) with $\beta = 1$. For the cone angle $\alpha \ll 1$ it is useful to define

$$\psi = (\pi/2 - \theta)/\alpha$$

as a new independent variable, so that in the flow we have $0 \le \psi \le 1$, and $\psi = 0$ on the plate, $\psi = 1$ on the cone. Analysis of the flow equations shows that for $\alpha \ll 1$, appropriate scalings are

$$\mathbf{q} = (v_r, v_\theta, v_\varphi) = (\alpha^2 u, \alpha^3 v, w)\,, \quad p = \hat{p}/\alpha^2\,, \quad \text{and}$$

$$\tau = \begin{pmatrix} \tau_{rr} & \tau_{r\theta} & \tau_{r\varphi} \\ \tau_{r\theta} & \tau_{\theta\theta} & \tau_{\theta\varphi} \\ \tau_{r\varphi} & \tau_{\theta\varphi} & \tau_{\varphi\varphi} \end{pmatrix} = \begin{pmatrix} \alpha^2\Sigma & \alpha\zeta & \gamma \\ \alpha\zeta & \alpha^2\Gamma & \Pi/\alpha \\ \gamma & \Pi/\alpha & \Delta/\alpha^2 \end{pmatrix} .$$

Note that these scalings are typical of a lubrication approximation.

The equations of motion are then, assuming axial symmetry (we drop the accent from $\hat{p}$ to simplify the notation),

$$\frac{\partial u}{\partial r} + \frac{2}{r}u - \frac{1}{4}\frac{\partial v}{\partial \psi} + \alpha\frac{v}{r}\tan(\alpha\psi) = 0 \tag{28}$$

$$\frac{\partial p}{\partial r} + \frac{\Delta}{r} = \alpha^2\left(\mathrm{Re}\,\frac{w^2}{r} - \frac{1}{r}\frac{\partial \zeta}{\partial \psi}\right) + o(\alpha^2) \tag{29}$$

$$\frac{1}{r}\frac{\partial p}{\partial \psi} = \alpha\tan(\alpha\psi)\frac{\Delta}{r} + O(\alpha^2) \tag{30}$$

$$\frac{1}{r}\frac{\partial \Pi}{\partial \psi} = 2\alpha\tan(\alpha\psi)\frac{\Pi}{r} + \alpha^2\left(\frac{\partial \gamma}{\partial r} + \frac{3}{r}\gamma - \mathrm{Re}\,\frac{\partial w}{\partial t}\right) \tag{31}$$

$$\mathrm{De}\,\frac{\partial \Sigma}{\partial t} + \Sigma = 2\frac{\partial u}{\partial r} - 2\mathrm{De}\,\frac{\zeta}{r}\frac{\partial u}{\partial \psi} + O(\alpha^2) \tag{32}$$

$$\mathrm{De}\,\frac{\partial \zeta}{\partial t} + \zeta = -\frac{1}{r}\frac{\partial u}{\partial \psi} + O(\alpha^2) \tag{33}$$

$$\mathrm{De}\,\frac{\partial \gamma}{\partial t} + \gamma = -\mathrm{De}\left(\frac{\Pi}{r}\frac{\partial u}{\partial \psi} + \frac{\zeta}{r}\frac{\partial w}{\partial \psi}\right) + \left(\frac{\partial w}{\partial r} - \frac{w}{r}\right) + O(\alpha^2) \tag{34}$$

$$\mathrm{De}\,\frac{\partial \Gamma}{\partial t} + \Gamma = 2\left(\frac{u}{r} - \frac{1}{r}\frac{\partial v}{\partial \psi}\right) + O(\alpha^2) \tag{35}$$

$$\begin{aligned}\mathrm{De}\,\frac{\partial \Pi}{\partial t} + \Pi \;=\; & \left(-\frac{\partial w}{\partial \psi} - \alpha\tan(\alpha\psi)\frac{w}{r}\right)\\ & +\alpha^2\mathrm{De}\Bigg(-\frac{\Gamma}{r}\frac{\partial w}{\partial \psi} + \frac{v}{r}\frac{\partial \Pi}{\partial \psi} - \frac{\Pi}{r}\frac{\partial v}{\partial \psi} - u\frac{\partial \Pi}{\partial r}\\ & -\frac{w}{r}\zeta + 2\frac{u}{r}\Pi + \zeta\frac{\partial w}{\partial r}\Bigg) + o(\alpha^2)\end{aligned} \tag{36}$$

and

$$\begin{aligned}\mathrm{De}\,\frac{\partial \Delta}{\partial t} + \Delta \;=\; & -2\mathrm{De}\,\frac{\Pi}{r}\frac{\partial w}{\partial \psi} + \alpha^2\mathrm{De}\Bigg(\frac{v}{r}\frac{\partial \Delta}{\partial \psi} - u\frac{\partial \Delta}{\partial r} + 2\frac{u}{r}\Delta\\ & -2\frac{\tan(\alpha\psi)}{\alpha}\frac{w}{r}\Pi\Bigg) + 2\alpha^2\mathrm{De}\,\gamma\left(\frac{\partial w}{\partial r} - \frac{w}{r}\right) + o(\alpha^2)\,.\end{aligned} \tag{37}$$

We look for solutions of the form

$$f(r,\psi;\mathrm{Re},\mathrm{De},\alpha) = f_0(r,\psi;\mathrm{Re},\mathrm{De}) + \alpha f_1(r,\psi;\mathrm{Re},\mathrm{De}) + \cdots$$

and find successive boundary value problems for the corresponding orders. Note that any nonzero u_0, v_0 means there is a secondary flow.

To summarize the successive approximations, we note that to zeroth order we can determine the variables only in terms of the (yet unknown) u_0, v_0 as follows:

$$\begin{aligned}
w_0 &= r\psi \quad \text{(the primary flow)},\\
p_0 &= -\Delta_0 \ln r\\
\zeta_0 &= \frac{1}{r}\frac{\partial u_0}{\partial \psi}\\
\Sigma_0 &= 2\frac{\partial u_0}{\partial r} + 2\mathrm{De}\,\frac{\zeta_0}{r}\frac{\partial u_0}{\partial \psi}\\
\gamma_0 &= -2\mathrm{De}\,\frac{\Pi_0}{r}\frac{\partial u_0}{\partial \psi}\\
\Gamma_0 &= 2\left(\frac{u_0}{r} - \frac{1}{r}\frac{\partial v_0}{\partial \psi}\right)\\
\Pi_0 &= -1\\
\Delta_0 &= 2\mathrm{De}\,.
\end{aligned}$$

To determine u_0, v_0, we must actually go to $O(\alpha^2)$ terms (for details see [4]). At that order a stream function can be introduced,

$$\frac{\partial \chi}{\partial \psi} = r^2 u_0\,, \quad \frac{\partial \chi}{\partial r} = r v_0,$$

which is governed by the nonhomogeneous fourth-order partial differential equation

$$\mathcal{L}\chi \equiv \left\{\frac{\partial^4}{\partial \psi^4} + 2\mathrm{De}^2\left(r\frac{\partial^3}{\partial r \partial \psi^2} + 2\frac{\partial^2}{\partial \psi^2}\right)\right\}\chi = 2\psi(4\mathrm{De}\, r^3 - \mathrm{Re}\, r^5)\,,$$

and the boundary conditions

$$\frac{1}{r}\frac{\partial \chi}{\partial r}, \quad \frac{1}{r^2}\frac{\partial \chi}{\partial \psi} \to 0 \quad \text{as} \quad r \to 0\,,$$

$$\frac{\partial \chi}{\partial \psi} = \frac{\partial \chi}{\partial r} = 0 \quad \text{on} \quad \psi = 0, 1\,.$$

In addition, we prescribe the following conditions at infinity:

$$\chi = O(r^3) \quad \text{for} \quad r \gg 1 \quad \text{if} \quad \mathrm{Re} = 0\,,$$

$$\chi = O(r^5) \quad \text{for} \quad r \gg 1 \quad \text{if} \quad \mathrm{Re} \neq 0 \, .$$

Since this is an inhomogeneous problem, any two solutions differ by a solution of the associated homogeneous problem. One such solution is $\chi =$ constant. Since these solutions yield only the trivial solution $u_0 = v_0 = 0$, we shall look only for nonconstant solutions. For De = 0 (Newtonian case) the unique solution is

$$\chi = \frac{\mathrm{Re}}{60} r^5 \psi^2 (3\psi - \psi^3 - 2) \, .$$

For the non-Newtonian case, De $\neq 0$, the associated homogeneous problem has an infinite sequence of eigenvalues. These eigenvalues are given by

$$\mathrm{De}^* = \delta\sqrt{\frac{2}{5}} , \quad \text{if} \quad \mathrm{Re} = 0 \text{ and, in addition, } \mathrm{De}^* = \delta\sqrt{\frac{2}{7}} , \quad \text{if} \quad \mathrm{Re} \neq 0 \, ,$$

where $\delta = n\pi$ (n a positive integer) or $\delta = \tan\delta$. The smallest of these eigenvalues is

$$\mathrm{De}_c = \pi\sqrt{\frac{2}{5}} , \quad \text{if} \quad \mathrm{Re} = 0 \quad \text{and} \quad \mathrm{De}_c = \pi\sqrt{\frac{2}{7}} , \quad \text{if} \quad \mathrm{Re} \neq 0 \, .$$

For De = De* the nonhomogeneous problem may have either no solution or infinitely many solutions. For further discussion see [4]. If De $\neq$ De*, one solution is

$$\chi = r^3 f_3(\psi) + r^5 f_5(\psi),$$

where

$$f_j(\psi) = C_j[A_j(\mu_j\psi - \sin\mu_j\psi) + B_j(1 - \cos\mu_j\psi) + \psi^3] \, .$$

The constants C_j, A_j and B_j ($j = 3, 5$) are

$$C_3 = \frac{4}{5\mathrm{De}} , \quad C_5 = \frac{-\mathrm{Re}}{7\mathrm{De}^2} , \quad A_j = \frac{h_1(\mu_j)}{h_3(\mu_j)} , \quad B_j = \frac{h_2(\mu_j)}{h_3(\mu_j)} ,$$

and $\mu_3 = \mathrm{De}\sqrt{10}, \mu_5 = \mathrm{De}\sqrt{14}$. The functions h_1, h_2, and h_3 are defined as

$$h_1(x) = 3\cos x + x\sin x - 3 , \quad h_2(x) = x\cos x + 2x - 3\sin x , \quad \text{and}$$

$$h_3(x) = x(2 - 2\cos x - x\sin x) \, .$$

For De < De_c, this solution is unique, and typical secondary flow streamlines are displayed in Fig. 3. If we consider these flows at a fixed point and consider De $\to$ De_c, we find the solutions are singular there; that is, the limiting value does not exist. This is related to the so-called high Weissenberg number problem.

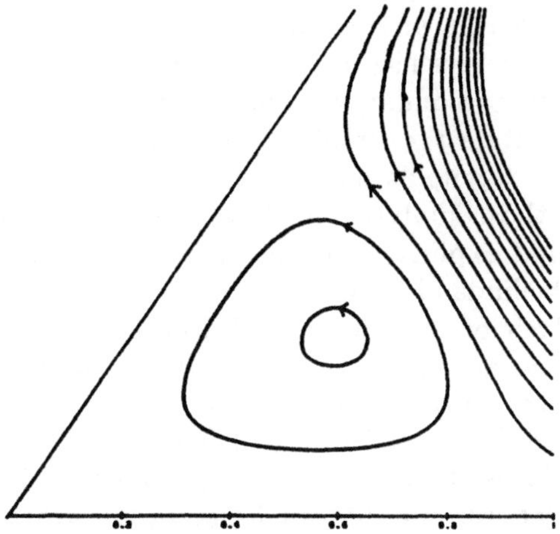

Fig. 3. **Secondary-flow streamlines for a Maxwell fluid:** De = 0.1 **and** Re = 0.5

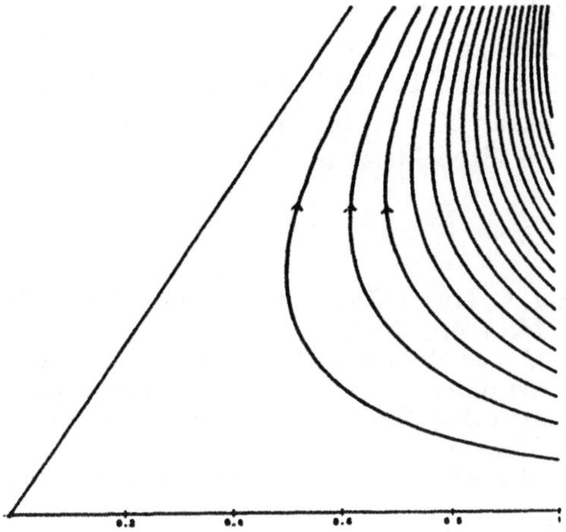

Fig. 4. **Secondary-flow streamlines for a Newtonian fluid:** De = 0 **and** Re = 1.0

Note that for De $= 0$ (Re $\neq 0$), there is a secondary flow (Fig. 4) that is inward along the stationary plate and outward along the rotating cone. As expected, inertial effects fling out fluid. As De increases, a secondary vortex appears near the vertex and grows with De. The elastically induced vortex rotates in the opposite sense to the inertially induced one.

Previous analyses of this secondary flow have been carried out for example by Walters and Waters [8] on the simple fluid model of Coleman and Noll, where a perturbation on Reynolds number was carried out, and by others for coaxial disc flow. In those analyses the flows were computed numerically. Here we were able to find explicit solution representations, and we can do so in fact for Oldroyd-B fluids also [4].

Linear stability analysis of the flows has also been carried out [5]. It was shown that De_c is exactly the value where the flow first becomes unstable to axisymmetric disturbance. Two critical values of the Deborah number were found, one for Re $= 0$ and the other for Re $\neq 0$. The critical Deborah number for creeping flow, $\pi\sqrt{2/5}$, is the same as that obtained by Phan-Thien [6]. Note that the critical Deborah number for Re $\neq 0$ (and order one), $\pi\sqrt{2/7}$, is less than that for creeping flow, thus showing that inertia is destabilizing. Recent experimental evidence, however, suggests that the first instability is to nonaxisymmetric time-dependent disturbances.

5. Summary

An overview of some problems and difficulties in viscoelastic fluid flow has been presented. Details and generalization are in referenced papers. In particular, various new phenomena, both physical and mathematical, arise because of the new critical parameter De (or We) that characterizes the amount of "elasticity" present in a viscoelastic fluid.

References

[1] Cook, L. P., and Schleiniger, G. (1991) 'The inlet layer in the flow of viscoelastic fluids', *J. Non-Newtonian Fluid Mech.* **40**, 307–321.

[2] Cook, L. P., Schleiniger, G. and Weinacht, R. J. (1990) 'Composite type, change of type, and degeneracy in first order systems with applications to viscoelastic flows', in B. L. Keyfitz and M. Shearer (eds.) *Nonlinear Evolution Equations That Change Type*, Springer-Verlag, New York, 32–46.

[3] Joseph, D. D. (1990) *Fluid Mechanics of Viscoelastic Liquids*, Springer-Verlag, New York.

[4] Olagunju, D. O. and Cook, L. P. (1992) 'Secondary flows in cone and plate flow of an Oldroyd-B fluid', *J. Non-Newtonian Fluid Mech.*, to appear.

[5] Olagunju, D. O. and Cook, L. P. (1992) 'Linear stability analysis of cone and plate flow of an Oldroyd-B fluid', preprint.

[6] Phan-Thien, N. (1985) 'Cone-and-plate flow of the Oldroyd-B fluid is unstable', *J. Non-Newtonian Fluid Mech.*, **17**, 37–44.

[7] Renardy, M. (1991) 'Compatibility conditions at corners between walls and inflow boundary conditions for fluids of Maxwell type', *Z. Angew. Math. Mech.*, **71**, 37.

[8] Walters, K. and Waters, N. D. (1968) 'On the use of a rheogoniometer part I — steady shear', in R. E. Wetton and R. W. Whorlow (eds.), *Polymer Systems*, MacMillan, London.

EXPONENTIALLY SMALL SPLITTING OF HOMOCLINIC ORBITS IN A WATER-WAVE PROBLEM

Wiktor Eckhaus
Mathematisch Instituut
Rijksuniversiteit Utrecht
Utrecht
the Netherlands

ABSTRACT. We describe the results of asymptotic analysis for a class of singularly perturbed ODEs that arise in the theory of water waves. The emphasis is on the question of existence of solitary waves and the phenomenon of exponentially small splitting of homoclinic orbits.

1. Introduction

The mathematical problem under consideration in this paper comes from the theory of water waves in the presence of surface tension. The task is to describe nonlinear surface waves of permanent form traveling with a constant speed on an inviscid fluid layer of constant depth with capillary forces acting on the free surface. Kirchgässner, in an impressive survey paper [7], shows how to obtain a center-manifold-type reduction leading to a finite system of nonlinear ODEs. Two nondimensional parameters govern the description of the water waves. They are the Froude number, $F_r = c/\sqrt{gh}$, and the Bond number, $b = T/\rho hc^2$, where c is the wave speed, g is the acceleration of gravity, h is the mean height of the fluid layer, T is the coefficient of surface tension, and ρ is the mass density. Extensive results over a wide range of these parameters are given in [7]. However, Kirchgässer remarks that when the Froude number is slightly higher than one and the Bond number less than one-third, "the situation...is more complicated and essentially unsolved." Kirchgässner does provide the center-manifold reduction for this case, which results in a system of four coupled ODEs.

In connection with this difficulty, Amick and Kirchgässner [2] have proposed a model problem of extreme simplicity. Consider $y(x)$, $x \in (-\infty, \infty)$,

H. G. Kaper and M. Garbey (eds.),
Asymptotic and Numerical Methods for Partial Differential Equations with Critical Parameters 207–214.

which satisfy the equation

$$\varepsilon^2 \frac{d^4y}{dx^4} + \frac{d^2y}{dx^2} - y + y^3 = 0, \tag{1}$$

where ε is a small parameter. Do there exist any trivial solutions such that

$$y(x) \to 0 \quad \text{as} \quad |x| \to \infty\ ?$$

Analysis of this simple problem turned out to be extremely difficult (see [1], [3], [4], [5]). The deep reason for the difficulty was shown in [3]. The problem appears to be linked to the phenomenon of "exponentially small splitting of separatrices" (see [6] and [8] and references quoted there). The phenomenon is typically represented by the breakup of homoclinic orbits under influence of rapid external oscillations. The analysis is highly nontrivial, because the quantity that one attempts to determine, the "splitting distance," is exponentially small. Poincaré was already familiar with this problem (see the quotation in the introduction of [6]), but very little progress had been made until recently.

We remark that Amick and Kirchgässner [2] do not give any relation between the model problem (1) and the relevant center manifold reduction presented in [7]. This relation was derived in [4]. It turns out that the equations of the center manifold can be further reduced to a *perturbed* version of the model problem (1) (i.e., the equation (1) with a small nonzero right-hand side).

Full asymptotic analysis of the (perturbed) model problem (1) was given in [3] and [4]. The main results can be stated as follows. The model problem (1) does not admit solitary-wave solution (i.e., $y(x) \to 0$ as $|x| \to \infty$). Perturbed model problems also, in general, do not have solitary wave solutions. However, there exist (a great many) perturbed problems for which solitary waves do exist. They are obtained, for any given problem, by modification of the right-hand side of (1) by terms of the order of magnitude ε^{2m}, where m is an arbitrarily large number. Furthermore, for any problems without solitary waves, one can construct "quasi-solitary waves," that is, solutions that over very large distances look like solitary waves. In what follows we describe and summarize the analysis and the results of [3] and [4].

2. Construction of Homoclinic Half-Orbits

The model problem (1), for $\varepsilon = 0$, has an integral given by

$$\left(\frac{dy}{dx}\right)^2 = y^2 - \frac{2}{3}y^3 + c. \tag{2}$$

Taking $c = 0$, one can define two half-orbits $y_+(x), y_-(x)$:

$$y'_\pm = \pm\sqrt{y^2 - \frac{2}{3}y^3}, \tag{3}$$

where prime denotes differentiation. The union of $y_+(x)$ and $y_-(x)$ is a homoclinic orbit sketched in the phase plane in Fig. 1.

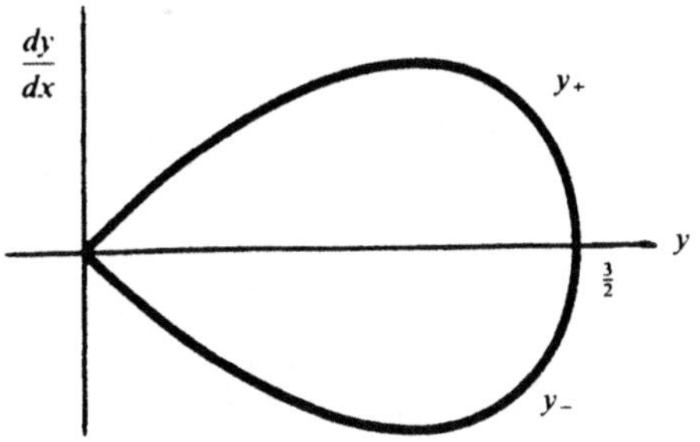

Fig. 1. **Homoclinic orbit in the phase plane**

One easily finds that $y_+ \cup y_-$, as a function of x, is twice continuously differentiable for $x \in (-\infty, \infty)$ and tends to zero as $|x| \to \infty$. This is the solitary wave solution of (1) for $\varepsilon = 0$.

For $\varepsilon \neq 0$, the equation (1) still has an integral, given by

$$\varepsilon^2 \left\{ \frac{dy}{dx}\frac{d^3y}{dx^3} - \frac{1}{2}\left(\frac{d^2y}{dy}\right)^2 \right\} = -\frac{1}{2}\left(\frac{dy}{dx}\right)^2 + \frac{1}{2}y^2 - \frac{1}{3}y^3 + c. \tag{4}$$

Define now

$$z := \left(\frac{dy}{dx}\right)^2, \tag{5}$$

and introduce y as an independent variable. One gets, as a differential equation for the trajectories in the phase plane, the equation

$$\varepsilon^2 \left\{ z\frac{d^2z}{dy^2} = -\frac{1}{4}\left(\frac{dz}{dy}\right)^2 \right\} + z = y^2 - \frac{2}{3}y^3 + c. \tag{6}$$

In [3] the equation (6), with $c = 0$, was taken as the starting point of asymptotic analysis. Formal asymptotic expansions can be constructed by two different methods: straightforward iteration and construction of a power series in y that at the same time is an asymptotic expansion in ε. We collect here some relevant results of the first method. We introduce a formal expansion with a remainder term as follows:

$$z(y;\varepsilon) = \Phi_m(y;\varepsilon) + \varepsilon^{2m}\rho_m(y;\varepsilon) \tag{7}$$

$$\Phi_m(y;\varepsilon) = \sum_{n=0}^{m-1} \varepsilon^{2n} z_n(y) \tag{8}$$

$$z_0 = y^2\left(1 - \frac{2}{3}y\right) \tag{9}$$

$$z_n =$$

$$\left\{\Phi_n \frac{d^2 z_{n-1}}{dy^2} - \frac{1}{2}\frac{d\Phi_{n-1}}{dy}\frac{dz_{n-1}}{dy} + \frac{d^2\Phi_{n-1}}{dy^2} z_{n-1} - \frac{1}{4}\varepsilon^{2(n-1)}\left(\frac{dz_{n-1}}{dy}\right)^2\right\} \tag{10}$$

For the remainder term ρ_m one gets the equation

$$\begin{aligned}\varepsilon^2 \Bigg\{ \left(\Phi_m + \varepsilon^{2m}\rho_m\right)\frac{d^2\rho_m}{dy^2} - \frac{1}{2}\frac{d\Phi_m}{dy}\frac{d\rho_m}{dy} + \frac{d^2\Phi_m}{dy^2}\rho_m - \\ -\frac{1}{4}\varepsilon^{2m}\left(\frac{d\rho_m}{dy}\right)^2\Bigg\} + \rho_m = f_m,\end{aligned} \tag{11}$$

where f_m is defined in a formal way as z_m in (10). It is important to note that f_m is a polynomial in y.

The main body of [3] is devoted to the analysis of the problem for the remainder ρ_m as solution of (11) within the space of functions that behave as y^2 for $y \to 0$. This is done by converting the problem to a suitable integral equation which permits a contraction-mapping argument. The main result is as follows.

Within the space of functions $z(y)$ which behave as y^2 for $y \downarrow 0$, there exists a unique solution of (6) on the interval $[0, y_0(\varepsilon)]$ such that $z(y) > 0$ for $y \in (0, y_0(\varepsilon))$, $z(y_0) = 0$, and $\left(\frac{dz}{dy}\right)_{y=y_0} < 0$. Furthermore,

$$z = \Phi_m + o(\varepsilon^{2m}), \quad y \in [0, y_0(\varepsilon)].$$

This result is further extended in [3] to a large class of perturbed problems of the structure

$$\varepsilon^2\frac{d^4y}{dx^4} + \frac{d^2y}{dx^2} - y + y^2 = \varepsilon^2\mathcal{P}\left(y, \frac{dy}{dx}, \frac{d^2y}{dx^2}; \varepsilon\right), \tag{12}$$

where

$$\begin{aligned}\mathcal{P} = X_0(y) + X_1(y)\left(\frac{dy}{dx}\right)^2 + \left[yX_2(y) + X_3(y)\left(\frac{dy}{dx}\right)^2\right]\frac{d^2y}{dx^2} \\ + X_4(y)\left(\frac{d^2y}{dx^2}\right)^2.\end{aligned} \tag{13}$$

The functions $X_i(y)$, $i = 0, 1, ..., 4$, are polynomials in y which may further depend on ε but in a regular way.

With these results one can define again the two half orbits $y_+(x), y_-(x)$ and study their union.

3. The (Non)-Existence of Solitary Waves

Given the existence of the solution $z(y)$ as specified in the preceding section, we define two half-orbits $y_+(x)$ and $y_-(x)$ through the equations

$$\frac{dy_+}{dx} = +\sqrt{z(y_+)}, \quad \frac{dy}{dx} = -\sqrt{z(y_-)}. \tag{14}$$

Next we investigate the smoothness of $y_+ \cup y_-$ at $y = y_0$. We find

$$\frac{d^2 y_\pm}{dx^2} = \frac{1}{2}\frac{dz}{dy} \quad \text{continuous at} \quad y = y_0 \tag{15}$$

$$\frac{d^3 y_\pm}{dx^3} = (\pm)\frac{1}{2}\sqrt{z}\frac{d^2 z}{dy^2} \quad \text{equals zero if} \quad \frac{d^2 z}{dy^2} \quad \text{exists for} \quad y \to y_0. \tag{16}$$

$$\frac{d^4 y_\pm}{dx^2} \quad \text{continuous at} \quad y = y_0, \quad \text{using (1.1) and (3.2).} \tag{17}$$

Let us interpose an investigation of half-orbits defined by asymptotic approximations $\Phi_m(y;\varepsilon)$ of $z(y;\varepsilon)$. Thus we define

$$\frac{dy_\pm^{(m)}}{dx} = \pm\sqrt{\Phi_m(y;\varepsilon)}. \tag{18}$$

It follows that

$$\frac{d^3 y_\pm^{(m)}}{dx^3} = (\pm)\frac{1}{2}\sqrt{\Phi_m}\frac{d^2\Phi_m}{dy^2} = 0 \quad \text{for} \quad y = y_0^{(m)}(\varepsilon), \tag{19}$$

where $y_0^{(m)}(\varepsilon)$ is the m-th asymptotic approximation of $y_0(\varepsilon)$, in other words, $\Phi_m(y_0^{(m)}(\varepsilon),\varepsilon) = 0$. Result (19) holds because $\Phi_m(y;\varepsilon)$ is a polynomial in y and hence a C^∞ function. Since this is true for any m, it follows that $\frac{d^3 y_\pm}{dx^3}$ at $y = y_0(\varepsilon)$, if not zero, must in absolute value be smaller than any power of ε. One is thus confronted with the notoriously difficult problem of "exponentially small splitting of separatrices" [6,7]. Our approach in [3] was as follows.

From (1) one can "solve" for $\frac{d^2 y}{dx^2}$ along $y_+(x)$ and obtain for $x \le 0$:

$$\frac{d^2 y_+}{dx^2}(x) = -\frac{1}{\varepsilon}\int_{-\infty}^{x} \sin\frac{1}{\varepsilon}(\xi - x)\left[y_+(\xi) - y_+^2(\xi)\right]\, d\xi. \tag{20}$$

The x-axis is parametrized in such a way that $y_+(0) = y_0(\varepsilon)$; that is, $\frac{dy}{dx} = 0$ at $x = 0$.

Next, after differentiation, one gets

$$I_\varepsilon := \frac{d^3 y_+}{dx^3}(0) = \frac{1}{\varepsilon^2} \int_{-\infty}^{0} \cos \frac{1}{\varepsilon}\xi \cdot \left[y_+(\xi) - (y_+(\xi))^2 \right] d\xi. \tag{21}$$

We have computed explicitly

$$I_\varepsilon^{(m)} := \frac{1}{\varepsilon^2} \int_{-\infty}^{0} \cos \frac{1}{\varepsilon}\xi \cdot \left[y_+^{(m)}(\varepsilon) - \left(y_+^{(m)}(\varepsilon) \right)^2 \right] d\xi \tag{22}$$

for $m = 1$ and $m = 2$. The results are surprising and illustrate how delicate the problem is. We found

$$I_\varepsilon^{(m)} = -\frac{\pi}{\varepsilon^5} c_m \, e^{-\frac{\pi}{\varepsilon}} [1 + o(1)] \tag{23}$$

$$c_1 = 6, \quad c_2 \cong 3.6 c_1. \tag{24}$$

Hence, a small correction of the trajectory does not result in a relatively small correction of $I_\varepsilon^{(m)}$. Computing the integral for $m > 2$ poses serious difficulties. However, we were able to prove that for all m

$$I_\varepsilon^{(m)} < 0. \tag{25}$$

These results can be seen as complementary to [1] and [5]. Amick and McLeod [1] prove that

$$\frac{d^3 y_-}{dx^3}(0) > 0. \tag{26}$$

They do not give any estimate of the order of magnitude. The main result of Hammersley and Mazzarino [5] (after change of their notation to ours) is

$$\frac{d^3 y_-}{dx^3}(0) = \Omega(\varepsilon) \frac{(1 + \varepsilon^2)^{\frac{7}{4}}}{\varepsilon^5} e^{-\frac{\pi}{\varepsilon}} \tag{27}$$

Concerning $\Omega(\varepsilon)$, two results are given: an estimate obtained "at the end of a long trail of inequalities,"

$$\Omega(\varepsilon) > 2^{-15}, \quad \varepsilon > 0, \tag{28}$$

and an extensive table of $\Omega(\varepsilon)$, obtained by *numerical* computation. For ε small $\Omega(\varepsilon)$ is of the order of magnitude of 10^2. We note that in orders of magnitude (in ε) (23) is in perfect agreement with (27). One can expect that c_m, in (23), for large m, will settle to a definite value, but analytic determination seems impossible.

Let us now turn to the perturbed model problem

$$\varepsilon^2 \frac{d^4y}{dx^2} + \frac{d^2y}{dx^2} = y - y^2 + \varepsilon^2 \mathcal{P}\left(y, \frac{dy}{dx}, \frac{d^2y}{dx^2}; \varepsilon\right). \tag{29}$$

The methods of [1] and [5] do not seem to be applicable. On the other hand, in [3] it was shown that for large classes of perturbation $\mathcal{P}$ given in (13), the asymptotic expansions (7), (8) can be computed explicitly.

Furthermore, for the perturbed problems, one gets

$$\begin{aligned} I_\varepsilon := \frac{d^3y}{dx^3}(0) = \frac{1}{\varepsilon^2} \int_{-\infty}^{0} \cos \frac{1}{\varepsilon} \xi \\ \times \left[y(\xi) - y^2(\xi) + \varepsilon^2 \mathcal{P}\left(y(\xi), \frac{dy}{dx}(\xi), \frac{d^2y}{dx^2}(\varepsilon); \varepsilon\right)\right] d\xi, \end{aligned} \tag{30}$$

and one can proceed to computations of the integral, replacing $y(x)$ by its asymptotic approximation $y^{(m)}(x)$. At this stage, however, the following observation should be made: *There are a great many perturbed problems for which solitary waves exist (i.e., $I_\varepsilon = 0$).* The observation is nearly trivial. Just take

$$\varepsilon^2 \mathcal{P} = -\varepsilon^{2m} \frac{d}{dy} f_m(y; \varepsilon), \tag{31}$$

where f_m is the polynomial that one gets by computing in (11) the error in the differential equation, obtained by replacing $z(y)$ by its asymptotic expansion Φ_m. Then Φ_m is an exact solution of the perturbed equation, and (19) holds. Note that the perturbation needed to get a solitary wave can be arbitrarily small in terms of powers of ε.

It is not difficult to set up an industry producing larger families of perturbed problems with solitary waves. Consider

$$\Phi(y) = y^2 - \frac{2}{3} y^3 + \varepsilon^2 \sum_{n=2}^{N} C_n y^n, \tag{32}$$

where c_n's are arbitrary but regular for $\varepsilon \to 0$. $\Phi(y)$ is the exact solution of a perturbed problem with

$$\mathcal{P} = \Phi(y) \frac{d^2\Phi}{dy^2} - \frac{1}{4}\left(\frac{d\Phi}{dy}\right)^2 + \sum_{n=2}^{N} C_n y^n. \tag{33}$$

In spite of the profusion of perturbed problems admitting solitary waves, it should be expected that they are structurally unstable; that is, in an arbitrarily small "neighborhood" of a problem admitting a solitary wave, there are problems that do not admit such waves (i.e., $I_\varepsilon \neq 0$).

With these considerations in mind, the problem of (non)-existence of solitary waves for perturbed model equations becomes almost a philosophical issue.

First, given the fact that the equation is known only with accuracy of ε^N, for some $N > 0$, is it consistent to worry about exponentially small jumps in the third derivative? (To this question, one should add the observation that in the basic equations of Kirchgässner [7] viscosity has been neglected.)

Second, as we have seen, there exist perturbations of the problem of order ε^M with M arbitrarily large, which ensure existence of solitary waves.

The way out is to show that even if $d^3y/dx^3(0) \neq 0$ the effect is not noticeable over large distances for $x > 0$. This has been done in [4]. In that paper the continuation of $y_+(x)$ for $x > 0$, denoted by $\tilde{y}_-(x)$, was studied as solution of the initial value problem for the model equation (1) with appropriate initial conditions such that $y_+ \cup \tilde{y}_-$ is smooth. It was shown that $\tilde{y}_-(x) - y_-(x)$ is exponentially small for $x \in [0, x]$, where x is at least of the order of magnitude of $\ln\left(\frac{1}{\varepsilon}\right)$.

References

[1] Amick, C. J., and McLeod, J. B. (1991) 'A singular perturbation problem in water waves', *Stab. and Appl. Anal. of Cont. Media,* **1**, no. 2.

[2] Amick, C. J., and Kirchgässner, K. (1989) 'A theory of solitary water-waves in the presence of surface tension', *Arch. Rat. Mech. & Analysis.*

[3] Eckhaus, W. (1992) 'Singular perturbations of homoclinic orbits in $\mathbf{R}^4$', *SIAM J. Math. Anal.,* to appear.

[4] Eckhaus, W. (1992) 'On water waves at Froude number slightly higher than one and Bond number less than $\frac{1}{3}$', *ZAMP,* **43**.

[5] Hammersley, H. M., and Mazzarino, G. (1989) 'Computational aspects of some autonomous differential equations', *Proc. Roy. Soc. London,* **A 424**.

[6] Holmes, P., Marsden, J., and Scheurle, J. (1988) 'Exponentially small splittings of separatrices with applications to KAM theory and degenerate bifurcations', *Contemporary Math.,* **81**.

[7] Kirchgässner, K. (1988) 'Nonlinearly resonant surface waves and homoclinic bifurcations', *Advanc. Appl. Mech.,* **26**.

[8] Scheurle, J. (1989) 'Chaos in a rapidly forced pendulum equation', *Contemporary Math.,* **97**.

SUPERSENSITIVE BOUNDARY VALUE PROBLEMS

Jacques G. Laforgue and Robert E. O'Malley, Jr.
Department of Applied Mathematics FS-20
University of Washington
Seattle WA 98195

ABSTRACT. We consider the interesting, but relatively uncommon, situation when the boundary value problem for the steady-state Burgers' equation has an interior shock layer. An expansion procedure utilizing exponentially small terms shows that a variety of asymptotically exponentially small perturbations can change the shock location by order one when the diffusion coefficient tends to zero. Such *supersensitivity* is expected for solutions to far more general boundary value problems. This suggests why computation of shocks is so challenging.

1. Introduction

Burgers' equation

$$v_t = \epsilon v_{xx} + v v_x \tag{1}$$

is a physically significant nonlinear partial differential equation which is exactly solvable by the Cole-Hopf transformation (cf. Whitham [14] and Kevorkian [6]). For $\epsilon > 0$ small, the convergence of the solution $v(x,t)$ of the initial-boundary value problem to its steady state $u(x)$ is extremely slow in the case of an interior shock, in contrast to the situation when there is an endpoint boundary layer. Kreiss and Kreiss [9] attribute this sluggishness to the occurrence of an asymptotically exponentially small eigenvalue $\lambda = O(e^{-1/\epsilon})$ for the linearized steady-state problem, with the corresponding eigenvector also being exponentially small away from the shock region (cf. also Kreiss [8] and Kreiss and Lorenz [10]).

The steady-state Burgers' equation

$$\epsilon u'' + u u' = 0 \tag{2}$$

can be integrated to yield $\epsilon u' + \frac{1}{2}u^2 =$ constant. Separating variables may then provide the solution in terms of the hyperbolic tangent function. By the uniqueness theorem for initial value problems, no nonconstant solution can

H. G. Kaper and M. Garbey (eds.),
Asymptotic and Numerical Methods for Partial Differential Equations with Critical Parameters 215–223.

have a zero of u'. Thus, if we consider the two-point problem on $-1 \leq x \leq 1$ with

$$u(\pm 1) = \pm 1, \tag{3}$$

the solution $u(x, \epsilon)$ must increase monotonically. If we, instead, had $u(1) < 1 (> 1)$, the limiting solution (as $\epsilon \to 0$) would be $U_{\text{LO}}(x) = -1$ ($U_{\text{RO}} = u(1)$) for all $x < 1$ (> -1). The ± 1 boundary values were selected so that the classical Rankine-Hugoniot condition $\int_{-1}^{1} s ds = 0$ fails to uniquely determine the location $x^*(\epsilon)$ of the shock layer which joins $U_{\text{LO}} = -1$ for $-1 \leq x < x^*$ to $U_{\text{RO}} = 1$ for $x^* < x \leq 1$ (cf. O'Malley [11, 12]). Symmetry assures us that $x^* = 0$. Numerical experiments by the late John Allen showed that the shock location x^* could be moved by $O(1)$ by letting the end value $u(1)$ vary with ϵ and change by an asymptotically exponentially small amount. Such *supersensitivity* suggests that extreme difficulties will be faced by most numerical codes which attempt to solve (2)–(3) for small values of ϵ. Our inclusion of transcendentally small terms in the upcoming analysis is, of course, only one of many recent confirmations of their critical importance for a wide variety of asymptotic problems. Readers should also note the work of Bohé [2], which considers equations of the form $\epsilon u'' = g(u) f(u')$ (with a Nagumo condition on f) through use of nonstandard analysis. It shows how determining the shock location x^* is truly a "halo effect." Further, as she observes, an example with a linear function f occurs as a model for compressible fluid flow in nozzles (cf. Crocco [3], Pearson [13], and Howes [5]).

Since the general solution of the steady-state Burgers' equation (2) is given by

$$u(x, \epsilon) = A \tanh\left(\frac{A}{2\epsilon}(x - x^*)\right) = A\left(\frac{1 - e^{-A(x-x^*)/\epsilon}}{1 + e^{-A(x-x^*)/\epsilon}}\right), \tag{4}$$

the boundary conditions $u(\pm 1, \epsilon) = \pm 1$ imply that

$$e^{-A(1 \mp x^*)/\epsilon} = \frac{A - 1}{A + 1} \tag{5}$$

for some $A > 1$. Thus, $x^* \equiv 0$ (as symmetry also implies) and $A(\epsilon)$ must satisfy the transcendental equation

$$A = 1 + (A + 1) e^{-A/\epsilon}. \tag{6}$$

We can solve it for ϵ small by using successive approximations. Setting $A_0 = 1$ and $A_{j+1} = 1 + (A_j + 1) e^{-A_j/\epsilon}$ for each $j \geq 0$, we have $A_1 = 1 + 2e^{-1/\epsilon}$ and $A_2 = 1 + 2(1 + e^{-1/\epsilon}) e^{-1/\epsilon} e^{-(2/\epsilon) e^{-(1/\epsilon)}} = 1 + 2(1 + e^{-1/\epsilon}) e^{-1/\epsilon}(1 - \frac{2}{\epsilon} e^{-1/\epsilon} + \frac{2}{\epsilon^2} e^{-2/\epsilon} + \ldots)$. Thus, we obtain

$$A(\epsilon) = 1 + 2e^{-1/\epsilon} - \frac{2}{\epsilon}(2 - \epsilon) e^{-2/\epsilon} + O\left(\frac{1}{\epsilon^2} e^{-3/\epsilon}\right). \tag{7}$$

Clearly, the asymptotic sequence involved is more complicated than a simple power series in the transcendentally small $e^{-1/\epsilon}$. Knowing $A(\epsilon)$ approximately, we get a corresponding asymptotic expansion of the solution $u(x,\epsilon)$ as a function of the stretched variable $\tau = x/2\epsilon$, namely,

$$\begin{aligned} u(x,\epsilon) &= \tanh\tau + 2e^{-1/\epsilon}\left[\tanh\tau + \tau(1-\tanh^2\tau)\right] \\ &\quad -2e^{-2/\epsilon}[2\tau^2\tanh\tau(1-\tanh^2\tau) \\ &\quad +\left(\frac{2}{\epsilon}-3\right)\tau(1-\tanh^2\tau) + \left(\frac{2}{\epsilon}-1\right)\tanh\tau] + O\left(\frac{1}{\epsilon^2}e^{-3/\epsilon}\right). \end{aligned} \tag{8}$$

Taking account of exponentially small terms in this way certainly provides a more precise approximation than the classical one which simply takes $A \sim 1$ and $u(x,\epsilon) \sim \tanh(x/2\epsilon)$. We should expect (8) to be more accurate for larger values of ϵ. For small ϵ values, observe that the boundary value problem has been recast as one on a $O(1/\epsilon)$ expanding τ interval, rather than the infinite one usually used.

2. The Perturbed Steady-State Problem

(i) Following Allen and Bohé, consider the slightly perturbed boundary value problem

$$\begin{aligned} &\epsilon u'' + uu' = 0, \ -1 \le x \le 1 \\ &u(-1) = -1, \ u(1) = 1 + 2be^{-a/\epsilon} \text{ for some fixed } a \text{ in } (0,1). \end{aligned} \tag{9}$$

Let us seek an asymptotic solution in the form

$$u(x,\epsilon) \sim u_0(\tau) + e^{-a/\epsilon}u_1(\tau,\epsilon) + \dots \tag{10}$$

using the stretched variable

$$\tau = (x - x^*)/2\epsilon. \tag{11}$$

We shall take the shock layer location $x^*(\epsilon)$ (i.e., $\tau = 0$) to be the zero of $u(x,\epsilon)$ and shall also require u_0 and u_1 to be bounded for $\tau = O(1/\epsilon)$. As a function of τ, u must satisfy

$$\begin{aligned} &\ddot{u} + 2u\dot{u} = 0 \\ &u\left(\frac{-1-x^*}{2\epsilon}\right) = -1, \ u\left(\frac{1-x^*}{2\epsilon}\right) = 1 + 2be^{-a/\epsilon} \end{aligned} \tag{12}$$

where the dots represent derivatives with respect to τ. Since $u(\tau,\epsilon) = A\tanh(A\tau)$ satisfies the unperturbed problem for $A = 1 + O(e^{-a/\epsilon})$, it is natural to take

$$u_0(\tau) = \tanh\tau. \tag{13}$$

Then, since

$$u_0((-1-x^*)/2\epsilon) = -1 + 2e^{-(1+x^*)/\epsilon} + O(e^{-2(1+x^*)/\epsilon}) \tag{14}$$

and

$$u_0((1-x^*)/2\epsilon) = 1 - 2e^{-(1-x^*)/\epsilon} + O(e^{-2(1-x^*)/\epsilon}), \tag{15}$$

we ask that u_1 satisfy the linearized problem

$$\begin{aligned} &\ddot{u}_1 + (2\tanh\tau)\dot{u}_1 + 2(1-\tanh^2\tau)u_1 = 0 \\ &u_1((-1-x^*)/2\epsilon) = -2e^{-(1+x^*-a)/\epsilon}, \\ &u_1((1-x^*)/2\epsilon) = 2b + 2e^{-(1-x^*-a)/\epsilon}, \end{aligned} \tag{16}$$

anticipating that $e^{-(1\pm x^*-a)/\epsilon}$ remains bounded as $\epsilon \to 0$. Note that the general solution of the differential equation is given by

$$u_1(\tau,\epsilon) = \alpha(\epsilon)(1-\tanh^2\tau) + \beta(\epsilon)(\tanh\tau + \tau(1-\tanh^2\tau)). \tag{17}$$

Because $u_1(0,\epsilon) = 0$ implies that $\alpha(\epsilon) \equiv 0$, the asymptotically negligible solution $1-\tanh^2\tau$ (as $\tau \to \pm\infty$) is eliminated from u_1. The boundary conditions then require that

$$-\beta = -2e^{-(1+x^*-a)/\epsilon} + O(\frac{1}{\epsilon}e^{-(1+x^*)/\epsilon})$$

and

$$\beta = 2b + 2e^{-(1-x^*-a)/\epsilon} + O\left(\frac{1}{\epsilon}e^{-(1-x^*)/\epsilon}\right). \tag{18}$$

Thus, $\beta > 0$, $\beta > 2b$, and $\beta(\beta-2b) = 4e^{-2(1-a)/\epsilon} + O\left(\frac{1}{\epsilon}e^{-(2-a)/\epsilon}\right)$ imply that

$$\beta(\epsilon) \sim b + \sqrt{b^2 + 4e^{-2(1-a)/\epsilon}}. \tag{19}$$

Thus, for $b > 0$, $\beta(0) = 2b$ and (18) implies that

$$x^*(\epsilon) = -1 + a - \epsilon\ \ln(\beta/2) + O(e^{-a/\epsilon}). \tag{20}$$

As anticipated, $e^{-1(\pm x^*-a)/\epsilon} = O(1)$. Note that the shock layer moves toward -1 as $a \to 0$ and toward 0 as $a \to 1$. If $b < 0$, we instead have the positive shock location

$$x^*(\epsilon) = 1 - a + \epsilon\ \ln((\beta-2b)/2) + O(e^{-a/\epsilon}). \tag{21}$$

(ii) As in stability theory generally, it is natural to consider the perturbation of the shock layer location x^* resulting from small changes in the coefficients of the differential equation as well as in the boundary values. Let us, then, consider a second perturbed problem

$$\epsilon u'' + (u + b_1 e^{-a/\epsilon})u' + b_2 e^{-a/\epsilon} = 0$$

$$u(-1) = -1 + 2b_3 e^{-a/\epsilon}, u(1) = 1 + 2b_4 e^{-a/\epsilon} \tag{22}$$

with exponentially small perturbations $b_i e^{-a/\epsilon}$ for some a in $(0, 1)$. We shall again use the stretched variable $\tau = (x - x^*)/2\epsilon$ and thereby transform (22) to the boundary value problem

$$\ddot{u} + 2(u + b_1 e^{-a/\epsilon})\dot{u} + 4\epsilon b_2 e^{-a/\epsilon} = 0$$

$$u((-1 - x^*)/2\epsilon) = -1 + 2b_3 e^{-a/\epsilon}, u((1 - x^*)/2\epsilon) = 1 + 2b_4 e^{-a/\epsilon} \tag{23}$$

for a solution $u(\tau, \epsilon)$ which satisfies $u(0, \epsilon) = 0$ and is defined on an $O\left(\frac{1}{\epsilon}\right)$ τ interval. Our experience suggests that we might seek an asymptotic solution through the Ansatz

$$u(\tau, \epsilon) \sim \sum_{n=0}^{\infty} u_n(\tau, \epsilon) e^{-na/\epsilon} \tag{24}$$

where we shall limit the growth of later terms by asking that $u_{n+1} = O(1/\epsilon^n)$ for $\tau = O(1/\epsilon)$ and all n. We shall first ask u_0 to satisfy the unperturbed (i.e., $\epsilon = 0$) problem, so take

$$u_0(\tau, \epsilon) \equiv \tanh \tau. \tag{25}$$

Next, equating coefficients of $e^{-a/\epsilon}$, we ask u_1 to satisfy the linearized problem

$$\ddot{u}_1 + (2 \tanh \tau)\dot{u}_1 + 2(1 - \tanh^2 \tau)(u_1 + b_1) + 4\epsilon b_2 = 0$$

$$u_1((-1 - x^*)/2\epsilon) = 2b_3 - 2e^{-(1+x^*-a)/\epsilon}$$

$$u_1((1 - x^*)/2\epsilon) = 2b_4 + 2e^{-(1-x^*-a)/\epsilon} \tag{26}$$

because, for example, $u_0((-1 - x^*)/2\epsilon) = -1 + 2e^{-(1+x^*)/\epsilon} + O(e^{-2(1+x^*)/\epsilon})$. The general solution of the differential equation is given by

$$\begin{aligned} u_1(\tau, \epsilon) &= \alpha_1(\epsilon)(1 - \tanh^2 \tau) + \beta_1(\epsilon)\left[\tau(1 - \tanh^2 \tau) + \tanh \tau\right] \\ &\quad -b_1 - \epsilon b_2 \left[\tau^2(1 - \tanh^2 \tau) + 2\tau \tanh \tau - 1\right]. \end{aligned} \tag{27}$$

Note, first, that $u_1(0, \epsilon) = 0$ implies that $\alpha_1(\epsilon) = b_1 - \epsilon b_2$ and, second, that the boundary conditions then require that

$$-\beta_1(\epsilon) - b_1 - b_2(1 + x^* - \epsilon) \sim 2b_3 - 2e^{-(1+x^*-a)/\epsilon}$$

and

$$\beta_1(\epsilon) - b_1 - b_2(1 - x^* - \epsilon) \sim 2b_4 + 2e^{-(1-x^*-a)/\epsilon} \tag{28}$$

(neglecting asymptotically exponentially small terms). Adding, we obtain

$$b_1 + b_2(1-\epsilon) + b_3 + b_4 \sim e^{-(1-a)/\epsilon}(e^{-x^*/\epsilon} - e^{x^*/\epsilon}). \tag{29}$$

Thus, if $\sum_{i=1}^4 b_i > 0$, we must have $1 - a + x^* \sim -\epsilon \ln(\sum_{i=1}^4 b_i)$, that is,

$$x^*(\epsilon) \sim -1 + a - \epsilon \ln\left(\sum_{i=1}^{4} b_i\right), \tag{30}$$

while

$$x^*(\epsilon) \sim 1 - a + \epsilon \ln\left|\sum_{i=1}^{4} b_i\right| \tag{31}$$

if $\sum_{i=1}^4 b_i < 0$. A more careful analysis provides later exponentially small terms in the expansions for $x^*(\epsilon)$ and for $\beta_1(\epsilon)$. For $a \geq 1$, another careful analysis shows that $x^*(\epsilon) \sim 0$.

(iii) To indicate what happens when different transcendentally small perturbations occur, consider the third perturbed problem

$$\epsilon u'' + uu' = 0$$

$$u(-1) = -1 + 2be^{-1/2\epsilon}, u(1) = 1 + 2ce^{-1/3\epsilon} \text{ for } c \neq 0. \tag{32}$$

As **(i)** would suggest, the shock layer will be located near $-2/3$ for $c > 0$ and near $2/3$ for $c < 0$. The solution will be given by

$$u(x,\epsilon) = A \tanh(A(x - x^*)/2\epsilon) \tag{33}$$

where A and x^* must satisfy

$$A\left[\frac{1 - e^{-A(1+x^*)/\epsilon}}{1 + e^{-A(1+x^*)/\epsilon}}\right] = 1 - 2be^{-1/2\epsilon}$$

and

$$A\left[\frac{1 - e^{-A(1-x^*)/\epsilon}}{1 + e^{-A(1-x^*)/\epsilon}}\right] = 1 + 2ce^{-1/3\epsilon}. \tag{34}$$

It is simplest to eliminate $e^{-Ax^*/\epsilon}$ and then formally seek a solution $A(\epsilon)$ in the form

$$A(\epsilon) \sim 1 + a_1 e^{-1/3\epsilon} + a_2 e^{-1/2\epsilon} + a_3 e^{-2/3\epsilon} + \dots$$

by using undetermined coefficients. This approach provides the approximate solution

$$A(\epsilon) \sim 1 + 2ce^{-1/3\epsilon} + \frac{2}{c}e^{-5/3\epsilon} + \dots \tag{35}$$

if $c > 0$, which corresponds to the shock location

$$x^*(\epsilon) = -\frac{2}{3} - \epsilon \ln c - \epsilon\frac{b}{c}e^{-1/6\epsilon} + O(e^{-1/3\epsilon}). \tag{36}$$

Likewise, for $c < 0$, we obtain

$$A(\epsilon) \sim 1 - 2be^{-1/2\epsilon} - \frac{2}{c}e^{-5/3\epsilon} + \dots$$

and

$$x^*(\epsilon) = 2/3 + \epsilon \ln(-c) + \epsilon\frac{b}{c}e^{-1/6\epsilon} + O(e^{-1/3\epsilon}). \tag{37}$$

Of course, when $c = 0$, the shock occurs near either $\pm 1/2$ depending on the sign of b.
(iv) It is possible to evaluate the appropriately normed solution of the linearized steady-state Burgers' equation at the shock location to show that these boundary value problems have an asymptotically exponentially large conditioning constant (cf. the discussions of well-conditioning by de Hoog and Mattheij [4], Ascher, Mattheij, and Russell [1], and Kramer [7]). This provides another effective method of evaluating *supersensitivity*.

3. A Less Sensitive Example

While an asymptotically exponentially small perturbation will move the shock location for the quasilinear Burgers' equation by $O(1)$, the situation is different for the quadratically nonlinear equation $\epsilon u'' + 2u(u')^2 = 0$. Note, first, that it can be integrated after division by u' and that the monotonic inverse function $x(u, \epsilon)$ can then be obtained after separating variables. Specifically, for the boundary value problem

$$\epsilon u'' + (2u + b_1\epsilon)(u')^2 = 0, \ -1 \le x \le 1$$

$$u(-1) = -1 + b_2\epsilon, \ u(1) = 1 + b_3\epsilon, \tag{38}$$

the inverse function will be given by

$$x(u, \epsilon) = \frac{I_L(u, \epsilon) - I_R(u, \epsilon)}{I_L(u, \epsilon) + I_R(u, \epsilon)} \tag{39}$$

where

$$I_L(u,\epsilon) = \int_{-1+\epsilon(b_1/2+b_2)}^{u+b_1\epsilon/2} e^{(s^2-1)/\epsilon} ds$$

and

$$I_R(u,\epsilon) = \int_{u+\epsilon b_1/2}^{1+\epsilon(b_1/2+b_3)} e^{(s^2-1)/\epsilon} ds.$$

For u near zero, the major contributions to the integrals I_L and I_R come from near $s = -1$ or 1, respectively. Using Laplace's method (cf. Wong [15]), we find that $I_L(0,\epsilon) = \frac{\epsilon}{2}e^{-(b_1+2b_2)} + O(\epsilon^2)$ and $I_R(0,\epsilon) = \frac{\epsilon}{2}e^{b_1+2b_3} + O(\epsilon^2)$, so the shock location for (38) is given by

$$x^*(\epsilon) = \left[\frac{e^{-(b_1+2b_2)} - e^{b_1+2b_3}}{e^{-(b_1+2b_2)} + e^{b_1+2b_3}}\right] \frac{e^{b_2-b_3}}{e^{b_2-b_3}} + O(\epsilon),$$

that is,

$$x^*(\epsilon) = \tanh\,(-\sum_{i=1}^{3} b_i) + O(\epsilon). \tag{40}$$

This shows that the shock layer moves toward -1 as $\sum_{i=1}^{3} b_i \to \infty$ and toward $+1$ as $\sum_{i=1}^{3} b_i \to -\infty$. We note that the conditioning constant for (38) is also asymptotically exponentially large.

Acknowledgements

The authors thank Professor Robert M. M. Mattheij and, especially, Professor Leonid V. Kalachev for their interest and assistance in this work, which is currently being generalized in a number of significant ways. This work was supported by the National Science Foundation under grant number DMS 9107196.

References

[1] Ascher, U. M., Mattheij, R. M. M., and Russell, R. D. (1988) *Numerical Solution of Boundary Value Problems for Ordinary Differential Equations*, Prentice-Hall, Englewood Cliffs, N.J.

[2] Bohé, A. (1990) 'Free layers in a singularly perturbed boundary value problem', *SIAM J. Math. Anal.*, **21**, 1264–1280.

[3] Crocco, L. (1965) 'A suggestion for the numerical solution of the steady Navier-Stokes equations', *AIAA J.* **3**, 1824–1832.

[4] de Hoog, F., and Mattheij, R. M. M. (1987) 'On dichotomy and well-conditioning in BVP', *SIAM J. Num. Anal.* **24**, 89–105.

[5] Howes, F. A. (1984) 'An analytical treatment of the formation of one-dimensional steady shock waves in uniform and diverging ducts', *J. Comp. Appl. Math.* **10**, 195–201.

[6] Kevorkian, J. (1990) *Partial Differential Equations*, Wadsworth and Brooks/Cole, Pacific Grove, Cal.

[7] Kramer, M. E. (1992) *Aspects of Solving Non-linear Boundary Value Problems Numerically*, Doctoral dissertation, Technische Universiteit Eindhoven.

[8] Kreiss, G. (1991) 'Convergence to steady state of solutions of viscous conservation laws', in H. G. Kaper and M. Garbey, eds., *Asymptotic Analysis and the Numerical Solution of Partial Differential Equations*, Marcel Dekker, New York, 225–237.

[9] Kreiss, G., and Kreiss, H.-O. (1986) 'Convergence to steady state of solutions of Burgers' equation', *Appl. Num. Math.* **2**, 161–179.

[10] Kreiss, H.-O., and Lorenz, J. (1989) *Initial-Boundary Value Problems and the Navier-Stokes Equations*, Academic Press, San Diego.

[11] O'Malley, Jr., R. E. (1991) 'Singular perturbations, asymptotic evaluation of integrals, and computational challenges', in H. G. Kaper and M. Garbey, eds., *Asymptotic Analysis and the Numerical Solution of Partial Differential Equations*, Marcel Dekker, New York, 3–16.

[12] O'Malley, Jr., R. E. (1991) *Singular Perturbation Methods for Ordinary Differential Equations*, Springer-Verlag, New York.

[13] Pearson, C. E. (1968) 'On non-linear ordinary differential equations of boundary layer type', *J. Math. and Physics* **47**, 351–358.

[14] Whitham, G. B. (1974) *Linear and Nonlinear Waves*, Wiley-Interscience, New York.

[15] Wong, R. (1989) *Asymptotic Approximations of Integrals*, Academic Press, San Diego.

ASYMPTOTICS OF THE THUAL-FAUVE PULSE

Piero de Mottoni
Matematica
2a Università di Roma
Tor Vergata
Italy

and

Michelle Schatzman
Laboratoire d'Analyse Numérique
Université Claude Bernard-Lyon 1
France

ABSTRACT. We consider the complex quintic Ginzburg-Landau equation

$$u_t = u_{xx} + \mu u + (1 + i\alpha)|u|^2 u - |u|^4 u,$$

where α is real, and $\mu \in (-3/16, 0)$. We prove that for small α this equation possesses pulse-like solutions that decay exponentially at infinity. The branch of solutions crosses a subcritical turning point; past that critical value, a small perturbation of the pulse evolves into the same pulse, up to translations in space and in phase.

1. Introduction

Thual and Fauve [8] have proposed the complex quintic Ginzburg-Landau equation

$$u_t = (d_r + id_i)u'' + \mu u + (\alpha_r + i\alpha_i)|u|^2 u(\beta_r + i\beta_i)|u|^4 u \tag{1}$$

as a model for localized structures that are stabilized by a nonvariational effect. In particular, they propose this model as an explanation of the observation of stable stationary interfaces between regions of convection and conduction in binary fluid mixtures, within a finite interval range of the control parameter.

The choice of parameters that make this equation interesting is as follows: the real part of the diffusion coefficient, d_r, has to be positive; and in order to have a difference between a quintic and a cubic Ginzburg-Landau equation,

H. G. Kaper and M. Garbey (eds.),
Asymptotic and Numerical Methods for Partial Differential Equations with Critical Parameters 225–239.

we demand that the cubic term be destabilizing (i.e., $\alpha_r > 0$) and that the quintic term be stabilizing (i.e., $\beta_r < 0$).

To keep matters simple, we will restrict our choice of parameters to

$$d_r = 1, \quad d_i = 0, \quad \alpha_r = 1, \quad \alpha_i = \alpha, \quad \beta_r = -1, \quad \beta_i = 0. \tag{2}$$

Under these restrictions, our equation becomes

$$u_t = u'' + \mu u + (1 + i\alpha)|u|^2 u - |u|^4 u. \tag{3}$$

In this article, for $-3/16 < \mu < 0$, we will give mathematical evidence for the existence of solutions of (3) of the form

$$u(x,t) = e^{i\omega t + i\phi(x)} r(x), \tag{4}$$

where, at infinity, r decays exponentially and $\phi(x) \sim -\bar{\theta}_0|x|$. Moreover, we will prove that the nonvariational term has a stabilizing effect, as was conjectured by Thual and Fauve [8]. Specifically, there exists for all small enough positive values of $\mu + (3/16)$ a subcritical turning point on the branch of solutions (u, α). Beyond this turning point, the solutions are linearly stable up to symmetries of the problem. The spectrum of the linearization is entirely in the complex half-plane $\Re(z) \geq \kappa > 0$, save for a double zero eigenvalue corresponding to translation modes in space and phase. According to Exercise 6, page 108, of Henry [5], this fact implies that a small initial perturbation of a pulse evolves into the same pulse, possibly translated in phase and in space.

Complete proofs will be given elsewhere.

Malomed and Nepiomastchy [6] published an interesting article on the same question, in which they state that they obtained a stable pulse for (1). There is no doubt that they obtained a pulse, using scaling and matched asymptotic expansions. However, they do not define stability, and it is probably not the same kind of stability as that considered here. Since their point of view is both intuitive and stimulating, it will be discussed below, so as to shed some light on the phenomena studied in the present article.

2. The Case $\alpha = 0$

In this article we are interested in the case where α is small. This suggests a continuation method, starting from $\alpha = 0$. Therefore, the case $\alpha = 0$ has to be studied precisely.

For $\alpha = 0$, a pulse is a solution of

$$r'' - \mu r - r^3 + r^5 = 0, \tag{5}$$

which decays exponentially to 0 at infinity. Let

$$\Phi(u) = \frac{1}{3}|u|^6 - \frac{1}{2}|u|^4 - \mu|u|^2.$$

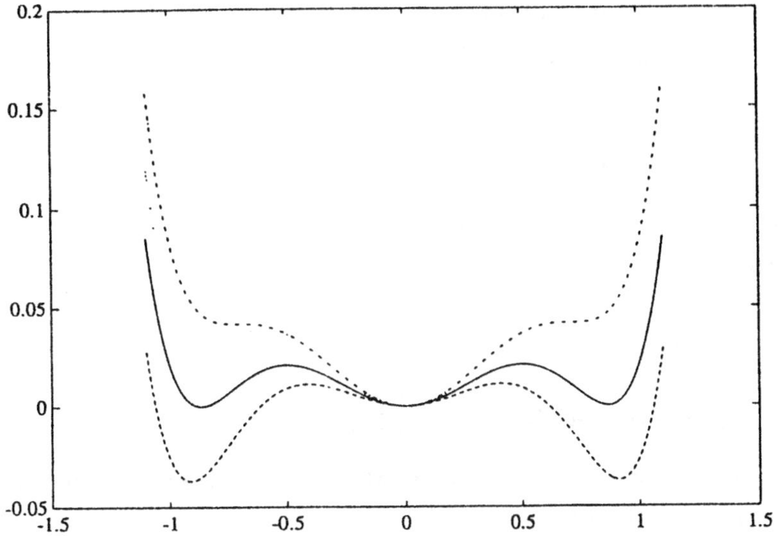

Fig. 1. **The function Φ for $\mu = -3/16$, $\mu = -1/4$, $\mu \in (-3/16, 0)$ and $\mu \in (-1/4, -3/16)$.**

A first integral of (5) is

$$r'^2 - \Phi(r).$$

The function Φ is graphed for different values of μ in Fig. 1.

It is easy to describe the phase portrait of (5) according to the values of μ; we observe that there is an orbit starting at 0 and returning at 0, that is, a homoclinic orbit iff $-(3/16) < \mu < 0$; therefore, it is possible to have a pulse only for $-(3/16) < \mu < 0$. In this case there is another homoclinic orbit through 0, corresponding to the symmetry $r \mapsto -r$, and there are two heteroclinic orbits; see Fig. 2a.

Moreover, the special case $\mu = -3/16$ corresponds to a "Maxwellian" situation where Φ has three minima equal to 0, and there are four heteroclinic orbits; see Fig. 2b. These are the limits of travelling waves for $\mu \neq -(3/16)$: there is a travelling wave $u(x,t) = K(x - ct)$ joining 0 at $-\infty$ to $\tilde{r}$ at $+\infty$, where $\tilde{r}$ is the abscissa of the minimum of Φ over $\{x > 0\}$. If $\mu \in (-(3/16), 0)$, $\Phi(\tilde{r}) < 0$, then the velocity c of this travelling wave is negative; if $\mu \in (-(1/4), -(3/16))$, the velocity is positive, which is the case in Fig. 2c. Of course, there are three other travelling waves corresponding to the symmetries of the problem, and their respective limits as μ tends to $-(3/16)$ are the four heteroclinic orbits appearing for $\mu = -(3/16)$.

Henceforth, we will perform the change of space variable $x \mapsto x|\mu|^{-1/2}$, which enables us to write (3) as

$$u_t = -\mu u'' + \mu u + (1+i\alpha)|u|^2 u - |u|^4 u \tag{6}$$

and (5) as

$$-\mu r'' - \mu r - r^3 + r^5 = 0. \tag{7}$$

We define a new parameter ν by

$$\nu = \frac{16\mu}{3} \Longleftrightarrow \mu = \frac{3}{16}(1-\nu).$$

There is an explicit expression of r_0, the solution of (7):

$$r_0(x) = \sqrt{R_0(x)}, \quad R_0(x) = \frac{3}{4}\frac{1-\nu}{1-\sqrt{\nu}\cosh 2x}. \tag{8}$$

See Fig. 3 for a plot of r_0 for different values of ν. This figure suggests that r_0 has simple asymptotics.

Indeed, let

$$L = \frac{1}{4}\ln\frac{9}{4\nu} \Longleftrightarrow e^{-4L} = \frac{4\nu}{9}. \tag{9}$$

Define

$$\tilde{R}(x) = \frac{e^{-2x}}{1 + 4e^{-2x}/3}. \tag{10}$$

Then, it is possible to prove that

$$|R_0(x+L) - \tilde{R}(x)| \le \nu e^{-2x}. \tag{11}$$

We will denote

$$\tilde{r}(x) = \sqrt{\tilde{R}(x)}. \tag{12}$$

3. A First Approach to the Existence of the Pulse

Let us first recall the argument in [6] for the existence of a pulse-like solution. A pulse would be a solution of (6) which leaves 0, goes to $\tilde{r}$, following essentially K (the kink), stays in the state $\tilde{r}$ on some space interval of length L, and goes back to 0 following the mirror image $\check{K}$ of K (the antikink); hence the name kink-antikink solution.

Under which conditions can this approximate solution be close to equilibrium? On the one hand, for $\mu \in (-(3/16), 0)$, the velocity of K is positive, so

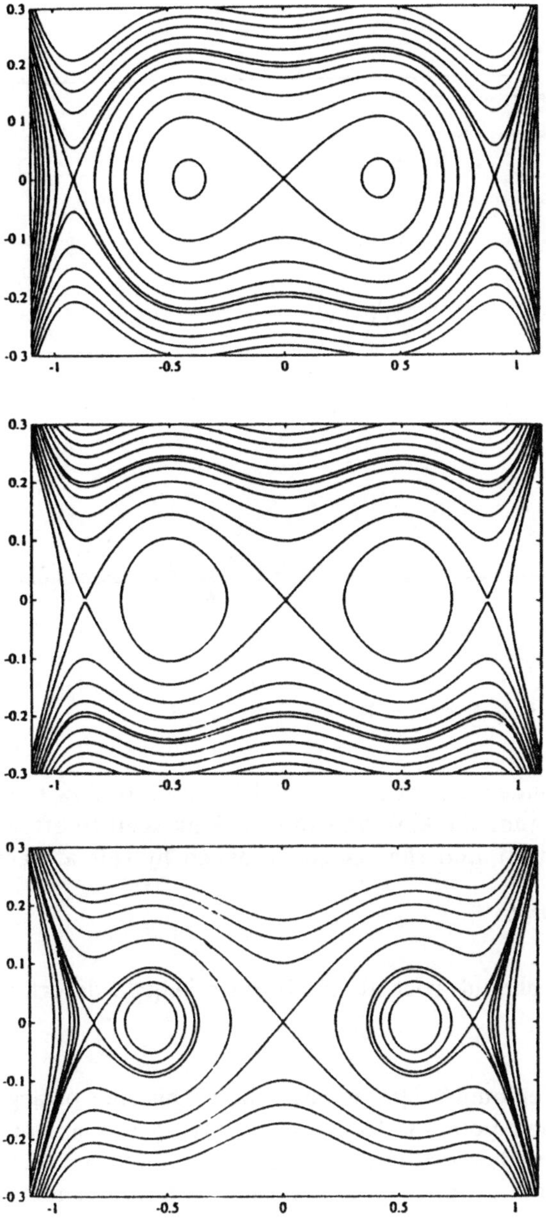

Fig. 2. **Phase portrait of (5): (a)** $\mu \in (-3/16, 0)$; **(b)** $\mu = -3/16$; **(c)** $\mu \in (-1/4, -3/16)$

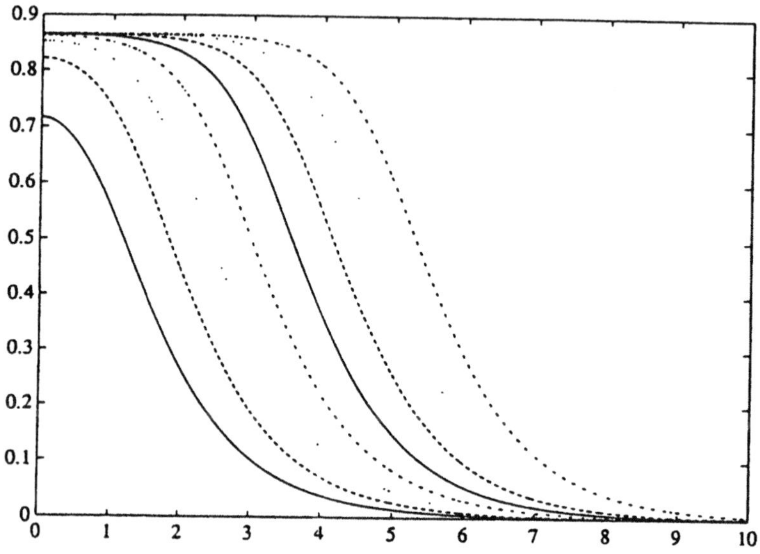

Fig. 3. **Graph of r_0 for $\nu = 10^{-j}, j = 1, \ldots, 8$**

that there is a tendency for the kink and antikink to flow out at the velocity c; on the other hand, the kink and the antikink tend to attract one another, as in [1], [3], and [4], and the velocity induced by this attraction is

$$\frac{dL}{dt} = -C_1 e^{-C_2 L}. \tag{13}$$

Therefore, there should exist an equilibrium length, determined by

$$c = C_1 e^{-C_2 L}. \tag{14}$$

Malomed and Nepiomastchy use essentially this line of argument in their paper. One of their points is to prove, by asymptotic expansions, that (13) holds in the nonvariational case. It must be pointed out that they consider a more complicated problem than (6): they keep all the generality of parameters d, α, and β in [6].

An estimate such as (13) is at the heart of [1] and [4] on the slow motion of interfaces for Allen-Cahn equation; to our knowledge, it has not been mathematically validated in more complicated situations, such as the present one. It would be of interest to prove the attractive behavior of interfaces in a situation of some generality.

We will follow a more mathematical approach, which will lead to *bona fide* proofs. Let

$$\mathcal{F}(u,\alpha,\omega,\nu) = i\omega u + \mu u'' - \mu u - |u|^2 u(1+i\alpha) + |u|^4 u. \tag{15}$$

We know already that

$$\mathcal{F}(r_0, 0, 0, \nu) = 0. \tag{16}$$

Hence, it would be nice to apply an implicit function theorem; the differential of $\mathcal{F}$ with respect to u is given by

$$D_u\mathcal{F}(r_0,0,0,\nu) = \begin{pmatrix} A_0 & 0 \\ 0 & \hat{A}_0 \end{pmatrix}, \tag{17}$$

where A_0 and $\hat{A}_0$ are Schrödinger operators given by

$$A_0 v = \mu\frac{d^2}{dx^2} + V_0, \quad \hat{A}_0 v = \mu\frac{d^2}{dx^2} + \hat{V}_0, \tag{18}$$

and the potentials V_0 and $\hat{V}_0$ are given by

$$V_0 = 5R_0^2 - 3R_0 - \mu, \quad \hat{V}_0 = R_0^2 - R_0 - \mu. \tag{19}$$

See Fig. 4 for a plot of V_0 and $\hat{V}_0$.

Classical results on the continuous spectrum of such operators in $L^2(\mathbf{R})$ imply that the continuous spectrum of A_0 and $\hat{A}_0$ is included in $[|\mu|, +\infty]$. Therefore, the intersection of the spectrum of A_0 or $\hat{A}_0$ contains only isolated eigenvalues; since we are in dimension 1, these eigenvalues are simple.

The function r_0' (resp. r_0) is an eigenvector corresponding to the eigenvalue 0 (bound state of zero energy) for V_0 (resp. $\hat{V}_0$), namely,

$$A_0 r_0' = 0, \quad \hat{A}_0 r_0 = 0.$$

Since r_0' changes sign, 0 cannot be the lowest eigenvalue of A_0: the lowest eigenvalue of A_0 is some negative number λ_-. On the other hand, r_0 does not change sign, so that 0 is the infimum of the spectrum of $\hat{A}_0$.

Let us give now precise spectral information on A_0 and $\hat{A}_0$. Since the two potential wells of V_0 are far removed from one another, the energy levels are duplicated, or, in another words, all the eigenvalues of A_0 come in pairs. In particular, λ_- can be estimated very precisely, and we obtain

$$\lambda_- \sim -\frac{3\nu}{4}. \tag{20}$$

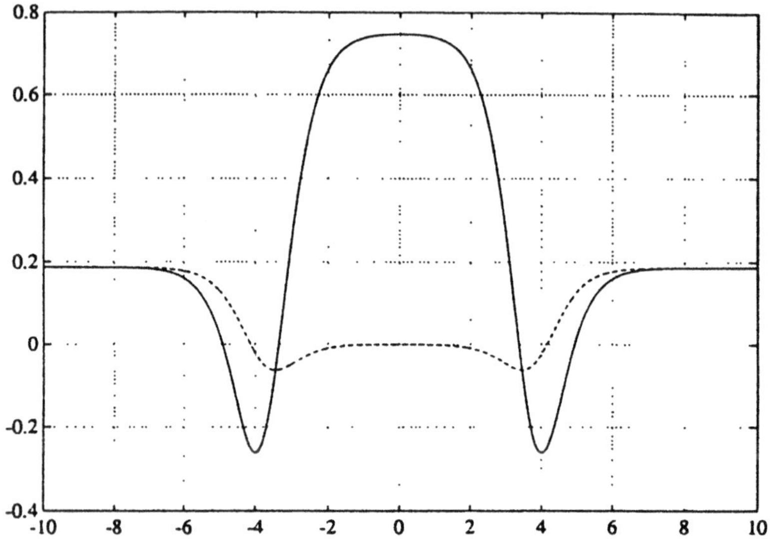

Fig. 4. **Graph of V_0 and $\hat{V}_0$**

The technique of proof of this result is similar to proofs written in [7]. The cluster of eigenvalues $\{\lambda_-, 0\}$ is bounded away from the remainder of the spectrum, uniformly in ν. The eigenfunction s_- spanning the eigenspace corresponding to λ_- can be taken strictly positive. It is possible to choose a proportionality factor in the definition of s_- so that

$$|s_- + \tilde{r}'(\cdot - L) + \tilde{r}'(- \cdot - L)|_{H^2(\mathbf{R})} = O(\nu\sqrt{L}). \tag{21}$$

From the definition of r_0 and (11), we obtain

$$|s_- + r_0'(x)|_{H^2(\mathrm{R}^+)} = O(\nu\sqrt{L}). \tag{22}$$

The derivative of $\mathcal{F}$ with respect to ω is

$$D_\omega\mathcal{F}(r_0, 0, 0, \nu) = \begin{pmatrix} 0 \\ r_0 \end{pmatrix}. \tag{23}$$

Define functional spaces X and Y by

$$\begin{aligned} X &= \{u \in (L^2(\mathbf{R}))^2 : u \text{ is even}\} \\ Y &= X \cap H^2(\mathbf{R}), \end{aligned} \tag{24}$$

where $H^2(\mathbf{R})$ is the Sobolev space of functions that are square integrable, as well as their first and second derivatives.

To apply the implicit function theorem, we have to verify that some adequately chosen linear mapping is an isomorphism. We choose to consider the mapping

$$\begin{aligned} & Y \times \mathbf{R} \to X \\ & \begin{pmatrix} v \\ \psi \end{pmatrix} \mapsto D_u F(r_0, 0, 0, \nu) v + D_\omega F(r_0, 0, 0, \nu) \psi. \end{aligned} \tag{25}$$

Given $\chi = \chi_r + i\chi_i$ in X, we have to solve

$$A_0 v_r = \chi_r \tag{26}$$

$$\hat{A}_0 v_i + \omega r_0 = \chi_i. \tag{27}$$

Clearly the first equation admits a unique solution, since r_0' is odd. If we choose ω so that $\chi_i - \omega r_0$ is the projection on $r_0^\perp$ of χ_i, we can see that the second equation has a unique solution. From this point on, we just have to verify the hypotheses of the implicit function theorem; it is rather straightforward and will be presented in a more complete publication. We obtain the following theorem.

Theorem 1. *For all $\nu \in (-(3/16), 0)$, there exists $\rho(\nu) > 0$ such that for each $\alpha \in [-\rho(\nu), \rho(\nu)]$, the problem*

$$u \in Y, \omega \in \mathbf{R}, \max(|u|, \omega) \leq \rho(\nu), \quad \mathcal{F}(u, \alpha, \omega, \nu) = 0$$

possesses a unique even solution. Moreover, v decreases exponentially to 0 as $|x|$ tends to $\pm\infty$, and there is a positive number $\bar{\theta}_0$ such that $\phi(x)$ is equivalent to $-|x|\bar{\theta}_0$ as x tends to $\pm\infty$.

The only statements that are not immediate are the qualitative ones. From Theorem 8.1 of Chapter 3 of Coddington and Levinson [2], we know that at $\pm\infty$,

$$u = \gamma e^{-\zeta|x|} + O(e^{-3\Re(\zeta)|x|}),$$

where ζ is the square root of $i\omega + |\mu|$. Therefore, $r \sim |\gamma| e^{-|x|\Re\zeta}$ and

$$\arg u(x) = \arg\gamma - |x|\Im(\zeta) + O(e^{-2\Re\zeta|x|}).$$

The remainder of the proof is straightforward.

4. Stability and Subcriticality

Our aim is to prove that the branch of solutions crosses a turning point beyond which the solution is stabilized for large enough α. Our previous attempts to prove this result used equivalents of the derivatives of higher order of $\mathcal{F}$ at $\alpha = \nu = 0$. Of course, done directly, the task looks impossible, since the symmetry $\alpha \mapsto -\alpha$ maps u into its conjugate; this would lead to consider a derivative of order 4, a highly technical endeavor. However, we can work on a simpler formulation where only even powers of α appear, thanks to the symmetries of the problem. We define a new function $F(\xi, \eta, \epsilon, \tau, \nu)$ by

$$F(\xi, \eta, \alpha^2, \tau, \nu) = \mathcal{F}_1(\xi + i\alpha\eta, \alpha, \alpha\tau, \nu) + \frac{i}{\alpha}\mathcal{F}_2(\xi + i\alpha\eta, \alpha, \alpha\tau, \nu). \tag{28}$$

More explicitly, we have

$$\begin{aligned} F_1(\xi, \eta, \epsilon, \tau, \nu) &= -\epsilon\tau\eta - |\mu|\xi'' - \mu\xi \\ &\quad - (\xi^2 + \epsilon\eta^2)(\xi - \epsilon\eta) + (\xi^2 + \epsilon\eta^2)^2\xi \end{aligned} \tag{29}$$

$$\begin{aligned} F_2(\xi, \eta, \epsilon, \tau, \nu) &= \tau\xi - |\mu|\eta'' - \mu\eta \\ &\quad - (\xi^2 + \epsilon\eta^2)(\xi + \eta) + (\xi^2 + \epsilon\eta^2)^2\eta). \end{aligned} \tag{30}$$

We are looking for a description of the solutions of

$$F(\xi, \eta, \epsilon, \tau, \nu) = 0, \tag{31}$$

in a neighborhood of a solution (ξ_0, η_0) of

$$F(\xi_0, \eta_0, 0, \tau_0, \nu) = 0, \tag{32}$$

which we will now determine.

Equation (32) can be written

$$-|\mu|\xi_0'' - \mu\xi_0 - \xi_0^3 + \xi_0^5 = 0 \tag{33}$$

$$\tau_0\xi_0 - |\mu|\eta_0'' - \mu\eta_0 - \xi_0^2(\xi_0 + \eta_0) + \xi_0^4\eta_0 = 0. \tag{34}$$

Therefore, we have to determine first ξ_0 and then η_0. We have already constructed a ξ_0: it is the function r_0 which has been determined previously. The second equation (34) can be written as

$$\hat{A}_0\eta_0 = r_0^3 - \tau_0 r_0.$$

This problem can have a square integrable solution if and only if

$$r_0^3 - \tau_0 r_0 \perp \ker A_0 = \mathbf{R} r_0.$$

This determines τ_0 by

$$\tau_0 = \frac{\int r_0^4 \, dx}{\int r_0^2 \, dx}. \tag{35}$$

Thus η_0 is determined uniquely provided that it is orthogonal to r_0. However, there is a more precise description of η_0. Let us look for a solution of (34) of the form

$$\eta_0 = r_0\phi_0.$$

We define for convenience

$$\theta_0 = \phi_0'.$$

Substituting in (34) and multiplying by r_0, we obtain

$$|\mu|(\theta_0 r_0^2)' = \tau_0 r_0^2 - r_0^4;$$

hence,

$$\theta_0(x) = \frac{1}{|\mu| r_0^2(x)} \int_{-\infty}^{x} [\tau_0 r_0^2(y) - r_0^4(y)]\, dy.$$

In particular, we can see that θ_0 is odd, thanks to the definition (35) of τ_0, and the limit of $\theta_0(x)$ as x tends to $\mp\infty$ is equal to $\pm\bar{\theta}_0$, where

$$\bar{\theta}_0 = \frac{\tau_0}{2|\mu|}.$$

Explicit formulas can be obtained for θ_0. It is also possible to perform an asymptotic analysis: from (11) and (35), we obtain

$$\tau_0 = \frac{3}{4} + O\left(\frac{1}{L}\right), \quad \bar{\theta}_0 \sim 2.$$

The asymptotics of θ_0 can be described rather precisely. We have

$$\theta_0(x) \sim \begin{cases} \bar{\theta}_0 & \text{if } x \leq -L \\ -x\bar{\theta}_0/L & \text{if } |x| \leq L, \\ -\bar{\theta}_0 & \text{if } x \geq L. \end{cases} \tag{36}$$

Therefore,

$$\phi_0(x) = \int_0^x \theta_0(y) + C,$$

where C should be determined so that $r_0\phi_0$ is orthogonal to r_0. This necessitates an asymptotic for $\int_0^x \theta_0(y)\, dy$:

$$\int_0^x \theta_0(y)\, dy \sim \begin{cases} -\bar{\theta}_0(-x - L/2) & \text{if } x \leq -L, \\ -x^2\bar{\theta}_0/(2L) & \text{if } |x| \leq L, \\ -\bar{\theta}_0(x - L/2) & \text{if } x \geq L. \end{cases}$$

Therefore,

$$C \sim \bar{\theta}_0 L/6.$$

Hence, we obtain that for $\nu \ll 1$,

$$\phi_0(0) \sim \frac{L}{3}, \quad \phi_0(L) \sim -\frac{2L}{3}.$$

If we keep with the implicit function theorem approach of Section 3, we can linearize F around $(r_0, \eta_0, 0, \tau_0)$. Denoting $\xi_1, \eta_1, \epsilon_1$ and τ_1 as the increments of the respective variables ξ, η, ϵ, and τ, we can write

$$D_\xi F_0\, \xi_1 + D_\eta F_0\, \eta_1 + D_\epsilon F_0\, \epsilon_1 + D_\tau F_0\, \tau_1 = 0. \tag{37}$$

Here, the index 0 on the derivatives of F means that the values are taken at $(r_0, \eta_0, 0, \tau_0)$ The first-order derivatives are given by

$$D_\xi F_0 = \begin{pmatrix} A_0 \\ \tilde{A}_0 \end{pmatrix}, \quad D_\eta F_0 = \begin{pmatrix} 0 \\ \hat{A}_0 \end{pmatrix}, \tag{38}$$

where

$$\tilde{A}_0 v = (\tau_0 - 3r_0^2 - 2r_0\eta_0 + 4r_0^3\eta_0)v, \tag{39}$$

and

$$D_\epsilon F_0 = \begin{pmatrix} -\tau_0\eta_0 - \eta_0^2 r_0 + \eta_0 r_0^2 + 2\eta_0^2 r_0^3 \\ -\eta_0^2 r_0 - \eta_0^3 + 2\eta_0^3 r_0^2 \end{pmatrix}, \quad D_\tau F_0 = \begin{pmatrix} 0 \\ r_0 \end{pmatrix}. \tag{40}$$

Now, interesting information can be extracted from (37). We perform the scalar product of the first component of (37) with s_-, and we obtain

$$\lambda_-(\xi_1, s_-) + \epsilon_1(-\tau_0\eta_0 - \eta_0^2 r_0 + \eta_0 r_0^2 + 2\eta_0^2 r_0^3, s_-) = 0. \tag{41}$$

Let us evaluate

$$a_0 = (-\tau_0\eta_0 - \eta_0^2 r_0 + \eta_0 r_0^2 + 2\eta_0^2 r_0^3, s_-).$$

We have

$$\begin{aligned} a_0 &= -2\int_0^\infty (-\tau_0\eta_0 - \eta_0^2 r_0 + \eta_0 r_0^2 + 2\eta_0^2 r_0^3) r_0'\, dx \\ &\quad + 2\int_0^\infty (-\tau_0\eta_0 - \eta_0^2 r_0 + \eta_0 r_0^2 + 2\eta_0^2 r_0^3)(s_- + r_0')\, dx. \end{aligned}$$

The second term is an $O(\sqrt{\nu}L^3)$, since $|s_- + |r_0'||_{L^2(\mathbb{R})} = O(\sqrt{\nu}L^{1/2})$ and $|\eta_0^2|_{L^2(\mathbb{R})} = O(L^{5/2})$; therefore, it suffices to estimate

$$\begin{aligned} a_1 &= -2\int_0^\infty (-\tau_0\eta_0 - \eta_0^2 r_0 + \eta_0 r_0^2 + 2\eta_0^2 r_0^3) r_0'\, dx \\ &= 2\int_0^\infty [(\tau_0 - r_0^2)\eta_0 + (r_0 - 2r_0^3)\eta_0^2] r_0'\, dx \\ &= 2\int_0^\infty (\tau_0 - r_0^2)\eta_0 r_0'\, dx - \int_0^\infty \hat{V}_0' \eta_0^2\, dx. \end{aligned}$$

We observe that

$$\begin{aligned} \int_0^\infty \hat{V}_0'\eta_0^2\, dx &= \hat{V}_0\eta_0^2 \mid_0^\infty - \int_0^\infty 2\hat{V}_0\eta_0\eta_0'\, dx \\ &= O(\sqrt{\nu}L^4) - 2\int_0^\infty (|\mu|\eta_0'' + (r_0^3 - \tau_0 r_0)\eta_0'\, dx \\ &= O(\sqrt{\nu}L^4) - 2\int_0^\infty (r_0^3 - \tau_0 r_0)\eta_0'\, dx. \end{aligned}$$

Therefore,

$$\begin{aligned} a_1 &= 2\int_0^\infty [(\tau_0 - r_0^2)\eta_0 r_0' + (r_0^3 - \tau_0 r_0)\eta_0']\, dx + O(\sqrt{\nu}L^4) \\ &= 2\int_0^\infty (\tau_0 - r_0^2)(\eta_0 r_0)'\, dx + O(\sqrt{\nu}L^4) \\ &= (\tau_0 - r_0^2)\eta_0 r_0 \mid_0^\infty - 2\int_0^\infty (-2r_0' r_0)\eta_0 r_0\, dx + +O(\sqrt{\nu}L^4). \end{aligned}$$

From the asymptotic behavior of τ_0 for $\nu \ll 1$, we know that

$$[(\tau_0 - r_0^2)\eta_0 r_0](0) = O(1).$$

Therefore

$$a = 4\int_0^\infty r_0^3 r_0' \phi_0\, dx + O(1) r_0^4 \phi_0 \mid_0^\infty - \int_0^\infty r_0^4 \theta_0\, dx.$$

From the asymptotic behavior of θ_0, we obtain

$$a_0 = \frac{3L}{8} + O(1). \tag{42}$$

Thus, we deduce from (20) and (42) that

$$\epsilon_1 \sim \frac{2\nu}{L}(\xi_1, s_-). \tag{43}$$

On the other hand, the projection of (37) on $s_-^\perp$ implies

$$A_0 \Pi_{s_-^\perp} \xi_1 = -\epsilon_1 \Pi_{s_-^\perp}(-\tau_0\eta_0 - \eta_0^2 r_0 + \eta_0 r_0^2 + 2\eta_0^2 r_0^3),$$

which shows that the component of ξ_1 along $s_-^\perp$ is very small with respect to its component along s_-.

Going back to the branch of solution of (31) out of $(r_0, \eta_0, 0, \tau_0, \nu)$, we observe that close enough to $(r_0, \eta_0, 0, \tau_0, \nu)$, we will have

$$\epsilon \sim \frac{3\nu}{4L}(\xi - r_0, s_-). \tag{44}$$

This means intuitively that the pulse is swollen by the nonvariational term: the distance between the transition zones increases with ϵ for small ϵ. Since we expect to find a turning point, it is appropriate to change coordinates. The new coordinate, denoted by y, will be the increase of the distance between the transition zones. Therefore, we come to our final ansatz: define

$$\xi^0 = r_0(1 - \nu e^{-4y}), \quad \eta^0 = \eta_0(1 - \nu e^{-4y}), \quad \tau^0 = \tau_0(1 - \nu e^{-4y}).$$

We look for a solution $(\xi, \eta, \epsilon, \tau)$ satisfying

$$\xi \sim \xi^0, \quad \eta \sim \eta^0, \quad \epsilon \sim \frac{\epsilon^0 \nu y}{L}, \quad \tau \sim \tau^0. \tag{45}$$

We determine ϵ^0 so that $F(\xi, \eta, \epsilon, \tau, \nu)$ will be orthogonal to $s_-^0 = s_-(1 - \nu e^{-4y})$, and we obtain

$$\epsilon^0 = \frac{3}{16}.$$

With the choice (45), $F(\xi, \eta, \epsilon, \tau, \nu)$ is so small that it is possible to apply the implicit function theorem and to obtain a branch of "swollen pulses" for $y \gg L^{-1}$. Then, we look at the derivative of F with respect to ξ and η on the branch of solutions. To show that the only negative eigenvalue λ_- crosses the imaginary axis for some value of y, we only have to show that

$$(D_\xi F_1(\xi, \eta, \epsilon, \tau, \nu)s_-^0, s_-^0)$$

crosses 0 for some value of y. This expression is equal to

$$\lambda_-^0 |s_-^0|_{L^2(\mathbb{R})} - \frac{3}{4} y\nu |s_-^0|_{H^2(\mathbb{R})} + \frac{\epsilon^0 \nu y}{L} \int [(6(\xi^0)^2 - 1)(\eta^0)^2 + 2\xi^0\eta^0](s_-^0)^2 \, dx + O(\nu^2 L^k),$$

where k is a bounded real number. The two norms of s_- tend to a constant as ν tends to 0. We can prove the following equivalence:

$$\int [(6(\xi^0)^2 - 1)(\eta^0)^2 + 2\xi^0\eta^0](s_-^0)^2 \, dx \sim C_2 L^2.$$

Therefore,

$$(D_\xi F_1(\xi, \eta, \epsilon, \tau, \nu)s_-^0, s_-^0) \sim -\frac{3}{4}\nu(1 - 4y) - C_0 \nu y + C_1 \epsilon^0 L \nu y. \tag{46}$$

This implies that for $y_c = O(1/L)$, the negative eigenvalue crosses the imaginary axis, and the pulse is stabilized.

A more intuitive way of describing this stabilization is the following. For $y < y_c$, the pulse responds to perturbations that are approximately proportional to s_-^0, either by the reciprocal annihilation of its transition zones or by the uncoupling of its transition zones, which flow away one from the other. For $y > y_c$, the coupling between the space translation mode and the phase translation mode is strong enough for the occurrence of a different phenomenon: a perturbation of the pulse that is approximately parallel to s_-^0 implies a change of phase in the pulse, which remains otherwise unchanged.

Acknowledgements

It is a pleasure to thank Stephan Fauve for introducing us to this problem and the stabilization conjecture, and Boris Malomed for pointing out most relevant work of his and promptly sending articles that are due to appear shortly. After the passing away of the first author of this article, the asymptotic show had to go on.

References

[1] Carr,J., and Pego, R. L. (1986) 'Metastable patterns in solutions of $u_t = \epsilon^2 u_{xx} - f(u)$', *Comm. Pure Appl. Math.*, **42**, 523–576.

[2] Coddington, E. A., and Levinson, N. (1955) *Theory of Differential Equations*, McGraw-Hill, New York.

[3] Fusco, G. (1990) 'A geometric approach to the dynamics of $u_t = \varepsilon^2 u_{xx} + f(u)$ for small ε', in K. Kirchgässner (ed.), *Lecture Notes in Physics, Vol. 359*, Springer-Verlag, Berlin, 53–73.

[4] Fusco, G., and Hale, J. K. (1989) ' Slow-motion manifolds, dormant instability, and singular perturbations', *J. Dynam. Differ. Equat.*, **1**, 75–94.

[5] Henry,D. (1981) *Geometric Theory of Semilinear Parabolic Equations, Lecture Notes in Mathematics, Vol. 840*, Springer-Verlag.

[6] Malomed, B. A., and Nepiomastchy, A. (1990) 'Kinks and solutions in the generalized Ginzburg-Landau equation', *Phys. Rev.*, **A 42**, 6009–6014.

[7] de Mottoni, P., and Schatzman, P. (1992) 'Evolution of developed interfaces in $\mathbf{R}^N$', *Trans. Amer. Math. Soc.*, to appear.

[8] Thual,O., and Fauve, S. (1988) 'Localized structures generated by subcritical instabilities', *J. Phys. France*, **49**, 1829–1833.

EXPONENTIAL CONVERGENCE OF THE GALERKIN APPROXIMATION FOR THE GINZBURG-LANDAU EQUATION

Arjen Doelman
Center for Applied Mathematics
Cornell University
Ithaca NY 14853-6201

and

Mathematical Institute
Rijksuniversiteit Utrecht
P.O. Box 80.010
3508 TA Utrecht
the Netherlands

and

Edriss S. Titi
Center for Applied Mathematics
Cornell University
Ithaca NY 14853-6201

and

Department of Mathematics
University of California
Irvine CA 92717

ABSTRACT. We show that the solutions of the Ginzburg-Landau equation on a periodic interval are of Gevrey-class regularity. Specifically, we prove that the modes of the Fourier decomposition of a solution u decay exponentially. Using this preliminary result, we establish that the N-dimensional linear Galerkin approximation of u converges exponentially fast to u as N goes to infinity. We discuss the influence of the parameters in the Ginzburg-Landau equation on the decay rate of the Fourier modes and on the rate of convergence of the Galerkin approximations.

1. Introduction

In this paper we study the Ginzburg-Landau equation

$$\frac{\partial u}{\partial t} = (r - (1 + ib)|u|^2)u + (1 + ia)\frac{\partial^2 u}{\partial x^2} \tag{1}$$

H. G. Kaper and M. Garbey (eds.),
Asymptotic and Numerical Methods for Partial Differential Equations with Critical Parameters 241–252.

on the interval $(0, 1)$ with periodic boundary conditions; $r > 0$ is the bifurcation parameter (related to the control parameter of the underlying physical system); a and b are real parameters. The equation appears as a modulation equation governing the nonlinear evolution of disturbances at near-critical conditions in very general, nonlinear stability problems, including the classical hydrodynamical examples such as Rayleigh-Bénard convection and Poiseuille flow. Parameters a and b can be computed for a given physical process; for details on the derivation of (1) we refer to [5], [19]. The Ginzburg-Landau equation has recently been the object of a heterogeneous collection of studies addressing, for instance, the dimension of the attractor [9], [12], various kinds of special solutions [2], [6], [15], its mathematical validity as modulation equation [13], [16], and numerical simulations [17], [18], [21]. As in most papers on this subject, we consider the equation with a cubic nonlinearity with negative real part (which occurs in most applications); in Remark 3 we briefly consider the complementary case.

Since we are considering (1) with periodic boundary conditions, it is natural to introduce the Fourier decomposition of $u(x, t)$,

$$u(x,t) = \sum_{n=-\infty}^{\infty} U_n(t) e^{2\pi i n x}. \tag{2}$$

It has been observed in numerical simulations that there is a fast exponential decay of the relative energy per mode U_n as $|n| \to \infty$ for solutions on the attractor of (1) [17, 18, 21] and that (1) can be approximated by very low dimensional Galerkin approximations [7, 21]. We wish to understand this behavior analytically. In [7] it was shown that the decay of $\limsup_{t\to\infty} |U_n(t)|$ is faster than algebraic, as a first step towards understanding the numerical observations. In this paper we improve this result: first, we present a result on the regularity of the solutions of (1) with periodic boundary conditions and initial condition $u(x, 0) = u_0(x) \in H^1_{per}((0, 1))$; then, we prove that there exist constants $K = K(r, a, b) > 0$ and $\sigma = \sigma(r, a, b) > 0$ such that

$$\limsup_{t\to\infty} |U_n(t)| < K e^{-\sigma|n|} \tag{3}$$

uniformly in $|n|$. Thus, as $t \to \infty$, the maximal modulus of a mode $U_n(t)$ decays exponentially as a function of $|n|$; in other words, the solutions of (1) with periodic boundary conditions are of Gevrey-class regularity (see [10]). Hence we have extended the result of [7]. This preliminary result can also be interpreted as follows: the finite-dimensional attractor for the flow induced by (1) (see [9], [12]) lies exponentially close to the linear space spanned by the first $2N + 1$ Fourier modes. Indeed, based on (3), we can understand the low-dimensional behavior mentioned above and obtain our main result: we show that for $0 \leq t \leq T$ a solution $u(x,t)$ can be approximated exponentially accurately by the Galerkin approximation, $u_N(x,t)$:

$$\|u(t) - u_N(t)\|_{H^1} \leq C(T)e^{-\sigma(N+1)}, \tag{4}$$

where the evolution of $u_N(x,t)$ is governed by

$$\frac{\partial u_N}{\partial t} = ru_N - (1+ib)P_N(|u_N|^2 u_N) + (1+ia)\frac{\partial^2 u_N}{\partial x^2}. \tag{5}$$

Here, P_N is the projection operator on the finite-dimensional space spanned by the first $2N+1$ Fourier modes $\{e^{2\pi inx}\}_{n=-N,\ldots,N}$. The initial condition for $u_N(x,t)$ is given by $u_N(x,0) = P_N u(x,0)$. The termm $C(T)$ in (4) is a positive constant which depends on the coefficients a, b, and r of (1) (see Section 4), σ in (4) is as in (3). Thus, the semi-discrete (linear) Galerkin approximation based on the Fourier decomposition (2) converges exponentially fast to the solution of (1) as the number of modes goes to infinity.

The proof of the first regularity result (i.e., (3)) is based on the methods of Foias and Temam [10] (who proved Gevrey-class regularity for solutions of the Navier-Stokes equations). As in [10], we derive *a priori* estimates of the solutions to prove the Gevrey-class regularity without using Galerkin approximations; to make our proof completely rigorous, one should apply the classical process in which one first considers a Galerkin approximation of (1), derives the *a priori* estimates, and then passes to the limit (see, for instance, [3], [23]). The convergence result (4) is based on the work of Devulder et al. [4] in which a method has been developed to estimate the rate of convergence of (in general, nonlinear) Galerkin approximations.

Remark 1. Independently, Promislow [20] has shown recently that the solutions of the Ginzburg-Landau equation are time analytic with values in the Gevrey class of functions. These results are then applied to the construction of approximate inertial manifolds. Our estimate (4) yields that the approximations u_N of u using the trivial approximate inertial manifold (i.e., the $2N+1$-dimensional linear space spanned by $\{e^{2\pi inx}\}_{n=-N,\ldots,N}$) converge already exponentially fast.

Remark 2. This paper can be considered as an illustration of a forthcoming paper [8], in which results similar to the ones described above are obtained for the n-dimensional Ginzburg-Landau equation on the periodic domain $(0,1)^n$ $(n = 1,2,3)$. We refer to that paper for more details on our methods; especially in the section on the approximation of u by u_N, we will refer frequently to [8].

2. Regularity of the Solutions of Equation (1)

In this section we present an outline of the proof of the Gevrey-class regularity result. To facilitate the presentation, we will omit, as far as possible, the technical derivations of the *a priori* estimates needed for the proof. Those will be given in Section 3. After formulating the main theorem, we will discuss some consequences of it.

Consider the selfadjoint, nonnegative operator $A = -\frac{\partial^2}{\partial x^2}$ with periodic boundary conditions:

$$Au(x,t) = \sum_{n=-\infty}^{\infty} 4\pi^2 n^2 U_n(t) e^{2\pi i n x}$$

(using (2)). We define $v(x,t) = e^{tA^{\frac{1}{2}}} u(x,t)$, that is,

$$v(x,t) = \sum_{n=-\infty}^{\infty} e^{2\pi |n| t} U_n(t) e^{2\pi i n x}.$$

Note that if we show that $\|v\|_2$, the L_2-norm of v, is bounded for all time, we have proved the exponential decay of $\limsup_{t\to\infty} |U_n(t)|$. However, as in [7], we prefer to derive estimates for $\|A^{\frac{1}{2}} v\|_2$ instead of $\|v\|_2$.

Lemma 1. $\|A^{\frac{1}{2}} v\|_2$ *satisfies the equation*

$$\frac{1}{2}\frac{d}{dt}\|A^{\frac{1}{2}} v\|_2^2$$

$$= \|A^{\frac{3}{4}} v\|_2^2 + r\|A^{\frac{1}{2}} v\|_2^2 - \|Av\|_2^2 - Re(< A^{\frac{1}{2}}(1+ib)e^{tA^{\frac{1}{2}}}|u|^2 u, A^{\frac{1}{2}} e^{tA^{\frac{1}{2}}} u >) \quad (6)$$

with $<,>$ *the scalar product of* $L_2((0,1))$.

Proof. One can derive an equation for v using (1). Substitution of this equation into $\frac{\partial}{\partial t}\|A^{\frac{1}{2}} v\|_2^2 = \int_0^1 (\frac{\partial v}{\partial t} A v^* + v A \frac{\partial v^*}{\partial t}) dx$ yields the result.

The regularity result is based on the following technical lemma. We will postpone its proof until Section 3.

Lemma 2. *There are positive constants* $\hat{C}_1(r,b)$, $\hat{C}_2(r,b)$, *and* $\hat{C}_3(r,b)$ *such that*

$$\frac{d}{dt}\|A^{\frac{1}{2}} v\|_2^2 \leq \hat{C}_1(r,b)\|A^{\frac{1}{2}} v\|_2^2 + \hat{C}_2(r,b)\|A^{\frac{1}{2}} v\|_2^4 + \hat{C}_3(r,b)\|A^{\frac{1}{2}} v\|_2^6, \quad (7)$$

where $\hat{C}_i(r,b) \to \infty$ *for* $r \to \infty$ *and* $\hat{C}_i(r,b) \to \infty$ *for* $|b| \to \infty$.

We now define $E(t) = 1 + \|A^{\frac{1}{2}}v\|_2^2$. It is easily checked from (7) that there is a positive constant $\hat{K} = \hat{K}(r,b)$ such that

$$\frac{dE}{dt} \leq \hat{K}(r,b)E^3, \tag{8}$$

and, if we consider $u(x,0) = u_0(x)$ as the initial condition for (1),

$$E(0) = 1 + \|A^{\frac{1}{2}}u_0\|_2^2$$

(since $v(x,0) = u(x,0)$). Hence, if $\|A^{\frac{1}{2}}u_0\|_2$ is finite—i.e., if $u_0 \in H^1_{per}((0,1))$—then $E(0) < \infty$ and therefore by (8) there exists a $T_0 > 0$ such that $E(t) \leq 2E(0)$ for $t \in [0, T_0]$, or

$$1 + \|A^{\frac{1}{2}}e^{tA^{\frac{1}{2}}}u\|_2^2 \leq 2 + 2M_0,$$

with $\sqrt{M_0} = \|u_0\|_{H^1}$. This yields an exponential decay of the Fourier modes U_n, for $|n| \to \infty$, for $t \in [\delta, T_0]$, for any $\delta > 0$. This result can be extended to unbounded t by means of the global upper bound for $\|A^{\frac{1}{2}}u\|_2$ established in [9] for any solution $u(x,t)$ of (1) with $u(x,0) \in H^1_{per}((0,1))$:

$$\|A^{\frac{1}{2}}u\|_2^2 \leq M_1 = M_1(a,r,b,M_0)$$

($M_1 \geq M_0$). Hence, at any time t_1 we can start the above process, choosing $u(x,t_1)$ as initial condition and using $\|A^{\frac{1}{2}}u(t=t_1)\|_2^2 \leq M_1$ (thus $E(0) \leq 1 + M_1$). We then find a $T_1 > 0$, T_1 independent of t_1, such that

$$E(T_1) = 1 + \|A^{\frac{1}{2}}e^{T_1A^{\frac{1}{2}}}u(t = t_1 + T_1)\|_2^2 \leq 2 + 2M_1.$$

Thus, for any time $t_0 > T_0$ ($T_0 \geq T_1$) we obtain the estimate

$$\|A^{\frac{1}{2}}e^{T_1A^{\frac{1}{2}}}u(t = t_0)\|_2 \leq \sqrt{1 + 2M_1},$$

which can be translated into

$$\sum_{n=-\infty}^{\infty} 4\pi^2 n^2 e^{2T_1|n|}|U_n(t)|^2 \leq 1 + 2M_1$$

for all $t > T_0$. We can now formulate the following theorem.

Theorem 1. *Consider the equation (1) with periodic boundary conditions and initial condition $u(x,0) = u_0(x) \in H^1_{per}((0,1))$. Then there exist constants $\sigma = \sigma(r,a,b)$ and $K = K(r,a,b)$ such that*

$$\sum_{n=-\infty}^{\infty} n^2 e^{2\sigma|n|}|U_n(t)|^2 \leq K^2$$

for all t larger than some T_0, with $U_n(t)$ defined by (2). Therefore (for $n \neq 0$)

$$\limsup_{t\to\infty} |U_n(t)| \leq \frac{1}{n} K e^{-\sigma |n|}$$

(see (3)). The proof of this theorem follows from the preceding analysis; σ and K can easily be expressed in terms of T_1 and M_1. This theorem states that $u(x,t)$ is of Gevrey-class regularity (see [10]). There is no reason to expect that the constants σ and K are sharp. However, by applying the properties of the constants $\hat{C}_i$ $(i = 1,2,3)$ and M_1 as functions of a, b, and r (Lemma 2 and [9]), one observes that

$$\sigma \downarrow 0 \text{ if } |a|,\ |b| \text{ or } r \to \infty \quad \text{and} \quad K \to \infty \text{ if } |a|,\ |b| \text{ or } r \to \infty$$

(from [9] it follows that $M_1 \to \infty$ for $|a|$, $|b|$, or $r \to \infty$; hence, although constant $\hat{K}$ in (8) does not depend on a, we find that K and σ in Theorem 1 depend on a). This result is in agreement with the numerical observations in [17], [18]; a similar behavior has been found in [7]. However, it should be noted that the constants that measure the decay of the modes in [7]—where it was shown that this decay is faster than algebraic—are independent of a.

Remark 3. As in most applications, we have chosen a negative real part for the coefficient of the cubic nonlinearity in (1). However, in some circumstances (for instance, in the case of Poiseuille flow), this real part is positive [22]. It has been observed that in that case there are solutions of the Ginzburg-Landau equation that blow up in finite time [14]. Yet, some parts of our results are still valid, since the method of proof is based "only" on the global *a priori* estimates of $\|u\|_2$ and $\|A^{\frac{1}{2}}u\|_2$; the sign of the real part of the nonlinearity has no influence on the other parts of the proof. Hence, one can show that solutions of this equation will be of Gevrey-class regularity for $t \in [\delta, T]$ for any $\delta > 0$ and $T(> 0)$ such that $\|u\|_2$ and $\|A^{\frac{1}{2}}u\|_2$ are (still) bounded for $t \in [0,T]$.

Remark 4. In [13] it has been proven that, if a solution of the Ginzburg-Landau equation is of Gevrey-class regularity for a certain time t_0, then it will remain so as long as it exists.

3. Proof of Lemma 2

The main part of the proof of Lemma 2 consists of the derivation of an upper bound for the nonlinearity in (6):

$$\left| -Re\left(< (1+ib)e^{tA^{\frac{1}{2}}} |u|^2 u, e^{tA^{\frac{1}{2}}} u > \right) \right|$$

$$\leq \sqrt{1+b^2}\left|< e^{tA^{\frac{1}{2}}}|u|^2u, e^{tA^{\frac{1}{2}}}u >\right|$$

$$= \sqrt{1+b^2}\left|\sum_{M=-\infty}^{\infty} 4\pi^2 M^2 U_M^* e^{2\pi t|M|}\Big(\sum_{k+l-m=M} e^{2\pi t|M|} U_k U_l U_m^*\Big)\right|$$

by substitution of (2) into the scalar product; this expression is bounded from above by

$$\sqrt{1+b^2}\sum_{M=-\infty}^{\infty} 4\pi^2 M^2 |U_M| e^{2\pi t|M|}\left(\sum_{k+l-m=M} e^{2\pi t|M|}|U_k||U_l||U_m|\right)$$

$$\leq \sqrt{1+b^2}\sum_{M=-\infty}^{\infty} 4\pi^2 M^2 |U_M| e^{2\pi t|M|}$$

$$\left(\sum_{k+l-m=M} e^{2\pi t|k|}|U_k| e^{2\pi t|l|}|U_l| e^{2\pi t|m|}|U_m|\right)$$

$$\left(\sum_{k+l-m=M} e^{2\pi t|M|}|U_k||U_l||U_m|\right)$$

$$= \sqrt{1+b^2} < |\hat{v}|^2\hat{v}, A\hat{v} >$$

with $\hat{v}(x,t) = \sum_{n=-\infty}^{\infty} e^{2\pi|n|t}|U_n|e^{2\pi i n x}$. (Note that, in Fourier decomposition, $v(x,t) = \sum_{n=-\infty}^{\infty} e^{2\pi|n|t}U_n e^{2\pi i n x}$.) Furthermore, using $\|.\|_p$ to denote the norm in L_p, we have

$$\begin{aligned} < |\hat{v}|^2\hat{v}, A\hat{v} > &\leq \int_0^1 |\hat{v}|^3|A\hat{v}|dx \leq \left(\int_0^1 |\hat{v}|^6 dx\right)^{\frac{1}{2}}\left(\int_0^1 |A\hat{v}|^2 dx\right)^{\frac{1}{2}} \\ &= \|\hat{v}\|_6^3\|A\hat{v}\|_2 \leq \tfrac{\alpha}{2}\|\hat{v}\|_6^6 + \tfrac{1}{2\alpha}\|A\hat{v}\|_2^2 \end{aligned}$$

for any $\alpha > 0$ (α will be specified later). Note that $\|\hat{v}\|_2 = \|v\|_2$, $\|A^{\frac{1}{2}}\hat{v}\|_2 = \|A^{\frac{1}{2}}v\|_2$ and $\|A\hat{v}\|_2 = \|Av\|_2$. By the interpolation (or Gagliardo-Nirenberg) inequality (see, for instance [11]) we find

$$\|\hat{v}\|_6 \leq C\|\hat{v}\|_2^{\frac{2}{3}}\|\hat{v}\|_{H^1}^{\frac{1}{3}} = C\|v\|_2^{\frac{2}{3}}(\|v\|_2^2 + \|A^{\frac{1}{2}}v\|_2^2)^{\frac{1}{6}}$$

for some positive constant C. Thus, all in all, we have

$$\begin{aligned} &\left|-Re\left(< (1+ib)e^{tA^{\frac{1}{2}}}|u|^2u, e^{tA^{\frac{1}{2}}}u >\right)\right| \\ &\leq \sqrt{1+b^2}\left(\tfrac{\alpha}{2}C\|v\|_2^4(\|v\|_2^2 + \|A^{\frac{1}{2}}v\|_2^2) + \tfrac{1}{2\alpha}\|Av\|_2^2\right) \\ &= \tfrac{1}{2}(1+b^2)C(\|v\|_2^6 + \|v\|_2^4\|A^{\frac{1}{2}}v\|_2^2) + \tfrac{1}{2}\|Av\|_2^2 \end{aligned}$$

by choosing $\alpha = \sqrt{1+b^2}$.

Next, we estimate the $\|A^{\frac{3}{4}}v\|_2^2$ term in (6):

$$\|A^{\frac{3}{4}}v\|_2 \leq \|Av\|_2\|A^{\frac{1}{2}}v\|_2 \leq \frac{1}{2}\|Av\|_2^2 + \frac{1}{2}\|A^{\frac{1}{2}}v\|_2^2.$$

Thus, upon substitution of these results into (6), we find

$$\frac{\partial}{\partial t}\|A^{\frac{1}{2}}v\|_2^2 \leq (2r+1)\|A^{\frac{1}{2}}v\|_2^2 + (1+b^2)C(\|v\|_2^6 + \|v\|_2^4\|A^{\frac{1}{2}}v\|_2^2). \tag{9}$$

Note that the "$-\|Av\|_2^2$" term has "killed" the two "$\frac{1}{2}\|Av\|_2^2$"-terms produced by the estimates. We now use that

$$\|v\|_2^2 \leq |U_0|^2 + \|A^{\frac{1}{2}}v\|_2^2$$

where U_0 is the zeroth Fourier mode of $u(x,t)$ (and of $v(x,t)$). In [7], [9] it was shown that $\|u\|_2 \leq \hat{C}$ for all time (with $\hat{C}^2 = \max(\|u_0\|_2^2, 2r)$. Hence, we substitute in (9)

$$\|v\|_2^2 \leq \hat{C}^2 + \|A^{\frac{1}{2}}v\|_2^2,$$

which yields that $\frac{\partial}{\partial t}\|A^{\frac{1}{2}}v\|_2^2$ is bounded from above by a cubic polynominal in $\|A^{\frac{1}{2}}v\|_2^2$, with positive coefficients; that is, we have completed the derivation of the inequality (7). Note that the asymptotic behavior of the constants $\hat{C}_i$, $i = 1,2,3$, stated in Lemma 2 follows immediately from the asymptotic properties of $\hat{C}$ and the coefficients in (9).

4. Convergence of u_N to u

In this section we present a (rough) sketch of the method by which the distance (in the H^1-norm) between solution $u(x,t)$ of (1) and its (linear) Galerkin approximation $u_N(x,t)$ (a solution of (5)) can be measured. We refer to [8] for a detailed description of the method and its proofs. Following [4] we introduce $p(x,t) = P_N u(x,t)$ and $q(X,t) = (I - P_N)u(x,t)$, where P_N is the projection operator defined in Section 1 and I is the identity; thus, $u(x,t) = p(x,t) + q(x,t)$. Furthermore we define $y(x,t) = u_N(x,t)$ and

$$\delta(x,t) = p(x,t) - y(x,t) \;\; (= P_N u(x,t) - u_N(x,t)).$$

As a consequence of Theorem 1 we have

$$\begin{aligned} \|q\|_{H^1}^2 &= \textstyle\sum_{|n|=N+1}^{\infty} |U_n|^2 + \sum_{|n|=N+1}^{\infty} 4\pi^2 n^2 |U_n|^2 \\ &\leq K^2 \left(\frac{1}{(N+1)^2} + 4\pi^2\right) e^{-2\sigma(N+1)}. \end{aligned} \tag{10}$$

Because

$$\|u - u_N\|_{H^1} = \|p + q - y\|_{H^1} \leq \|\delta\|_{H^1} + \|q\|_{H^1}, \tag{11}$$

one can obtain a result like (4) if one can estimate $\|\delta\|_{H^1}$ in terms of $\|q\|_{H^1}$. This estimation can be done in two steps. First, one derives an equation that governs the evolution of $\|\delta\|_2^2$ (i.e., the (square of the) L_2-norm of δ), and, based on this equation, one determines an upper bound on $\|\delta\|_2$ of the type (4). The second step is then similar to the first: based on an evolution equation for $\|A^{\frac{1}{2}}\delta\|_2^2$ (with A as defined in Section 2), one derives an upper bound for $\|A^{\frac{1}{2}}\delta\|_2$ using the same techniques as in the first step. We will only sketch the derivation of the first step.

Theorem 2. *Let $u(x,t) \in H^1_{per}((0,1))$ (as a function of x) be a solution of (1) with initial condition $u(x,0) = u_0(x) \in H^1_{per}((0,1))$, and let $u_N(x,t)$ be the Galerkin approximation of $u(x,t)$; that is, the evolution of $u_N(x,t)$ is governed by (5), and P_N is the projection operator onto the space spanned by $\{e^{2\pi inx}\}_{n=-N,\ldots,N}$, $u_N(x,0) = P_N(u_0(x))$. Then, for $t \in [0,T]$, there is a positive constant $C_2 = C_2(T)$ such that*

$$\|u - u_N\|_2 \le \frac{C_2}{N+1} e^{-\sigma(N+1)}. \tag{12}$$

For a proof of the more general result (4), we refer to [8]. The appearance of the extra term $N+1$ in the denominator of the right-hand side of (12) can be explained immediately: we will use $\|q\|_2$ instead of $\|q\|_{H^1}$ to estimate $\|\delta\|_2$; hence, by Theorem 1 and by (10) we observe that we will automatically get the extra factor $N+1$ in the denominator of (12).

If we use (1) and (5), it is straightforward to derive the evolution equation for $\|\delta\|_2^2$:

$$\begin{aligned}\tfrac{1}{2}\tfrac{d}{dt}\|\delta\|_2^2 = {} & r\|\delta\|_2^2 - \|A^{\frac{1}{2}}\delta\|_2^2 \\ & -\tfrac{1}{2}(1+ib) < P_N(|u|^2u - |y|^2y), \delta > \\ & -\tfrac{1}{2}(1-ib) < \delta, P_N(|u|^2u - |y|^2y) >,\end{aligned} \tag{13}$$

in which $<,>$ denotes the scalar product in $L_2((0,1))$ (see also Lemma 1, in which a similar equation is derived). As in the proof of Theorem 1, we will use the term $-\|A^{\frac{1}{2}}\delta\|_2^2$ to "kill" the "bad" terms appearing from the estimates of the nonlinearities in (13). Again, as in the proof of Theorem 1 (see Sections 2 and 3), the main ingredient of the proof of Theorem 2 is the derivation of a suitable upper bound on these nonlinearities. Define $\mathcal{N}(u,y)$ by

$$\begin{aligned}\mathcal{N}(u,y) = {} & -(1+ib) < P_N(|u|^2u - |y|^2y), \delta > \\ & -(1-ib) < \delta, P_N(|u|^2u - |y|^2y) > .\end{aligned}$$

Then, substituting $u = \delta + q + y$ and $y = u - \delta - q$ in some parts of $\mathcal{N}(u,y)$ and performing some manipulations with the scalar products (see [8]), one can show that

$$|\mathcal{N}(u,y)| \leq 2\sqrt{1+b^2}[|<|u|^2\delta,\delta>|+|<|u|^2q,\delta>|+|<|y|^2\delta,\delta>| \\ +|<|y|^2q,\delta>|+|<uy,\delta^2>|+|<uy,q\delta>|]. \tag{14}$$

Note that each of the six terms of (14) contains either a factor δ and a factor q or two factors δ. Next, one has to derive the "suitable" upper bounds for each of these terms. By "suitable" we mean an upper bound of the following structure:

$$L_1^j(\alpha_j)\|\delta\|_2^2 + \alpha_j L_2^j \|A^{\frac{1}{2}}\delta\|_2^2 + L_3^j\|q\|_2^2 \tag{15}$$

for $j = 1,2,...,6$ in which the positive parameters α_j can still be chosen. The "constants" L_i^j depend on the properties of u and $u_N = y$:

$$L_i^j = L_i^j(\|u\|_2, \|A^{\frac{1}{2}}u\|_2, \|y\|_2, \|A^{\frac{1}{2}}y\|_2)$$

(for $j = 1,...,6$ and $i = 1,2,3$). The derivations of these upper bounds are essentially based on the interpolation inequalities; the techniques are similar to those in Section 3. The appearance of the "free" parameters α_j are due to an application of Young's inequality. We again refer to [8] for the details of the derivations of the six upper bounds of type (15).

The dependence on u of factors L_i^j can be eliminated by the global upper bounds known for $\|u\|_2$ and $\|A^{\frac{1}{2}}u\|_2$ (see [9] and Sections 2 and 3). The finite-dimensional system that governs the behavior of $y = u_N$ has been studied in [7]. Based on the estimates on U_k^N (i.e., the kth Fourier mode of u_N, $|k| \leq N$) derived in [7], we have global upper bounds on $\|y\|_2$ and $\|A^{\frac{1}{2}}y\|_2$ that depend only on the coefficients of the Ginzburg-Landau equation (and the initial data). Thus, the factors L_i^j can be bounded from above by constants that depend only on the coefficients of (1) and the initial condition u_0. Hence, if we choose the free parameters α_j such that $\sum_{j=1}^{6}\alpha_j L_2^j \leq 1$, the term $-\|A^{\frac{1}{2}}\delta\|_2^2$ in the evolution equation (13) will eliminate the corresponding terms (but with the opposite signs) emerging from the estimates on the nonlinearities. Thus, we obtain

$$\frac{d}{dt}\|\delta\|_2^2 \leq (2r + 4\sqrt{1+b^2}\hat{L}_1)\|\delta\|_2^2 + 4\sqrt{1+b^2}\hat{L}_2\|q\|_2^2$$

in which the $\hat{L}_i$'s are (positive) constants, $\hat{L}_i = \hat{L}_i(a,b,r,M_0)$ for $i = 1,2$, with $M_0 = \|u_0\|_{H^1}^2$, as in Section 2. One can now apply the Gronwall lemma to derive

$$\|\delta\|_2^2 \leq \frac{\mathcal{L}_2}{\mathcal{L}_1}\|q\|_2^2(e^{\mathcal{L}_1 t} - 1),$$

with the obvious definitions for $\mathcal{L}_1$ and $\mathcal{L}_2$. Combining this result with estimates (10) and (11) (using L_2-norms instead of H_1-norms) yields (12) with

$$C_2(T) = K(1 + \sqrt{\frac{\mathcal{L}_2}{\mathcal{L}_1}(e^{\mathcal{L}_1 T} - 1)}), \tag{16}$$

where K is defined in Theorem 1.

Remark 5. By checking the influence of a, b, and r on the coefficients in (16) one finds that C_2 is an increasing function of $|a|$, $|b|$ and r in Theorem 2 (i.e., $C_2 \to \infty$ if $|a|$ or $|b|$ or $r \to \infty$; see [8] for more details). Since we already noted that $\sigma \downarrow 0$ if $|a|$ or $|b|$ or $r \to \infty$ (see Section 2), we observe that the (linear) Galerkin approximation u_N converges more slowly to u for $|a|$ or $|b|$ or r large.

Acknowledgements

The work of A. Doelman was supported by Air Force grant AFOSR-91-0329 and NSF grant DMS-90-24929; the work of E. S. Titi was supported in part by AFOSR and NSF grant DMS-89-15672.

References

[1] Bartuccelli, M., Constantin, P., Doering, C. R., Gibbon, J. D. and Gisselfält, M. (1990) 'On the possibility of soft and hard turbulence in the complex Ginzburg-Landau equation', *Phys.* **44D**, 421–444.

[2] Bernoff, A. J. (1988) 'Slowly varying fully nonlinear wavetrains in the Ginzburg-Landau equation', *Phys.*, **30D**, 363–381.

[3] Constantin, P., and Foias, C. (1988) *Navier-Stokes Equations*, University of Chicago Press, Chicago.

[4] Devulder, C., Marion, M., and Titi, E.S. (1992) 'On the rate of convergence of the nonlinear Galerkin methods', *Math. of Comp.*, to appear.

[5] DiPrima, R. C., Eckhaus, W., and Segel, L. A. (1971) 'Non-linear wave-number interaction in near-critical two-dimensional flows', *J. Fluid Mech.*, **49**, 705–744.

[6] Doelman, A. (1989) 'Slow time-periodic solutions of the Ginzburg-Landau equation', *Phys.*, **40D**, 156–172.

[7] Doelman, A. (1991) 'Finite dimensional models of the Ginzburg-Landau equation', *Nonlin.*, **4**, 231–250.

[8] Doelman, A., and Titi, E. S. (1992) 'Regularity of solutions and the convergence of the linear Galerkin method in the Ginzburg-Landau equation', in preparation.

[9] Doering, C. R., Gibbon, J. D., Holm, D. D., and Nicolaenko, B. (1988) 'Low dimensional behaviour in the complex Ginzburg-Landau equation', *Nonlin.*, **1**, 279–310.

[10] Foias, C., and Temam, R. (1989) 'Gevrey class regularity for the solutions of the Navier-Stokes equations', *J. Funct. An.*, **87**, 359–369.

[11] Friedman, A. (1976) *Partial Differential Equations*, R. Krieger Pub. Co., New York.

[12] Ghidaglia, J. M., and Héron, B. (1987) 'Dimension of the attractors associated to the Ginzburg-Landau equation', *Phys.*, **28D**, 282–304.

[13] van Harten, A. (1991) 'On the validity of the Ginzburg-Landau equation', *J. Nonl. Sc.*, **1**, 397–422.

[14] Hocking, L. M., Stewartson, K., and Stuart, J. T. (1972) 'A nonlinear stability burst in plane parallel flow', *J. Fluid Mech.*, **51**, 705–735.

[15] Holmes, P. (1986) 'Spatial structure of time-periodic of the Ginzburg-Landau equation', *Phys.*, **23D**, 84–90.
[16] Iooss, G., Mielke, A., and Demay, Y. (1989) 'Theory of steady Ginzburg-Landau equation, in hydrodynamic stability problems', *Eur. J. Mech. B/Fluids*, **8**, 229–268.
[17] Keefe, L. R. (1985) 'Dynamics of perturbed wavetrain solutions to the Ginzburg-Landau equation', *Stud. in Appl. Math.*, **73**, 91–153.
[18] Moon, H. T., Huerre, P., and Redekopp, L. G. (1983) 'Transitions to chaos in the Ginzburg-Landau equation', *Phys.*, **7D**, 135–150.
[19] Newell, A. C. (1974) 'Envelope equations' *Lect. Appl. Math.*, Vol. **15**, 157–163.
[20] Promislow, K. (1991) 'The development and numerical implementation of approximate inertial manifolds for the Ginzburg-Landau equation', Ph.D. thesis, Indiana University.
[21] Rodriguez, J. D., and Sirovich, L. (1990) 'Low-dimensional dynamics for the complex Ginzburg-Landau equation', *Phys.*, **43D**, 77–86.
[22] Stewartson, K., and Stuart, J. T. (1971) 'A non-linear instability theory for a wave system in plane Poiseuille flow', *J. Fluid Mech.*, **48**, 529–545.
[23] Temam, R. (1977) *Navier-Stokes Equations: Theory and Numerical Solution*, Stud. Math. Appl. Vol. 2, North-Holland, Amsterdam.

Part 5: Symbolic Manipulation Tools for Asymptotic Analysis

General Asymptotic Scales and Computer Algebra

Bruno Salvy

GENERAL ASYMPTOTIC SCALES AND COMPUTER ALGEBRA

Bruno Salvy
Algorithms Project
INRIA Rocquencourt
78153 Le Chesnay Cedex
France

ABSTRACT. In many applications, one encounters asymptotic expansions of a form more complicated than mere Puiseux series. Existing computer algebra systems lack good algorithms for handling such asymptotic expansions. We present tools that permit the representation and automatic handling of general exp-log asymptotic expansions.

1. Introduction

The process of computing an asymptotic expansion generally consists of two distinct parts: one is the application of a method (e.g., Laplace's method, saddle-point method, integration by parts, indeterminate coefficients), which usually requires some expertise, and the other part is a simple but tedious manipulation of asymptotic expansions (adding or multiplying them, computing their logarithm, and so on). The purpose of this paper is to describe the automation of this latter part, which must precede any attempt at automating the particular application considered. We shall focus on finding a good representation for asymptotic expansions and on describing algorithms to perform elementary operations (sum, product, power, exponential, and logarithm) on objects stored with this representation. The emphasis is put on asymptotic expansions as *objects*, as opposed to asymptotic expansion as a *process*.

Many more-or-less general definitions of asymptotic expansions have been given. We shall use the following.

Definition 1: Asymptotic Scales. *Let $\mathcal{E}$ be a set of points with a limit point $\mathcal{P}$, and let $\mathcal{S} = \{\phi_i\}_{i\in I}$ be a set of functions defined in a neighborhood of $\mathcal{P}$ taking their values in $\mathbb{C}\setminus\{0\}$. I is a totally ordered index set. Then $\mathcal{S}$*

H. G. Kaper and M. Garbey (eds.),
Asymptotic and Numerical Methods for Partial Differential Equations with Critical Parameters 255–266.

is an asymptotic scale *at* $\mathcal{P}$ *if and only if*

$$\forall (i,j) \in I \times I, \quad i < j \Longrightarrow \frac{|\phi_j(x)|}{|\phi_i(x)|} \to 0, \quad x \to \mathcal{P} \text{ in } \mathcal{E}.$$

Definition 2: Asymptotic Expansions. *Let* $\mathcal{S} = \{\phi_i\}_{i\in I}$ *be an asymptotic scale. An* asymptotic expansion *in* $\mathcal{S}$ *is a formal sum of the form*

$$a_1(x)\phi_{i_1}(x) + a_2(x)\phi_{i_2}(x) + \cdots + a_k(x)\phi_{i_k}(x) + r(x),$$

where $i_1 < i_2 < \cdots < i_k$; $a_i(x)$ *are functions from* $\mathcal{E}$ *to* $\mathbb{C}$ *whose modulus has a nonzero finite upper limit as* $x \to \mathcal{P}$; *and*

$$|r(x)|/|\phi_{i_k}(x)| \to 0, \quad \text{as } x \to \mathcal{P} \text{ in } \mathcal{E}.$$

When the functions a_i are constants, these expansions are *of Poincaré type*, but more general expansions are necessary, for instance, to express the asymptotic expansion of Bessel J functions. In this paper, "asymptotically" means "as $x \to \mathcal{P}$." In most of our examples, the set $\mathcal{E}$ is $\mathbb{C}$ and $\mathcal{P}$ is either 0 or ∞.

Examples of asymptotic scales at 0 include $\{x^k, k \in \mathbb{Z}\}$, which gives (truncated) formal power series as corresponding asymptotic expansions. A more general scale is $\{x^r, r \in \mathbb{Q}\}$, corresponding to Puiseux expansions. Both of these scales are important in computer algebra, since their manipulation reduces to operations on polynomials in one variable. Our aim in this paper is to algorithmically reduce manipulations of more general asymptotic expansions to manipulations of polynomials, tasks that are routinely handled by computer algebra systems. Currently, most of these systems reduce asymptotic expansions to formal power series manipulation. Maple is an exception, using a model of "hierarchical series" introduced in [5].

In Section 2, we separate computations on asymptotic expansions into operations on the coefficients and operations on the asymptotic scale. Section 3 deals with the former ones. The rest of the paper is concerned with operations on the asymptotic scale and on computing with the most general asymptotic scale. In Section 4, we study the case of a scale that is fixed *a priori*, thus generalizing computations with formal power series. This corresponds to restricting the logarithm and the exponential to expansions of particular valuations. Then, relaxing this limitation in Section 5, we give an algorithm to perform expansions in a scale that is determined dynamically during the computation.

2. Coefficients and Scale

The choice of a finite representation for mathematical objects depends heavily on the operations one wishes to perform on these objects. In the case of asymptotic expansions, the desired operations contain addition, product, inverse, exponential, and logarithm. Henceforth, we restrict ourselves to these operations, which we shall call *elementary*. We now spend some time on the machinery required to perform these elementary operations on asymptotic expansions, decomposing them into operations on the set of coefficients and operations on functions of the asymptotic scale.

Addition

Let A and B be the following asymptotic expansions:

$$\begin{aligned} A &= a_1\phi_{i_1} + a_2\phi_{i_2} + \cdots + a_k\phi_{i_k} + o(\phi_{i_k}), \\ B &= b_1\phi_{j_1} + b_2\phi_{j_2} + \cdots + b_k\phi_{j_p} + o(\phi_{j_p}). \end{aligned} \tag{1}$$

Then one computes $A + B$ as follows:

- Compare ϕ_{i_1} and ϕ_{j_1}.
- If one asymptotically dominates the other, then the corresponding term is output, and the same process is applied to the remainder of the expansions.
- Otherwise, $a_1 + b_1$ is compared to 0, and depending on the result, either one term or zero terms of the result have been found, and the same process is applied to the remainder of the expansions.
- When one of the $o()$ is reached, its argument is compared against the current ϕ_m in the other expansion, and a similar process applies.

The necessary operations are thus: ordering in the scale, addition of coefficients, and comparison of a coefficient to 0. When the coefficients are not constant, we mean by comparison to 0, the determination of the upper limit of the modulus and then comparison of it to 0. This test is necessary to insure that the resulting expansion still obeys the definition of asymptotic expansions.

Product

With the same notation,

- If $A = a_1\phi_{i_1}$, then provided the functions $\phi_{i_1} \cdot \phi_{j_k}$ are in the scale, the result is

  $$a_1 b_1 \phi_{i_1}\phi_{j_1} + \cdots + o(\phi_{i_1}\phi_{j_k}).$$
- Otherwise, expand the product to reduce the computation to the previous case and to addition.

This operation necessitates that the scale is closed by product, and the product of coefficients is needed too.

Inverse

With the same notation as above, in order to compute the expansion $1/A$, the functions $1/\phi_{i_1}$ and the constant function 1 have to belong to the scale, and it is necessary that the lower limit of $|a_1|$ is nonzero. Then the inverse is computed as

$$\frac{1}{a_1}\frac{1}{\phi_{i_1}} \cdot [1 - \psi + \psi^2 - \psi^3 + \cdots],$$

where

$$\psi = \frac{a_2}{a_1}\frac{\phi_{i_2}}{\phi_{i_1}} + \frac{a_3}{a_1}\frac{\phi_{i_3}}{\phi_{i_1}} + \cdots + o(\phi_{i_k}/\phi_{i_1}),$$

whose integral powers can be computed by successive products.

Besides those that have already been mentioned, the necessary operations in this case are taking an inverse in the scale, invertibility, and inverse of the coefficients.

Logarithm

We shall say that A *has valuation 0* when ϕ_{i_1} is the constant function 1. Then, by analogy with formal power series, we define the logarithm of an asymptotic expansion only if this asymptotic expansion has valuation 0, and in this case the result is

$$\log a_1 - \psi + \psi^2/2 - \psi^3/3 + \cdots,$$

with ψ as above. A generalization of this definition of the logarithm will be given in Section 5. With this definition, the new operations concern only the coefficients. These are logarithm and product by rational numbers.

Exponential

Several choices have to be made in order to define an exponential of asymptotic expansions. We take a minimal approach here and give a more general version in Section 5.

If ϕ_{i_1} tends to 0 (which is computed by comparing it to the constant function 1), we say that the expansion has *positive valuation*; the result is

$$1 + A + A^2/2! + A^3/3! + \cdots,$$

which is computed with operations that have already been defined. Once again, by analogy with formal power series, we shall define the exponential in this case only.

With the restrictions we make on logarithm and exponential, we get the important property that, *during the computation, the scale and the set of coefficients are handled independently from each another.* From a computational point of view, this means that one can adopt a modular way of programming, implementing first the part that concerns the coefficients, then the part that concerns the scale, and finally the algorithm just outlined to produce the asymptotic expansions. We have thus implemented a prototype program in Gauss,[1] whose input consists of a domain of coefficients and an asymptotic scale and whose output is the corresponding domain of asymptotic expansions (see [14] for more details).

Putting together all the operations we have found necessary during this description, we can be rather precise concerning the scale and the coefficients:

> *The asymptotic scale has to be an effective ordered Abelian group, and the set of coefficients has to be an effective* $\mathbb{Q}$*-algebra.*

By "effective ordered Abelian group," we mean a group whose law and order can be computed, as well as the inverse with respect to the law. Examples are $\mathbb{Z}$, corresponding to Laurent expansions; $\mathbb{Q}$, corresponding to Puiseux expansions; and $\mathbb{Z}^2$ ordered lexicographically, corresponding to expansions in the scale $\{x^p \log^q x\}$ (for instance).

Similarly, by "effective $\mathbb{Q}$-algebra," we mean an algebra whose operations can be computed, most notably inversibility, inverse, and product by rational numbers. Examples include $\mathbb{Q}$ itself, any algebraic extension of $\mathbb{Q}$ and, for instance, $\mathbb{Q}[\sin(x), \cos(x)]$.

3. Coefficient Domains

Since we are working with symbolic objects, most operations can be performed symbolically. For instance, the inverse of an object A is simply $1/A$. Working this way, the real difficulty has been moved to decision procedures: deciding whether a function of the scale is asymptotically larger than another one, or deciding whether a constant is 0. The latter problem is classically called *the equivalence problem.* It is known to be decidable in any algebraic extension of $\mathbb{Q}$ (see, for instance, [8]), and it is known to be undecidable for sufficiently rich classes of expressions involving a variable [10].

For classes of constants more general than algebraic extensions of $\mathbb{Q}$, the problem is difficult and related to conjectures in number theory, such as Schanuell's conjecture. Some heuristics are known (see [11] for a review of these problems), and we shall assume here the existence of an "oracle" that

[1] Gauss is a language developed by Michael Monagan [9] in Maple, which implements domains and categories *à la* Axiom.

is able to decide whether a constant is 0 or not in the class of constants obtained by closing $\mathbb{Q}$ under exponential and logarithm. In practice, such an oracle can often be obtained by a numerical evaluation routine and a given ϵ threshold.

Another coefficient domain of use in practice is $\mathbb{Q}[\sin x, \cos x]$, where equivalence is decidable. One can even replace $\mathbb{Q}$ by any algebraic extension of $\mathbb{Q}$.

4. Effective Scales

In this section we consider effective ordered groups given by a finite set of generators and the required operations. We show that they form the basis for handling very general asymptotic expansions purely automatically.

To fix the ideas, here are several examples of effective ordered groups:

- $\{x^\alpha; \alpha \in \mathbb{Q}\}$, which corresponds to Puiseux series;
- $\{x^r \log^k x; r \in \mathbb{Q}[\alpha], k \in \mathbb{N}\}$, which is the scale necessary to handle the expansions of singular solutions of linear differential equations in the neighborhood of a regular singular point;
- more complicated scales, for example, one involving Euler's Γ function:
$$\{\Gamma^p(\Gamma(x)) \exp(qe^x)\Gamma^r(x)x^s \log^t x; (p, q, r, s, t) \in \mathbb{Q}^5\}.$$

In each of these examples, the problem of comparing asymptotically two monomials in the scale (two elements of the group) is reduced to a lexicographic comparison of exponents. This will be the case in general.

4.1. SETS OF EXPONENTS

If we wish to handle scales that are as general as possible, the set of exponents must be made as large as possible. Ideally, one would wish to work with $\mathbb{R}$, and we have to content ourselves with an effective ordered subgroup of $\mathbb{R}$.

Algorithms exist that permit computations in real algebraic extensions of $\mathbb{Q}$ [2]. Like in the constant case, more general, ordered, real, constant structures are not yet completely effective and are related to long-standing conjectures in number theory. We shall therefore once again assume the existence of an oracle deciding the comparison of exp-log real constants, and we shall denote by $\mathcal{E}$ any ordered group given with an implementation.

4.2. FINITELY GENERATED SCALES

The following remark makes possible very general automatic computations. Its simplest special case is formal power series.

Let $\{\phi_1, \ldots, \phi_k\}$ be a finite set of functions tending to infinity as $x \to \mathcal{P}$ such that

$$\forall i \in \{1, \ldots, k-1\}, \quad \log|\phi_{i+1}(x)| = o(\log|\phi_i(x)|), \quad \text{as } x \to \mathcal{P}.$$

Then

$$\{\phi_1^{\alpha_1}\phi_2^{\alpha_2}\cdots\phi_k^{\alpha_k}, \alpha_i \in \mathcal{E}\}$$

is an effective scale. It is clearly a group, and the order is computed by lexicographical comparison of the exponents.

An important special case is as follows: for any polynomial $P \in \mathbb{Q}[\alpha][X^{p/q}]$ with α an algebraic number and p and q integers, the scale given by the generators $\{\exp(P(x)), x, \log x\}$, which is sufficient to express asymptotic expansions of singular solutions of linear differential equations about an irregular singular point.

4.3. A GENERAL EXP-LOG SCALE

To simplify notations, suppose $\mathcal{P} = \infty$. As usual in this context, we denote by $e_n(x)$ the nth iterate of the exponential function, and likewise $l_n(x)$ for the logarithm. If $n < 0$ we define e_n to be l_{-n} and vice versa. These functions occur very frequently in asymptotic expansions. It is clear that any finite set of them can be handled as above. But more useful is to encode e_n as X^n and l_n as X^{-n}, and then use a group morphism from the scale into the additive group of Laurent polynomials in one variable. In other words, the monomial

$$e_{n_1}^{\alpha_1}(x)e_{n_2}^{\alpha_2}(x)\cdots e_{n_p}^{\alpha_p}(x)l_{m_1}^{\beta_1}(x)\cdots l_{m_k}^{\beta_k}(x)$$

is encoded as

$$\alpha_1 X^{n_1} + \cdots + \beta_k X^{-m_k}.$$

The order between two such polynomials is obtained by computing the sign of the leading coefficient of their difference.

This scale is very well suited to an implementation in computer algebra systems, whose polynomials are generally carefully implemented, as this is crucial to their overall efficiency.

5. Dynamical Scales

With the restrictions we have made on the definition of the logarithm and the exponential in Section 2, there is no way the scale can be extended during the computation. We now consider defining the exponential function of asymptotic expansions in some cases when the argument does not have a positive valuation. Then independence between the algebra of coefficients and the scale no longer holds, and the scale gets richer as the computation proceeds. Adding a few restrictions, we show how such an extended algorithm can be handled.

5.1. EXTENSION STEP

To control the relation between the scale and the algebra of coefficients, we restrict ourselves to cases where the algebra of coefficients is strictly ordered, so that it can also play the role of a set of exponents. A good example of an effective ordered $\mathbb{Q}$-algebra of coefficients is given by any real algebraic extension of $\mathbb{Q}$.

We now generalize the definition of the exponential of Section 2. Consider again the asymptotic expansion A defined in equation (1). If there exists $j \in \{1, \ldots, k\}$ such that ϕ_{i_j} has positive valuation, define ψ as

$$\psi = a_j \phi_{i_j} + \cdots + a_k \phi_{i_k} + o(\phi_{i_k}),$$

and extend the scale with the function

$$\Phi = \exp[a_1 \phi_{i_1} + \cdots + a_{j-1} \phi_{i_{j-1}}]. \tag{2}$$

The result is given by

$$\Phi \cdot (1 + \psi + \psi^2/2! + \psi^3/3! + \cdots),$$

this last product being computed by the same method as before. The remainder of this section addresses the extension of the scale.

In some cases, it may happen that the function Φ defined above already belongs to the scale. However, a famous theorem [4] states that, given any asymptotic scale, it is possible to construct a function that is not asymptotically equivalent to any function of the scale. This means that there is no "universal" asymptotic scale and that we probably cannot avoid the problem of extending the scale.

It can be seen that the only extra operation needed to handle this extension is the exponential in the ordered group of the scale. Once again, the problem is in the decision procedure. We have to find an ordered group closed under exponential, where the order is computable.

5.2. USING SHACKELL'S ALGORITHM

The class of exp-log functions (functions whose expression is obtained by repeated application of logarithm, exponential, and field operations starting from one variable x and rational numbers, with the restriction that any subexpression must be real for sufficiently large x) was studied from an asymptotic point of view in [7]. However, it was only in [3] that the comparison problem for exp-log functions was shown to be reducible to constant comparisons, and in [15] that an effective algorithm was produced to perform this reduction.

Although no complexity study of this comparison algorithm has been done, the algorithm is rather expensive in practice. We now describe a way to minimize the number of calls to this algorithm. The idea is to compute an extension of the exp-log scale we have described in Section 4, inserting rational powers of the variable X in the Laurent polynomials, so that eventually we work with Puiseux polynomials.

Suppose then that we start from a scale encoded by $\mathcal{S} = \mathbb{Z} \cup \{r_1, \ldots, r_k\}$ corresponding to a scale based on $\mathcal{B} = \{e_n(x), l_n(x); n \in \mathbb{N}\} \cup \{\psi_1, \ldots, \psi_k\}$, where each ψ_i is represented by X^{r_i} with $r_i \in \mathbb{Q} \setminus \mathbb{Z}$, such that if X^u and X^v are the representations of two functions ϕ_u and ϕ_v, then

$$u < v \Longrightarrow \log|\phi_u(x)| = o(\log|\phi_v(x)|), \quad \text{as } x \to \mathcal{P}. \tag{3}$$

Although the representation of the scale as an additive group of Puiseux polynomials is efficient computationally, to simplify the notation, we shall define e_r to be the member of $\mathcal{B}$ whose representative in $\mathcal{S}$ is r, and we shall denote the group of Puiseux polynomials multiplicatively. Thus a member of the scale is a function

$$e_{r_1}^{\alpha_1}(x) \cdot e_{r_2}^{\alpha_2}(x) \cdots e_{r_k}^{\alpha_k}(x).$$

When a new function Φ is added to the scale by computing the exponential of an asymptotic expansion of nonpositive valuation as in Equation (2), the algorithm first expands Equation (2) and associates a monomial to each $\exp(a_p\phi_{i_p})$, creating it if it does not belong to $\mathcal{B}$ yet. From Equation (3), this will be done by comparing logarithms of functions that are already in the scale to the logarithm of the function whose adjunction we are considering, this logarithm being simply $a_p\phi_{i_p}$, which is already a member of the scale. The following operations are performed:

1. From the representation of ϕ_{i_p} as a Puiseux polynomial
   $$e_{r_1}^{\alpha_1}(x) \cdot e_{r_2}^{\alpha_2}(x) \cdots e_{r_k}^{\alpha_k}(x),$$
   with $r_1 > r_2 > \cdots > r_k$ and $\alpha_1 > 0$, it is not difficult to see that a representation of $\exp(\phi_{i_p})$ as X^R must satisfy
   $$\lfloor r_1 \rfloor \leqslant R \leqslant \lceil r_1 \rceil + 1,$$
   these inequalities being strict when r_1 is not an integer.
2. Using Shackell's algorithm and a dichotomy among the *finite* set of rational numbers belonging to $(\lfloor r_1 \rfloor, \lceil r_1 \rceil + 1) \cap \mathcal{S}$, we reach a stage where either we have found two rational numbers r_m and r_{m+1} in $\mathcal{S}$ such that $r_m < R < r_{m+1}$, or we have found an index $r_m \in \mathcal{S}$ such that
   $$\lim_{x \to \mathcal{P}} \frac{\phi_{i_p}(x)}{\log \phi_{i_m}(x)} = c \in \mathbb{R} \setminus \{0\}. \tag{4}$$

3. In the first case, a new index $R = (r_m + r_{m+1})/2$ is inserted in $\mathcal{S}$; the corresponding function $\exp(\phi_{i_p})$ enters $\mathcal{B}$ and is represented as e_R.
4. In the second case, we claim that $\exp(\phi_{i_p})$ is not merely *equivalent* to $\phi_{i_m}^c$, but equal to it, and that $c = 1$. Then the representation of $\exp(\phi_{i_p})$ is e_{r_m}.

We sketch the proof of our last claim, which is an inductive proof based on the algorithm we have just outlined, and construct along the way $\mathcal{B}$. In Equation (4), since ϕ_{i_m} is represented by some r_m in $\mathcal{S}$, it has one of the types e_n or $\exp(U)$ with U a function of the scale. In both cases, this implies that $\log \phi_{i_m}$ is equal to a monomial

$$e_{i_1}^{\beta_1} e_{i_2}^{\beta_2} \cdots e_{i_s}^{\beta_s},$$

with the functions $e_{i_j} \in \mathcal{B}$. On the other hand, ϕ_{i_p} belonging to the scale, it is also equal to some monomial

$$e_{j_1}^{\alpha_1} e_{j_2}^{\alpha_2} \cdots e_{j_k}^{\beta_k},$$

with the functions e_{j_l} belonging to $\mathcal{B}$. From the property of $\mathcal{B}$ given in Equation (3), we deduce that to satisfy (4), these two monomials have to be equal.

Computation of the Logarithm. The computation of the logarithm has already been alluded to above. We can generalize the definition of the logarithm given in Section 2 to accommodate any asymptotic expansion. Consider the asymptotic expansion A defined in Equation (1). We can now define the logarithm in the case when the value of A is not zero. Provided we can compute $\log \phi_{i_1}$,

$$\log A = \log \phi_{i_1} + \log a_1 - \psi + \psi^2/2 - \psi^3/3 + \cdots,$$

with ψ as before. The computation of $\log \phi_{i_1}$ is straightforward in this scale. The representation of ϕ_{i_1} as a Puiseux polynomial

$$e_{r_1}^{\alpha_1}(x) \cdot e_{r_2}^{\alpha_2}(x) \cdots e_{r_k}^{\alpha_k}(x)$$

gives for the logarithm

$$\alpha_1 \log(e_{r_1}) + \cdots + \alpha_k \log(e_{r_k}),$$

so that the problem is reduced to computing logarithms of elements of $\mathcal{B}$. But this is evident since an element of $\mathcal{B}$ either is of the type e_p with $p \in \mathbb{Z}$ or has been constructed by a former exponentiation of a function of the scale.

Example. We show how our program finds the asymptotic expansion of

$$\log\left(x + \frac{2e^{x \log^{3/2} x}}{x^x}\right), \qquad \text{as } x \to \infty.$$

1. x is first translated into $e_0(x)$.
2. Then the second summand is considered, and the term $\exp(x\log^{3/2}x)$ is encountered. This is rewritten $\exp[e_0(x)e_{-1}^{3/2}(x)]$, which must lie between $e_1(x)$ and $e_2(x)$. As a consequence, this function enters the scale as $e_{3/2}(x)$.
3. The monomial $x^x = \exp(x\log x)$ is then handled. It is rewritten as $\exp[e_0(x)e_{-1}(x)]$. Thus it also has to lie between $e_1(x)$ and $e_2(x)$. Shackell's algorithm then helps, deciding that this function grows faster than the previous one, and this function enters the scale as $e_{5/4}(x)$.
4. We next have to compute $\log(2e_{3/2}(x)/e_{5/4}(x)+e_0(x))$, which is
$$\log[e_{3/2}(x)]-\log[e_{5/4}(x)]+\log 2-\sum_k\frac{(-1)^k}{k}\left(\frac{e_0(x)e_{5/4}(x)}{2e_{3/2}(x)}\right)^k.$$
The result is then obtained by recovering from their definitions the values of $e_{5/4}(x)$ and $e_{3/2}(x)$.

This is the actual output of our Maple program:

```
> gdev(log(x+2*exp(x*log(x)^(3/2))/x^x),x=infinity,5);
              3/2                                    x exp(x ln(x))
   (x ln(x)      ) + (- x ln(x)) + (ln(2)) + (1/2 ---------------)
                                                            3/2
                                                 exp(x ln(x)   )

                 2            2                 3    x 3
                x  exp(x ln(x))                x   (x )
      + (- 1/8 -----------------) + (O(----------------))
                          3/2 2                    3/2 3
                exp(x ln(x)    )       exp(x ln(x)    )
```

5.3. OTHER SCALES

In a yet-unpublished work, G. Gonnet and D. Gruntz [6] have devised an algorithm that performs asymptotic comparisons in very general scales. In one of their examples, they compare automatically

$$\exp(\exp(\exp(\Psi(\Psi(\Psi(x)))))) \text{ and } x,$$

where Ψ is the logarithmic derivative of Euler's Γ function. Such a comparison algorithm permits the construction of a very general asymptotic scale. However, it is not yet clear which class is handled by this new algorithm.

Another class of scales can be obtained by extending the exp-log scale we have presented, appealing to the theory of Hardy fields. Briefly, Hardy fields provide an algebraic setting to express and prove results about formal asymptotic expansions. Relevant work can be found in [1, 12, 13, and 16].

6. Conclusion

In this paper we have presented algorithms that perform manipulations of asymptotic expansions. The first step is to help make the program modular by reducing the operations on asymptotic expansions to operations on the set of coefficients and operations on the scale. The second step is to get as general an asymptotic scale as possible. Such a scale has been presented for exp-log functions. Its important features are that it is created dynamically and that it necessitates a minimal amount of comparisons (a costly operation that is handled by Shackell's algorithm).

Acknowledgements

This work was supported in part by the ESPRIT III Basic Research Action Programme of the European Community under contract ALCOM II (#7141).

References

[1] Bourbaki, N. (1961) *Éléments de Mathématiques*, Chap. V: Fonctions d'une variable réelle (appendice), 2nd ed., Hermann, Paris, 36–55.
[2] Coste, M., and Roy, M. F. (1988) 'Thom's lemma, the coding of real algebraic numbers and the topology of semi-algebraic sets', *J. Symbolic Computation*, **5**, 121–129.
[3] Dahn, B. I., and Göring, P. (1986) 'Note on exponential-logarithmic terms', *Fundamenta Mathematicae*, **127**, 45–50.
[4] du Bois-Reymond, P. (1875) 'Ueber asymptotische Werthe, infinitäre Approximationen und infinitäre Auflösung von Gleichungen', *Mathematische Annalen*, **VIII**, 363–414, 574–576.
[5] Geddes, K. O., and Gonnet, G. H. (1989) 'A new algorithm for computing symbolic limits using hierarchical series', in P. Gianni (ed.), *Symbolic and Algebraic Computation*, *Lecture Notes in Computer Science*, Vol. 358, Springer-Verlag, New York, 490–495.
[6] Gonnet, G., and Gruntz, D. (1992). Personal communication.
[7] Hardy, G. H. (1910) 'Orders of infinity', *Cambridge Tracts in Mathematics*, **12**.
[8] Loos, R. (1982) 'Computing in algebraic extensions', in B. Buchberger, G. E. Collins, and R. Loos (eds.), *Computer Algebra. Symbolic and Algebraic Computation, Computing Supplementum*, Vol. 4, Springer-Verlag, New York, 173–188.
[9] Monagan, M. B. (1989) *Signatures + Abstract Data Types = Computer Algebra – Intermediate Expression Swell*, Ph.D. thesis, University of Waterloo.
[10] Richardson, D. (1968) 'Some undecidable problems involving elementary functions of a real variable', *J. Symbolic Logic*, **33**, no. 4, 514–520.
[11] Richardson, D. (1992) 'The elementary constant problem', in P. Wang (ed.), *ISSAC'92*, ACM Press, to appear.
[12] Rosenlicht, M. (1983) 'Hardy fields', *J. Math. Anal. Appls.*, **93**, 297–311.
[13] Rosenlicht, M. (1987) 'Growth properties of functions in Hardy fields', *Trans. AMS*, **299**, no. 1, 261–272.
[14] Salvy, B. (1991) 'Asymptotique automatique et fonctions génératrices', Ph.D. thesis, École Polytechnique.
[15] Shackell, J. (1990) 'Growth estimates for exp-log functions', *J. Symbolic Computation*, **10**, 611–632.
[16] Shackell, J. (1991). 'Rosenlicht fields', *Trans. AMS*, to appear.

Part 6: Numerical Methods, Algorithms, and Architectures

DOMAIN DECOMPOSITION BY THE MORTAR ELEMENT METHOD

C. Bernardi and Y. Maday
Laboratoire d'Analyse Numérique
CNRS and Université Pierre et Marie Curie
4 place Jussieu
75252 Paris Cedex 05
France

and

A. T. Patera
Department of Mechanical Engineering
Massachusetts Institute of Technology
77 Massachusetts Avenue
Cambridge MA 02139

ABSTRACT. The paper reviews recent results concerning the mortar element method, which allows for coupling variational discretizations of different types on nonoverlapping subdomains. The basic ideas and proofs are recalled on a model problem, and new extensions are presented.

1. Introduction

The mortar element method was born in 1987 from the difficulty of handling complex geometries with spectral discretizations. It has been implemented on plane-bounded domains in two different ways: using spectral elements with no restrictions on the decomposition into subdomains or the degrees of polynomials (see [2, 12, 21]), or coupling spectral methods with finite elements (see [10, 18]). However, its range is much larger, since the mortar technique allows for coupling any discretization of variational type on each subdomain of a partition without overlapping, with no restriction on the decomposition. It leads to (generally nonconforming) approximations of elliptic or parabolic equations, and the error estimate is the sum of terms that involve only the local regularity of the exact solution associated with the local discretization parameters.

H. G. Kaper and M. Garbey (eds.),
Asymptotic and Numerical Methods for Partial Differential Equations with Critical Parameters 269–286.

The discretization relies on the variational formulation of the continuous problem: the discrete problem is obtained by the Galerkin method with or without numerical integration; that is, the variational formulation is applied with the exact space of functions replaced by a finite-dimensional discrete space, and, when necessary, the integrals are replaced by appropriate quadrature formulas. As a consequence, the most important part of the discretization is the construction of the discrete space: independent finite-dimensional spaces are given on each subdomain, and the so-called mortar function is defined on the intersection of the boundaries of the subdomains in order to enforce matching conditions. One of the main features of the method is to be in most cases nonconforming, in the sense that the discrete space is not included in the exact one. The construction of the discrete space is described in Section 2. The discrete problem is written in the simple case of a Poisson equation, and the analysis is performed in this case, only for the sake of brevity, since there is no restriction for applying this method to any elliptic second-order problem.

Once the discrete problem is written, the first question is the existence and uniqueness of the solution. It relies on the ellipticity of the associated bilinear form and, because of the nonconformity of the method, this cannot be derived from the ellipticity of the exact problem in a straightforward way. Hence, Section 3 is devoted to the proof of the ellipticity property, from which the well-posedness of the problem follows.

The next and final step of the numerical analysis is the derivation of error estimates. By Strang's lemma, the distance between the exact solution and the discrete one is bounded by a constant times the sum of the approximation error (i.e., the distance of the exact solution to the discrete space) and the consistency error resulting from the nonconformity of the method. These two terms are successively studied in Section 4, and the corresponding error estimates are written.

Because of the variational form, an appropriate choice of the basis of the discrete space allows for writing immediately the discrete problem as a global linear system, which is symmetric in most cases. Then, the system can be solved, for example, iteratively on subdomains after an appropriate preconditioning, as will be explained in Section 5.

The mortar element method has a number of natural extensions and important applications. The already existing results will be described in Section 6: they concern the discretization of the incompressible Stokes (and Navier–Stokes) equations, the treatment of domains with a moving part by "sliding" grids, and the extension to three-dimensional domains. Some other applications are currently under consideration—for instance, the discretization of fourth-order problems with a "double" mortar, the use of domain decomposition for adaptive refinements, and coupling operators. Some details will be given in Section 7.

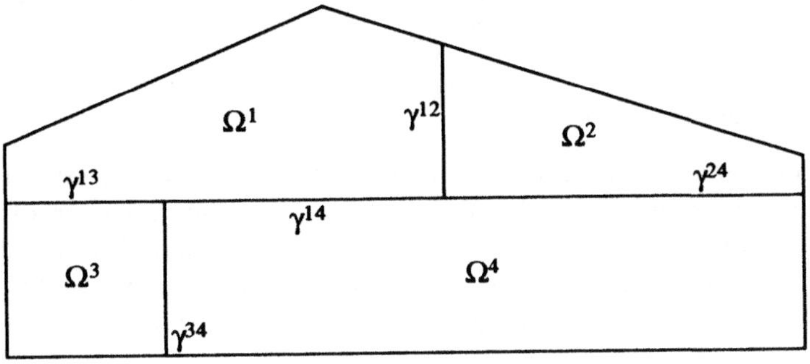

Fig. 1. **Geometry of the domain**

2. The Discretization

Three steps are necessary to present the discretization: first, some anatomy concerning the geometry of the domain is described; second, the discrete space is constructed—this is the crucial point; and third, the discrete problem is formulated.

2.1. THE GEOMETRY

Let Ω denote a bounded open set in $\mathbb{R}^2$, which we assume to be polygonal only for the sake of simplicity (the mortar element method has been used for handling curved boundaries; see, for instance, [3]. However, the description is too technical to be written here). We assume that there exists a partition of the domain Ω into nonoverlapping open polygons Ω^k, $1 \le k \le K$, that is,

$$\overline{\Omega} = \bigcup_{k=1}^{K} \overline{\Omega}^k \qquad \text{and} \qquad \Omega^k \cap \Omega^\ell = \emptyset, \quad 1 \le k < \ell \le K, \tag{1}$$

such that $\partial\Omega$ is the union of entire edges of subdomains. Note that there is no restriction on the geometry of the decomposition: the decomposition is said to be *conforming* if the intersection of two subdomains is empty or a vertex or a whole edge of both of them, but this restriction is not necessary for the mortar technique. We denote by $\Gamma^{k,j}$, $1 \le j \le J(k)$, the edges of Ω^k and by $\gamma^{k\ell}$, $1 \le k \ne \ell \le K$, the open segment such that $\overline{\gamma}^{k\ell} = \partial\Omega^k \cap \partial\Omega^\ell$. We also introduce the set $\mathcal{V}$ of all vertices of the Ω^k, $1 \le k \le K$, and the "skeleton" $\mathcal{S} = \cup_{k=1}^{K} \cup_{j=1}^{J(k)} \Gamma^{k,j}$. A representative example of geometry is presented in Fig. 1.

The basic point is to decompose the skeleton as the union of disjoint segments $\bar{\gamma}^m$, where m runs through a finite set M and each γ^m is an edge

$\Gamma^{k(m),j(m)}$ of an element $\Omega^{k(m)}$ (for the sake of generality, we assume that a polygon can have "virtual" vertices with angle π, i.e., consecutive edges on the same line). The segments γ^m are called *mortars*.

2.2. THE DISCRETE SPACE

Let X^* be a space of functions defined on Ω. We assume that the problem is set in a closed subspace X of X^* (essentially the subspace of functions in X^* that satisfy some essential boundary conditions). We denote by X^k, $1 \leq k \leq K$, the space of restrictions of functions of X^* to Ω^k. In most cases, there exist trace mappings T^k, $1 \leq k \leq K$, such that the following equivalence holds:

$$v \in X^* \iff \{ \, v_{|\Omega^k} \in X^k, \quad 1 \leq k \leq K,$$

$$T^k v_{|\gamma^{k\ell}} = T^\ell v_{|\gamma^{k\ell}}, \quad 1 \leq k < \ell \leq K.$$

For second-order problems, X is generally a closed subspace of $X^* = H^1(\Omega)$, for instance, $H_0^1(\Omega)$, and T^k is the standard first trace operator on Ω^k.

Now, let $\delta = (\delta_1, \ldots, \delta_K)$ be a discretization parameter, and let X_δ^k be a subspace of X^k, $1 \leq k \leq K$. For each j, $1 \leq j \leq K$, we define $W_\delta^{k,j}$ as the image of X_δ^k by T^k restricted to $\Gamma^{k,j}$, and we also introduce a subspace $\tilde{W}_\delta^{k,j}$ of $W_\delta^{k,j}$ with codimension equal to 2. The need for this last space will appear later on. We first give two basic examples.

Example 1: Finite Element Case. Let r_k be a fixed integer ≥ 1. Let $(\mathcal{T}_{h_k})_{h_k}$ be a regular family of triangulations of Ω^k, in the sense that $\overline{\Omega}^k$ is the union of triangles K in $\mathcal{T}_{h_k}$; the diameter of each K in $\mathcal{T}_{h_k}$ is less than h_k; its intersection with another triangle is empty, or one of its vertices; and the ratio of the inscribed circle in K to the diameter of K is bounded independently of h_k. We also assume that the virtual vertices of Ω^k, if any, are vertices of some K in $\mathcal{T}_{h_k}$. Then, setting $\delta_k = h_k$, we define X_δ^k as the space of continuous functions on $\overline{\Omega}^k$ such that their restrictions to any K in $\mathcal{T}_{h_k}$ is a polynomial of total degree $\leq r_k$. We denote by $\mathcal{P}_{r_k}(K)$ the space of such polynomials. Consequently, $W_\delta^{k,j}$ is the space of continuous functions on $\Gamma^{k,j}$ that are polynomial with degree $\leq r_k$ on each $K \cap \Gamma^{k,j}$ for each K in $\mathcal{T}_{h_k}$. We choose $\tilde{W}_\delta^{k,j}$ as the space of such functions which, moreover, have degree $\leq r_k - 1$ on the two $K \cap \overline{\Gamma}^{k,j}$ that contain the extremities of $\Gamma^{k,j}$.

Example 2: Spectral Case. Here, we assume that Ω^k is a rectangle. Setting $\delta_k = N_k^{-1}$ for an integer $N_k \geq 2$, we define X_δ^k as the space $\mathbb{P}_{N_k}(\Omega^k)$ of polynomials with degree $\leq N_k$ with respect to each variable. Thus, $W_\delta^{k,j}$ is the space $\mathbb{P}_{N_k}(\Gamma^{k,j})$ of polynomials with degree $\leq N_k$ on $\Gamma^{k,j}$, and we simply choose $\tilde{W}_\delta^{k,j}$ as the space $\mathbb{P}_{N_k-2}(\Gamma^{k,j})$.

Next, we define the space of *mortar functions* as

$$W_\delta = \{\varphi \text{ continuous on } \mathcal{S};\ \forall m \in M,\ \varphi_{|\gamma^m} \in W_\delta^{k(m),j(m)}\}. \qquad (2)$$

Finally, the discrete space X_δ is the space of functions v_δ such that
(i) for each k, $1 \le k \le K$, the restriction of v_δ to Ω^k belongs to X_δ^k;
(ii) v_δ satisfies the essential boundary conditions that are satisfied by the functions of X;
(iii) there exists a mortar function φ in W_δ such that, for $1 \le k \le K$,

$$\begin{cases} \forall \alpha \text{ vertex of } \Omega^k, & (T^k v_\delta)(\alpha) = \varphi(\alpha), \\ \forall \psi \in \tilde{W}_\delta^{k,j}, & \int_{\Gamma^{k,j}} (T^k v_\delta - \varphi)(\tau)\psi(\tau)\, d\tau = 0, \quad 1 \le j \le J(k). \end{cases} \qquad (3)$$

We conclude this definition with some remarks:
(i) In most cases, the discretization is nonconforming; that is, X_δ is not included in X. For instance, in the spectral case, conformity occurs when the decomposition is conforming and for an appropriate choice of the degrees of the mortar functions. There is never conformity when finite element and spectral discretizations are coupled.
(ii) The last conditions in (3) are of integral type. Indeed, it is proved in [11, 14] that pointwise matching conditions do not lead to optimal error estimates.
(iii) The number of conditions in (3) is equal to the dimension of $W_\delta^{k,j}$ (which justifies the codimension 2 of $\tilde{W}_\delta^{k,j}$). For appropriate choices of $\tilde{W}_\delta^{k,j}$, these conditions define an operator $\tilde{\pi}_\delta^{k,j}$: for any sufficiently smooth function φ on $\Gamma^{k,j}$, $\tilde{\pi}_\delta^{k,j}\varphi$ belongs to $W_\delta^{k,j}$ and satisfies

$$\begin{cases} \forall \alpha \text{ extremity of } \Gamma^{k,j}, & (\tilde{\pi}_\delta^{k,j}\varphi)(\alpha) = \varphi(\alpha), \\ \forall \varphi \in \tilde{W}_\delta^{k,j}, & \int_{\Gamma^{k,j}} (\varphi - \tilde{\pi}_\delta^{k,j}\varphi)(\tau)\psi(\tau)\, d\tau = 0. \end{cases} \qquad (4)$$

The stability and approximation properties of the $\tilde{\pi}_\delta^{k,j}$ will be important in the derivation of error estimates.

2.3. THE DISCRETE PROBLEM

We have to approximate a second-order partial differential equation that admits the variational formulation: *find u in X such that*

$$\forall v \in X, \quad \ell(u, v) = < f, v >, \qquad (5)$$

where $\ell(.,.)$ is a bilinear form on $X \times X$, f is given in the dual space of X' of X, and $< .,. >$ stands for the duality pairing between X' and X.

Then, for each value of δ, the discrete problem is written: *find u_δ in X_δ such that*

$$\forall v_\delta \in X_\delta, \quad \ell_\delta(u_\delta, v_\delta) = < f, v_\delta >_\delta . \tag{6}$$

The bilinear forms $\ell_\delta(.,.)$ and $< .,. >_\delta$ are constructed by writing the integrals over Ω which appear in $\ell(.,.)$ and $< .,. >$, respectively, as sums of integrals on the Ω^k and eventually replacing these integrals by appropriate quadrature formulas.

In the following sections, the numerical analysis will be performed in the simple case of a Poisson equation. So, setting $X^* = H^1(\Omega)$ and $X = H_0^1(\Omega)$, we consider the following variational problem, where f is given in $L^2(\Omega)$: *find u in $H_0^1(\Omega)$ such that*

$$\forall v \in H_0^1(\Omega), \quad \int_\Omega \mathbf{grad}\, u \,.\, \mathbf{grad}\, v \, dx = \int_\Omega f(\mathbf{x}) v(\mathbf{x}) \, d\mathbf{x}. \tag{7}$$

Then, with each Ω^k, we associate the space X_δ^k of Example 1 or 2, and we can write the discrete problem: *find u_δ in X_δ such that*

$$\forall v_\delta \in X_\delta, \quad \sum_{k=1}^{K} \int_{\Omega^k} \mathbf{grad}\, u_\delta \,.\, \mathbf{grad}\, v_\delta \, d\mathbf{x} = \sum_{k=1}^{K} \int_{\Omega^k} f(\mathbf{x}) v_\delta(\mathbf{x}) \, d\mathbf{x}. \tag{8}$$

Here, for the sake of brevity, we do not use numerical integration. It is straightforward to extend this discretization to more general second-order problems.

3. Well-Posedness

Because of the variational form of the problem, its well-posedness is a direct consequence of the ellipticity of the bilinear form $\ell_\delta(.,.)$ on X_δ, which is not trivial since X_δ is not contained in X. The complete proof is given in the Appendix of [13]; only a sketch of the proof is written here.

In the case of problem (8), the natural norm on the space X_δ is a broken $H^1(\Omega)$-norm defined by

$$\|v_\delta\|_{H^1_*(\Omega)} = \Big(\sum_{k=1}^{K} \|v_\delta\|^2_{H^1(\Omega^k)} \Big)^{\frac{1}{2}}. \tag{9}$$

Thus, the ellipticity property reduces to the inequality

$$\forall v_\delta \in X_\delta, \quad \sum_{k=1}^{K} |v_\delta|^2_{H^1(\Omega^k)} \geq c \sum_{k=1}^{K} \|v_\delta\|^2_{H^1(\Omega^k)}, \tag{10}$$

for a positive constant c. Let us fix a value δ of the parameter and assume that (10) does not hold; then, there exists a sequence $(v_{\delta n})_n$ in X_δ such that

$$\sum_{k=1}^{K} |v_{\delta n}|^2_{H^1(\Omega^k)} \leq \frac{1}{n} \quad \text{and} \quad \sum_{k=1}^{K} \|v_{\delta n}\|^2_{H^1(\Omega^k)} = 1. \tag{11}$$

There exists a subsequence converging towards a function v in X_δ. From the previous inequality, it is readily checked that v is constant on each Ω^k, and the conditions at the corners of the domains that appear in (3) yield that constants are equal. Hence, v is constant on Ω; and, since it belongs to $H^1_0(\Omega)$, it is equal to 0. This is not compatible with (11). As a consequence, property (10) holds for each value of δ.

However, for stability and error estimates, it is necessary to know whether the constant c in (10) can be bounded independently of δ. The idea here is to introduce a Banach space $\tilde{X}$ such that (10) holds for all functions v in $\tilde{X}$ and that contains (nearly) all the discrete spaces X_δ. The definition of this space is linked to the geometry of the decomposition in the following way: with each $\Gamma^{k,j}$, we associate a first space $\tilde{W}^{k,j}$ (1) in the finite element case, made of continuous functions on $\Gamma^{k,j}$ that are affine on each interval delimited by two consecutive points of $\mathcal{V} \cap \overline{\Gamma}^{k,j}$ (with a slight modification if $r_k = 1$); (2) in the spectral case, made of polynomials in $\mathbb{P}_{L-2}(\Gamma^{k,j})$, where L is the maximal number of points of $\mathcal{V}$ contained in the same $\Gamma^{k,j}$. And, finally, we introduce the space $\tilde{X}$ of functions v on Ω vanishing on $\delta\Omega$ such that their restriction to Ω^k belongs to $H^1(\Omega^k)$ and which satisfy

$$\forall \psi \in \tilde{W}^{k,j}, \quad \int_{\Gamma^{k,j}} (v_{|\Omega^k} - \varphi)(\tau)\psi(\tau)\, d\tau = 0, \tag{12}$$

where the mortar function φ is defined on each γ^m, $m \in M$, as the trace of $v_{|\Omega^{k(m)}}$. Thus, exactly the same arguments as in the preceding lines for X_δ prove that

$$\forall v \in \tilde{X}, \quad \sum_{k=1}^{K} |v|^2_{H^1(\Omega^k)} \geq c \sum_{k=1}^{K} \|v\|^2_{H^1(\Omega^k)}, \tag{13}$$

and the constant of (10) can be bounded independently of δ when X_δ is contained in $\tilde{X}$. The next theorem is now a simple application of the Lax–Milgram lemma.

Theorem 1. *Let us assume that, for each k, $1 \leq k \leq K$:*
(1) in the finite element case, the points of $\mathcal{V} \cap \Gamma^{k,j}$ are nodes of the triangulation $\mathcal{T}_{h_k}$;
(2) in the spectral case, N_k is $\geq L$.
Then, for any f in $L^2(\Omega)$, problem (8) has a unique solution u_δ in X_δ.

Moreover, this solution satisfies

$$\Big(\sum_{k=1}^{K} \|u_\delta\|^2_{H^1(\Omega^k)}\Big)^{\frac{1}{2}} \le c\,\|f\|_{L^2(\Omega)} \tag{14}$$

for a constant c independent of δ.

Of course, the ellipticity property is also the basis of the following error estimates.

4. Error Estimates

From Strang's lemma and if the assumptions of Theorem 1 hold, we derive from (9) and (10) the abstract estimate:

$$\|u - u_\delta\|_{H^1_*(\Omega)}$$

$$\le c\Big(\inf_{v_\delta \in X_\delta} \|u - v_\delta\|_{H^1_*(\Omega)} + \sup_{w_\delta \in X_\delta} \frac{\sum_{1 \le k < \ell \le K} \int_{\gamma^{k\ell}} (\frac{\partial u}{\partial n})(\tau)[w_\delta(\tau)]\,d\tau}{\|w_\delta\|_{H^1_*(\Omega)}}\Big), \tag{15}$$

where $[w_\delta]$ denotes the jump $w_{\delta\,|\Omega^\ell} - w_{\delta\,|^k}$ and $\frac{\partial}{\partial n}$ is the normal derivative operator on $\gamma^{k\ell}$ with the appropriate direction. The first term in the right-hand side is the approximation error; the second one is the consistency error. We shall establish bounds for both of them.

We begin with the consistency error. For any function w_δ in X_δ, if χ denotes the mortar function associated with w_δ, we have

$$\sum_{1 \le k < \ell \le K} \int_{\gamma^{k\ell}} (\frac{\partial u}{\partial n})(\tau)[w_\delta(\tau)]\,d\tau = \sum_{\Gamma^{k,j}} \int_{\Gamma^{k,j}} (\frac{\partial u}{\partial n})(\tau)(w_\delta - \chi)(\tau)\,d\tau,$$

where the last sum is taken on all the edges $\Gamma^{k,j}$ that are not contained in $\partial\Omega$. Next, using the integral matching conditions in (3), we derive that, for any $\psi^{k,j}$ in $\tilde{W}^{k,j}$,

$$\sum_{1 \le k < \ell \le K} \int_{\gamma^{k\ell}} (\frac{\partial u}{\partial n})(\tau)[w_\delta(\tau)]\,d\tau = \sum_{\Gamma^{k,j}} \int_{\Gamma^{k,j}} (\frac{\partial u}{\partial n} - \psi^{k,j})(w_\delta - \chi)(\tau)\,d\tau$$

$$\le \sum_{k=1}^{K} \sum_{j=1}^{J(k)} \|\frac{\partial u}{\partial n} - \psi^{k,j}\|_{(H^{\frac{1}{2}}(\Gamma^{k,j}))'} (\|w_\delta\|_{H^{\frac{1}{2}}(\Gamma^{k,j})} + \|\chi\|_{H^{\frac{1}{2}}(\Gamma^{k,j})}).$$

Finally, noting that χ on $\gamma^{k\ell}$ coincides either with the trace of $w_{\delta\,|\Omega^k}$ or with the trace of $w_{\delta\,|\Omega^\ell}$, we obtain the following conclusion:

$$\frac{\sum_{1 \le k < \ell \le K} \int_{\gamma^{k\ell}} (\frac{\partial u}{\partial n})(\tau)[w_\delta(\tau)]\,d\tau}{\|w_\delta\|_{H^1_*(\Omega)}}$$

$$\leq c, \sum_{k=1}^{K} \sum_{j=1}^{J(k)} \inf_{\psi^{k,j} \in \tilde{W}^{k,j}} \| \frac{\partial u}{\partial n} - \psi^{k,j} \|_{(H^{\frac{1}{2}}(\Gamma^{k,j}))'}. \tag{16}$$

To estimate the distance to $\tilde{W}^{k,j}$ in $(H^{\frac{1}{2}}(\Gamma^{k,j}))'$, we use the orthogonal projection operator $\pi_\delta^{k,j}$ from $L^2(\Gamma^{k,j})$ onto $\tilde{W}_\delta^{k,j}$. We recall the standard duality estimates, which hold for any function ψ in $H^\sigma(\Gamma^{k,j})$, $\sigma \geq 0$:

$$\|\psi - \pi_\delta^{k,j} \psi\|_{(H^{\frac{1}{2}}(\Gamma^{k,j}))'} \leq c \begin{cases} \delta_k^{\frac{1}{2}+\inf\{\sigma, r_k\}} \|\psi\|_{H^\sigma(\Gamma^{k,j})} & \text{(finite elements)}, \\ \delta_k^{\frac{1}{2}+\sigma} \|\psi\|_{H^\sigma(\Gamma^{k,j})} & \text{(spectral)}. \end{cases} \tag{17}$$

Estimating the approximation error is a bit more technical; it proceeds in three steps.

(1) Let us first introduce, for each k, an approximation v^k of $u_{|\Omega^k}$ in X_δ^k that vanishes on $\partial\Omega^k \cap \partial\Omega$: the most natural choice is to take the Lagrange interpolate of u at the nodes of the finite elements in the finite element case and on the grid of Gauss–Lobatto nodes (i.e., at the points whose coordinates are the (N_k+1)–nodes of a Gauss–Lobatto quadrature formula) in the spectral case. Then, we define v_* on Ω as being equal to v^k on each Ω^k, $1 \leq k \leq K$, so that

$$\|u - v_*\|_{H^1_*(\Omega)} = \Big(\sum_{k=1}^{K} \|u - v^k\|^2_{H^1(\Omega^k)} \Big)^{\frac{1}{2}}. \tag{18}$$

The function v_* is a good approximation of u, but it does not belong to X_δ.

(2) Next, in the spectral case, we enforce the matching conditions at the points of $\mathcal{V}$. With each point α in $\mathcal{V} \cap \Omega^k$ that is not a vertex of Ω^k, we associate a polynomial q_α of minimal degree on Ω^k that is equal to 1 in α and vanishes in any other point of $\mathcal{V} \cap \Omega^k$; we still denote by q_α its extension by 0 on Ω. Next, we set $v_{**} = \sum_\alpha [v_*(\alpha)] q_\alpha$, where $[v_*(\alpha)]$ denotes the jump of v_* in α and the summation is taken on all Ω^k, $1 \leq k \leq K$, and all the points of $\mathcal{V} \cap \Omega^k$ that are not vertices of Ω^k. We have

$$\|v_{**}\|_{H^1_*(\Omega)} \leq c \sum_{k=1}^{K} \sum_{\alpha \in \mathcal{V} \cap \partial\Omega^k} |(u - v_*)(\alpha)|,$$

so that, by the Gagliardo-Nirenberg inequality,

$$\|v_{**}\|_{H^1_*(\Omega)} \leq c \sum_{k=1}^{K} \sum_{j=1}^{J(k)} \|u - v_*\|^{\frac{1}{2}}_{L^2(\Gamma^{k,j})} \|u - v_*\|^{\frac{1}{2}}_{H^1(\Gamma^{k,j})}. \tag{19}$$

In the finite element case, we can take v_{**} equal to 0, because of the hypothesis made in Theorem 1.

(3) Finally, we note from the conditions (3) that the trace on $\Gamma^{k,j}$ of a function w_δ in X_δ with mortar function χ is equal to $\tilde{\pi}_\delta^{k,j}\chi$. Hence, for each m in M, we set

$$v^m = \begin{cases} R_\delta^{k(m),j(m)}\tilde{\pi}_\delta^{k(m),j(m)}[v_* + v_{**}] & \text{on } \Omega^{k(m)}, \\ 0 & \text{elsewhere,} \end{cases} \tag{20}$$

where $R_\delta^{k(m),j(m)}$ is a lifting operator from $W_\delta^{k,j} \cap H_0^1(\Gamma^{k,j})$ into X_δ^k: the trace of $R_\delta^{k(m),j(m)}\varphi$ coincides with φ on $\Gamma^{k,j}$ and vanishes on the other edges of Ω^k. The existence and stability properties of such operators are proven by P. E. Bjørstad and O. B. Widlund [15] in the finite element case when the family of triangulations $(\mathcal{T}_{h_k})_{h_k}$ is uniformly regular on $\Gamma^{k,j}$ (this hypothesis will be made in the following theorem) and in [9] in the spectral case:

$$\forall \varphi \in W_\delta^{k,j} \cap H_0^1(\Gamma^{k,j}), \quad \|R_\delta^{k,j}\varphi\|_{H^1(\Omega^k)} \le c\,\|\varphi\|_{H_{00}^{\frac{1}{2}}(\Gamma^{k,j})}, \tag{21}$$

for a constant c independent of δ. As a consequence,

$$\sum_{m\in M} \|v^m\|_{H_*^1(\Omega)} \le c \sum_{k=1}^{K}\sum_{j=1}^{J(k)} \Big(\|u - (v_* + v_{**})_{|\Omega^k}\|_{H_{00}^{\frac{1}{2}}(\Gamma^{k,j})} + \|(id - \tilde{\pi}_\delta^{k,j})(u - (v_* + v_{**})_{|\Omega^k})\|_{H_{00}^{\frac{1}{2}}(\Gamma^{k,j})}\Big). \tag{22}$$

The approximation properties of the operator $\tilde{\pi}_N^{k,j}$ are given in [13]: for any function ψ in $H^\sigma(\Gamma^{k,j})$ $\sigma \ge 1$,

$$\|\psi - \tilde{\pi}_\delta^{k,j}\psi\|_{H^{\frac{1}{2}}(\Gamma^{k,j})} \le c \begin{cases} \delta_k^{\inf\{\sigma, r_k+1\}-\frac{1}{2}}\,\|\psi\|_{H^\sigma(\Gamma^{k,j})} & \text{(finite elements),} \\ \delta_k^{\sigma-\frac{1}{2}}\,\|\psi\|_{H^\sigma(\Gamma^{k,j})} & \text{(spectral).} \end{cases} \tag{23}$$

As a conclusion, we choose

$$v_\delta = v_* + v_{**} + \sum_{m\in M} v^m. \tag{24}$$

The function v_δ belongs to X_δ, and the distance $\|u - v_\delta\|_{H_*^1(\Omega)}$ can be estimated from the previous inequalities.

Theorem 2. *Let us assume that the hypotheses of Theorem 1 hold and, moreover, in the finite element case that the length of each $K \cap \Gamma^{k,j}$, $K \in \mathcal{T}_{h_k}$, is larger than a constant times h_k. If the following regularity property is satisfied by the solution u of problem* (3),

$$\forall k, 1 \le k \le K, \quad u_{|\Omega^k} \in H^{s_k}(\Omega^k), \tag{25}$$

for real numbers $s_k > \frac{3}{2}$, the following estimate holds between u and the solution u_δ of problem (4):

$$\|u - u_\delta\|_{H^1_*(\Omega)} \leq c \sum_{k=1}^{K} \delta_k^{\inf\{s_k, r_k+1\}-1} \|u_{|\Omega^k}\|_{H^{s_k}(\Omega^k)}, \tag{26}$$

for a constant c independent of δ, with r_k equal to the degree of the polynomials in the finite element case and to $+\infty$ in the spectral case.

As a conclusion, the estimate depends only on the local regularity of the exact solution on each subdomain. Thus, for instance, the method allows for handling the singularities of the solution that come from the geometry of the domain in a very simple way.

5. Implementation

In the simple case that we study, the discrete problem (8) can be written as a linear system $\tilde{A}\tilde{U} = \tilde{F}$ in a straightforward way: the unknowns in the vector U are the values of u_δ at the nodes in the interior of each domain Ω^k, $1 \leq k \leq K$ (the nodes of the triangulation in the finite element case, the Gauss–Lobatto nodes in the spectral case) and the values of the associated mortar function φ at the nodes of $\mathcal{S}$. The matrix $\tilde{A}$ is symmetric and positive definite, so that the system is usually solved by a conjugate gradient algorithm (we refer to P. LeTallec and T. Sassi [20] for the use of Lagrangian-type algorithms for the mortar element discretization).

Note that the values of $u_{\delta\,|\Omega^k}$ on $\partial\Omega^k$ can be easily computed from the values of φ by the formula $u_{\delta\,|\Gamma^{k,j}} = \tilde{\pi}_\delta^{k,j}\varphi$, $1 \leq j \leq J(k)$. Hence, there exists a rectangular matrix Q such that the vector $Q\tilde{U}$ is made up of the vectors U^k of the values of $u_{\delta\,|\Omega^k}$ on all nodes of $\overline{\Omega}^k$, and the system can be written equivalently as $Q^T A Q\tilde{U} = Q^T F$. The corresponding matrix A is block-diagonal, each block A^k being the matrix associated with the discrete Laplace equation with Neumann conditions on Ω^k. Hence, the procedure for residual evaluation at each step of the conjugate gradient is the following: at the step ℓ, being given an intermediate vector $\tilde{W}_\ell$ of values at the nodes of $\mathcal{S}$, we compute, thanks to the matching conditions (3), the vectors W_ℓ^k of values at the (internal and boundary) nodes of $\overline{\Omega}^k$. Then, we evaluate $A^k W_\ell^k$ in each Ω^k, from which we derive a vector $W_{\ell+1}^k$ of values at the nodes of $\overline{\Omega}^k$. Finally, we compute $\tilde{W}_{\ell+1}$ by multiplying by Q^T. It can be observed that the largest part of the computation is performed separately on each Ω^k, which allows for an easy parallelization of the algorithm (see [19]).

The matrices A^k that are derived from the finite-element discretization are sparse, but those that are derived from the spectral discretization are full and ill conditioned (recall that, in our problem, the condition number is of the order of N_k^3). Even if these matrices are smaller than the finite element ones for equal accuracy, preconditioning the linear system is necessary. Two preconditioners currently are used: (i) a global one, which is generally the inverse of the diagonal of the matrix A, and (ii) a local one, in which each spectral matrix A^k is multiplied by the inverse of the corresponding finite-difference discretization.

Some second-order elliptic problems have now been solved via the mortar element method, and its efficiency is well established (see [1, 5, 17]).

6. First Extensions

We review here some generalizations of the mortar technique to more complicated problems or geometries. Only the main features of the discretization are presented, and references are given for the detailed study and the proofs.

6.1. THE STOKES EQUATIONS IN PRIMITIVE VARIABLES

The discretization of the following problem was analyzed and implemented by N. Debit [17], in the framework of coupling finite elements and spectral discretization. We keep the domain Ω and the decomposition described in Section 2.1, we recall that $L_0^2(\Omega)$ denotes the space of functions in $L^2(\Omega)$ with a null integral on Ω, and we consider the Stokes problem in saddle-point variational formulation, for a positive viscosity ν and a source term $\mathbf{f}$ given in $L^2(\Omega)^2$: *Find* $(\mathbf{u}, p)$ *in* $H_0^1(\Omega)^2 \times L_0^2(\Omega)$ *such that*

$$\begin{aligned} &\forall \mathbf{v} \in H_0^1(\Omega)^2, \\ &\qquad \nu \int_\Omega \mathbf{grad}\,\mathbf{u} \,.\, \mathbf{grad}\,\mathbf{v}\, d\mathbf{x} - \int_\Omega (\operatorname{div} \mathbf{v})(\mathbf{x}) p(\mathbf{x}) d\mathbf{x} = \int_\Omega \mathbf{f}(\mathbf{x}) \mathbf{v}(\mathbf{x})\, d\mathbf{x}, \\ &\forall q \in L_0^2(\Omega), \quad \int_\Omega (\operatorname{div} \mathbf{u})(\mathbf{x}) q(\mathbf{x})\, d\mathbf{x} = 0. \end{aligned} \tag{27}$$

Here, the unknowns are the velocity $\mathbf{u}$ and the pressure p. As previously, we consider a discrete problem that is derived by the Galerkin method: *Find* $(\mathbf{u}_\delta, p_\delta)$ *in* $(X_\delta)^2 \times M_\delta$ *such that*

$$\begin{aligned} \forall \mathbf{v}_\delta \in (X_\delta)^2, \quad & \nu \sum_{k=1}^K \int_{\Omega^k} \mathbf{grad}\,\mathbf{u}_\delta \,.\, \mathbf{grad}\,\mathbf{v}_\delta\, d\mathbf{x} - \int_{\Omega^k} (\operatorname{div} \mathbf{v}_\delta)(\mathbf{x}) p_\delta(\mathbf{x})\, d\mathbf{x} \\ & \qquad = \sum_{k=1}^K \int_{\Omega^k} \mathbf{f}(\mathbf{x}) \mathbf{v}_\delta(\mathbf{x})\, d\mathbf{x}, \\ \forall q_\delta \in M_\delta, \quad & \sum_{k=1}^K \int_{\Omega^k} (\operatorname{div} \mathbf{u}_\delta)(\mathbf{x}) q_\delta(\mathbf{x})\, d\mathbf{x} = 0, \end{aligned} \tag{28}$$

and the crucial point is to define compatible discrete spaces X_δ and M_δ.

Let us assume that we are given on each Ω^k two discrete spaces X_δ and M_δ^k such that the following Babuška–Brezzi inf-sup condition holds:

$$\forall q_\delta \in M_\delta^k \cap L_0^2(\Omega^k),$$
$$\sup_{\mathbf{v}_\delta \in (X_\delta^k \cap H_0^1(\Omega^k))^2} \frac{\sum_{k=1}^K \int_{\Omega^k} (\mathrm{div}\,\mathbf{v}_\delta)(\mathbf{x}) q_\delta(\mathbf{x})\, d\mathbf{x}}{\|\mathbf{v}_\delta\|_{H^1(\Omega^k)}^2} \geq \beta_\delta^k \, \|q_\delta^k\|_{L^2(\Omega^k)}, \quad (29)$$

for a positive constant β_δ^k possibly depending on δ_k. Such a condition is obtained
(1) for a number of well-known finite elements with a constant β_δ^k that is bounded independently of δ_k; and
(2) in spectral approximation, by taking, for instance, X_δ^k equal to $I\!P_{N_k}(\Omega^k)$ and M_δ^k equal to $I\!P_{N_k-2}(\Omega^k)$, the constant β_δ^k behaving as $N_k^{-\frac{1}{2}}$.
The corresponding spaces $\tilde{W}_\delta^{k,j}$ are chosen in the same way as in examples 1 and 2. Next, from these spaces, the global space X_δ is defined exactly as Section 2.2, while the space M_δ is the space of all functions in $L_0^2(\Omega)$ such that their restrictions to Ω^k, $1 \leq k \leq K$, belong to M_δ^k.

With these definitions, N. Debit and Y. Maday [18] prove a global inf-sup condition by an argument due to J. Boland and R. Nicolaides [16]:

$$\forall q_\delta \in M_\delta, \quad \sup_{\mathbf{v}_\delta \in (X_\delta)^2} \frac{\sum_{k=1}^K \int_{\Omega^k} (\mathrm{div}\,\mathbf{v}_\delta)(\mathbf{x}) q_\delta(\mathbf{x})\, d\mathbf{x}}{\|\mathbf{v}_\delta\|_{H_*^1(\Omega)^2}} \geq \beta_\delta \, \|q_\delta\|_{L^2(\Omega)}, \quad (30)$$

where the positive constant β_δ depends only on the β_δ^k. Then, the numerical analysis is performed exactly as in Sections 3 and 4. Exactly the same estimate as (26) is derived for the velocity, together with an estimate for the pressure in $L^2(\Omega)$.

6.2. SLIDING GRIDS

Let us for a while consider a time-dependent problem, in a domain such that part of a boundary is moving. Since defining new grids on the domain at each time step is very expensive, the idea is to introduce two different grids on the initial domain: a fixed one and a grid that follows the moving part of the boundary and "slides" on the fixed grid. Thus, it is necessary to enforce only the mortar matching conditions of Section 2.2 at each time step. This technique was implemented by G. Anagnostou [1] in the case of the flow around an impeller that rotates at a fixed speed, as shown in Fig. 2. The sliding grid is defined on the circle (which is itself divided in several elements), and the fixed grid is defined on the outside of the circle in the rectangle. This computation leads to optimal numerical results. The numerical analysis is made in [3].

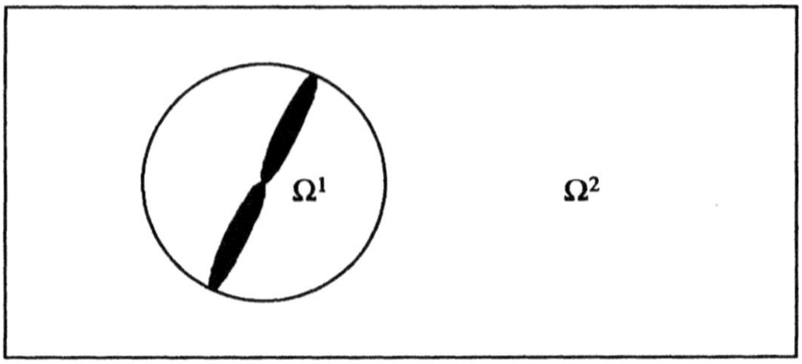

Fig. 2. **A domain with "sliding" grids**

6.3. THREE-DIMENSIONAL DOMAINS

The extension to three-dimensional domains is performed by F. Ben Belgacem [5] for a spectral discretization coupled with finite elements and for a spectral element discretization. Let us describe the method in simple cases of geometry. A polyhedron Σ is assumed to be the union of disjoint Σ^k, $1 \le k \le K$. The open faces of each Σ^k are denoted by $\Omega^{k,i}$, $1 \le i \le I(k)$, and its edges by $\Gamma^{k,j}$, $1 \le j \le J(k)$. We define $\mathcal{F}$ as the union of all $\Omega^{k,i}$ and $\mathcal{S}$ as the union of all $\Gamma^{k,j}$. As in Section 2.1, we also write $\mathcal{F}$ as the union of disjoint $\omega^n = \Omega^{k(n),i(n)}$, $n \in N$, and $\mathcal{S}$ as the union of disjoint $\gamma^m = \Gamma^{k(m),j(m)}$, $m \in M$.

Next, taking $Y = H^1(\Sigma)$, we are given discrete spaces Y_δ^k on each Σ^k, $1 \le k \le K$, contained in $H^1(\Sigma^k)$ (standard finite-element spaces on a regular family of triangulations of Σ^k in the finite element case, polynomials of degree $\le N_k$ with respect to each of the three variables in the spectral case). The trace of Y_δ^k on $\Omega^{k,i}$ is denoted $X_\delta^{k,i}$; its trace on $\Gamma^{k,j}$ is denoted by $W_\delta^{k,j}$. Then, the space $\tilde{W}_\delta^{k,j}$ is chosen as in Section 2.1, of codimension 2 in $W_\delta^{k,j}$, and a space $\tilde{X}_\delta^{k,i}$ is also introduced: in the spectral case and for triangulations made of rectangles in the finite element case, it is the tensorial product of spaces $\tilde{W}^{k,j}$ on two orthogonal edges of $\Omega^{k,i}$.

Then, W_δ is defined on $\mathcal{S}$ exactly as in (2), and X_δ is also defined as in Section 2.2 (with the X_δ^k replaced by the $X_\delta^{k(n),i(n)}$, $n \in N$). Finally, as might be expected, the discrete space Y_δ is the space of functions v_δ such that
(i) for each k, $1 \le k \le K$, the restriction of v_δ to Ω^k belongs to Y_δ^k;
(ii) v_δ satisfies the essential boundary conditions that are satisfied by the functions of Y;
(iii) there exist a mortar function φ in W_δ and a mortar function χ in X_δ

such that, for $1 \leq k \leq K$,

$$\begin{cases} \forall \alpha \text{ vertex of } \Omega^k, & (v_{\delta\,|\Omega^k})(\alpha) = \varphi(\alpha), \\ \forall \psi \in \tilde{W}_\delta^{k,j}, & \int_{\Gamma^{k,j}} (v_{\delta\,|\Omega^k} - \varphi)(\tau)\psi(\tau)\,d\tau = 0, \qquad 1 \leq j \leq J(k), \\ \forall \zeta \in \tilde{X}_\delta^{k,i}, & \int_{\Omega^{k,i}} (v_{\delta\,|\Omega^k} - \chi)(\mathbf{x})\zeta(\mathbf{x})\,d\mathbf{x} = 0, \qquad 1 \leq i \leq I(k). \end{cases} \tag{31}$$

Slight modifications in this definition are necessary for complex decompositions, as explained in [5].

The numerical analysis of the discrete problem for the Poisson equation, associated by the Galerkin method to this discrete space, is analyzed by F. Ben Belgacem and Y. Maday [6, 7] for spectral discretizations with or without finite elements, in special cases of geometry. It requires several interesting results, such as the construction of a polynomial lifting operator in a cube that satisfies a stability property analogous to (21) (see [5] for details). Note that the technique is more difficult to describe than to implement.

7. Extensions under Consideration

We present here the partial results of work with fourth-order problems, adaptive refinements of the mesh, and coupling operators.

7.1. FOURTH-ORDER PROBLEMS

The extension to fourth-order problems is performed by Z. Belhachmi [4] in the case of spectral discretizations. Thus, in a domain Ω that is the union of disjoint rectangles, the following model problem is considered: *find u in $H_0^2(\Omega)$ such that*

$$\forall v \in H_0^2(\Omega), \quad \int_\Omega (\Delta u)(\mathbf{x})(\Delta v)(\mathbf{x})\,d\mathbf{x} = \int_\Omega f(\mathbf{x})v(\mathbf{x})\,d\mathbf{x}, \tag{32}$$

where f is a given function of $L^2(\Omega)$. The discrete problem is constructed by a Galerkin method; hence a discrete space that approximates $H_0^2(\Omega)$ must be defined. Two mortar functions are needed here, one for the trace and one for the first normal derivative.

We take X_δ^k equal to $\mathbb{P}_{N_k}(\Omega^k)$, so that $W_\delta^{k,j}$ is equal to $\mathbb{P}_{N_k}(\Gamma^{k,j})$. Next, we define the space of mortar functions W_δ^* as the space of couples (φ_0, φ_1) defined on $\mathcal{S}$ such that their restrictions to any γ^m, $m \in \mathcal{M}$, belong to $W_\delta^{k(m),j(m)} \times W_\delta^{k(m),j(m)}$ and which satisfy the following compatibility property at each point α of $\mathcal{V}$: they define a unique value noted φ^α at α, a unique derivative φ_x^α with respect to x, a unique derivative φ_y^α with respect to y, and a unique derivative φ_{xy}^α with respect to x and y. This implies a series of compatibility conditions at each vertex α; however, they are natural: for instance, $\varphi_{0|\Gamma^{k,j}}(\alpha) = \varphi_{0|\Gamma^{k,j+1}}(\alpha)$, $\varphi_{1|\Gamma^{k,j}}(\alpha) = \pm\varphi'_{0|\Gamma^{k,j+1}}(\alpha)$.

The discrete space X_δ^* is the space of functions v_δ such that
(i) for each k, $1 \le k \le K$, the restriction of v_δ to Ω^k belongs to $I\!P_{N_k}(\Omega^k)$;
(ii) v_δ and its normal derivative vanish on $\partial\Omega$;
(iii) there exists a mortar couple (φ_0, φ_1) in W_δ^* such that, for $1 \le k \le K$,

$$\begin{cases} \forall\alpha \text{ vertex of } \Omega^k, & (v_{\delta|\Omega^k})(\alpha) = \varphi^\alpha, \quad (\frac{\partial(v_{\delta|\Omega^k})}{\partial x})(\alpha) = \varphi_x^\alpha, \\ & (\frac{\partial(v_{\delta|\Omega^k})}{\partial y})(\alpha) = \varphi_y^\alpha, \quad (\frac{\partial^2(v_{\delta|\Omega^k})}{\partial x \partial y})(\alpha) = \varphi_{xy}^\alpha, \\ \forall\psi \in I\!P_{N_k-4}(\Gamma^{k,j}), & \int_{\Gamma^{k,j}} (v_{\delta|\Omega^k} - \varphi_0)(\tau)\psi(\tau)\,d\tau = 0, \\ & \int_{\Gamma^{k,j}} (\frac{\partial(v_{\delta|\Omega^k})}{\partial n} - \varphi_1)(\tau)\psi(\tau)\,d\tau = 0, \\ & \qquad\qquad 1 \le j \le J(k). \end{cases} \tag{33}$$

This means that the operator $\tilde{\pi}_N^{k,j}$ has to be replaced by another one that takes into account the derivatives of the function at the extremities of the edge, which justifies the codimension 4 of $I\!P_{N_k-4}(\Gamma^{k,j})$ in $I\!P_{N_k}(\Gamma^{k,j})$.

The numerical analysis of this discretization is more difficult than the analysis for second-order problems, especially as far as the ellipticity property is concerned; it is being carried out by Z. Belhachmi [4], and numerical tests are currently being performed. The first application is the discretization of the Stokes problem in stream-function formulation.

7.2. ADAPTIVE REFINEMENTS OF THE MESH

In the finite element framework, the mortar element method is already used to define a much finer grid around corners, in order to take into account the singularities of the solution. The next idea is to determine at each step of the iteration and by appropriate estimators, in which part of the domain a finer grid is necessary. Then, this part is "cut out" as a new subdomain, the finer grid can be constructed completely independently of the previous grid, the mortar matching conditions are enforced between the two grids, and the computation goes on. This algorithm is being implemented by M. Ben Younes [8]. The refinement could be done in an automatic way.

7.3. COUPLING OPERATORS

The mortar element method is already used to couple the same differential operator on several subdomains with different coefficients on each subdomain. It could be a basic tool to couple significantly different operators in complex geometries: for instance, fluid and structure equations, Euler equations, and Navier-Stokes equations. Handling boundary layers by the mortar element method is under consideration [22].

References

[1] Anagnostou, G. (1991) 'Nonconforming sliding spectral element methods for the unsteady incompressible Navier–Stokes equations', Ph.D. thesis, Massachusetts Institute of Technology, Cambridge.

[2] Anagnostou, G., Maday, Y., Mavriplis, C., and Patera, A. T. (1990) 'On the mortar element method: Generalizations and implementation', in R. Glowinski (ed.), *Third International Conference on Domain Decomposition Methods for Partial Differential Equations*, SIAM, Philadelphia.

[3] Anagnostou, G., Maday, Y., and Patera, A. T. (1992) 'A sliding mesh method for partial differential equations in nonstationary geometries: Application to the incompressible Navier–Stokes equations', *SIAM J. Numer. Anal.*, to appear.

[4] Belhachmi, Z. (1992) Thesis, Université Pierre et Marie Curie, Paris, in preparation.

[5] Ben Belgacem, F. (1992) Thesis, Université Pierre et Marie Curie, Paris, in preparation.

[6] Ben Belgacem, F., and Maday, Y. (1992) 'Adaptation de la méthode des éléments avec joints au couplage spectral–éléments finis, étude de l'erreur pour l'équation de Poisson', Internal Report Électricité de France.

[7] Ben Belgacem, F., and Maday, Y. (1992) 'Extension of the mortar element method to three-dimensional domains: Spectral discretization', in preparation.

[8] Ben Younes, M. (1992) Thesis, Université Pierre et Marie Curie, Paris, in preparation.

[9] Bernardi, C., Dauge, M., and Maday, Y. (1992) 'Relèvement de traces préservant les polynômes', *Note aux C.R.A.S. Paris*, to appear.

[10] Bernardi, C., Debit, N., and Maday, Y. (1987) 'Couplage de méthodes spectrale et d'éléments finis: premiers résultats d'approximation', *Note aux C.R.A.S.* Série I **305**, 353–356.

[11] Bernardi, C., Debit, N., and Maday, Y. (1990) 'Coupling finite elements and spectral methods: First results', *Math. Comput.* **54**, 21–39.

[12] Bernardi, C., Maday, Y., Mavriplis, C., and Patera, A. T. (1989) 'The mortar element method applied to spectral discretizations', in T. J. Chung an G. R. Karr (eds.), *Finite Element Analysis in Fluids, Seventh International Conference on Finite Element Methods in Flow Problems*, UAH Press, Huntsville.

[13] Bernardi, C., Maday, Y., and Patera, A. T. (1992) 'A new nonconforming approach to domain decomposition: The mortar element method', in H. Brézis and J.-L. Lions (eds.), *Nonlinear Partial Differential Equations and Their Applications, Collège de France Seminar* **XI**.

[14] Bernardi, C., Maday, Y., and Sacchi-Landriani, G. (1989–90) 'Nonconforming matching conditions for coupling spectral and finite element methods', *Applied Numer. Math.* **6**, 65–84.

[15] Bjørstad, P. E., and Widlund, O. B. (1986) 'Iterative methods for the solution of elliptic problems on regions partitioned into substructures', *SIAM J. Numer. Anal.* **23**, 1097–1120.

[16] Boland, J., and Nicolaides, R. (1983) 'Stability of finite elements under divergence constraints', *SIAM J. Numer. Anal.* **20**, 722–731.

[17] Debit, N. (1991) 'La méthode des éléments avec joints dans le cas du couplage de méthodes spectrales et méthodes d'éléments finis: Résolution des équations de Navier–Stokes, Thesis, Université Pierre et Marie Curie, Paris.

[18] Debit, N., and Maday, Y. (1992) 'The coupling of spectral and finite element method for the approximation of the Stokes problem', preprint.

[19] Fisher, P. F., and Patera, A. T. (1991) 'Parallel spectral element solution of the Stokes problem', *J. Comp. Physics* **92**, 380–421.

[20] LeTallec, P., and Sassi, T. (1992) 'The domain decomposition with nonmatching grids', *SIAM J. Numer. Anal.*, to appear.

[21] Maday, Y., Mavriplis, C., and Patera, A. T. (1988) 'Nonconforming mortar element methods: Application to spectral discretizations,, in T. Chan, R. Glowinski, J. Périaux, and O. B. Widlund (eds.), *Second International Conference on Domain Decomposition*

Methods for Partial Differential Equations, SIAM, Philadelphia.
[22] Pironneau, O. (1992) Personal communication.

MULTILEVEL PROJECTION METHODS FOR PARTIAL DIFFERENTIAL EQUATIONS

Steve McCormick
Computational Mathematics Group
Campus Box 170
P.O. Box 173364
Denver CO 80217-3364

ABSTRACT. The multilevel projection method provides a framework for development of multilevel algorithms applied to any variational principle or equation that is amenable to discretization by subspace projection techniques. The objective of this methodology is to provide a systematic and consistent approach to design of multilevel algorithms for a wide range of applications. This paper briefly summarizes the basic aspects of this approach.

1. Introduction

This paper summarizes basic concepts of the newly developed multilevel projection method (PML [3]). The PML methodology provides a structure for development of multilevel algorithms that applies formally to any variational principle or equation that is itself amenable to discretization by a finite-element type of projection technique. For example, PML can be used to minimize differential functionals or solve partial differential equations (including their weak formulations). The basic idea behind PML is to pose the two essential multilevel components, relaxation and coarsening, as abstract processes that are related naturally to discretization. More specifically, the discretizations considered here involve projecting the target problem onto a subspace. Similarly, relaxation and coarsening can be posed a processes that involve correction of the current approximation by a quantity that is projected onto a set of subspaces and a coarse subspace, respectively. The central task for discretization is reduced to choosing a suitable subspace; similarly, the central tasks for relaxation and coarsening are reduced to choosing their suitable subspaces. All other essential aspects of multilevel processing (e.g., interlevel transfers, scaling, local refinement, and treatment of nonlinearities) are dictated by the PML framework, at least in principle.

This paper provides a brief summary of the core of the SIAM monograph

H. G. Kaper and M. Garbey (eds.),
Asymptotic and Numerical Methods for Partial Differential Equations with Critical Parameters 287–292.

on PML [3]. As such, it is mostly formal, it assumes several simplifications, it avoids many essential questions, and it includes little in the way of history, background, motivation, or concrete examples. The monograph and the references it cites should be consulted for some of these omissions. In other words, this paper is meant only as a digest or preview of the SIAM monograph.

We reemphasize that the following development is formal and is unconcerned with such important matters as proper posedness or specification of appropriate function spaces.

2. Abstract Problem Formulations

Here we pose three types of problem, based on three appropriate Hilbert spaces H, H_1, and H_2; a functional $R : H \rightarrow \Re$; an operator $K : H_1 \rightarrow H_2$; and a form $k : H_2 \times H_1 \rightarrow \Re$. The objective is to determine the element $u \in H$ or H_1 that satisfies one of the following problems:

$$\begin{array}{lll} \textit{variation}: & R(u) = \min_{v \in H} R(v), & u \in H. \\ \textit{equation}: & K(u) = 0, & u \in H_1. \\ \textit{weak form}: & k(v,u) = 0, & \forall v \in H_2, u \in H_1. \end{array}$$

For a concrete understanding of the principles developed here, it is useful to keep in mind specific cases of the abstract problems posed above. If the target problem is the Poisson equation

$$\begin{aligned} -\triangle u &= f \text{ in } \Omega \\ u &= 0 \text{ on } \partial\Omega, \end{aligned}$$

then the appropriate quantities are

$$R(u) = <\nabla u, \nabla u> -2 <u, f> \quad (\textit{energy functional}),$$

$$K(u) = -\triangle u - f,$$

and

$$k(v,u) = <\nabla v, \nabla u> - <v, f>,$$

where $<,>$ denotes the L_2 inner product. If the target is the eigenvalue problem

$$\begin{aligned} -\triangle u &= \lambda u \text{ in } \Omega \\ u &= 0 \text{ on } \partial\Omega, \end{aligned}$$

then the appropriate quantities are

$$R(u) = \frac{< \nabla u, \nabla u >}{< u, u >} \quad (\textit{Rayleigh quotient}),$$

$$K(u) = \; - < u, u > \; \Delta u - \; < \nabla u, \nabla u > u,$$

and

$$k(v, u) = < u, u > < \nabla v, \nabla u > \; - < \nabla u, \nabla u >< v, u > .$$

3. Discretization

The discretization of the abstract formulation is based on a subspace S^h of either H or H_1 and a subspace T^h of H_2, together with their orthogonal projections $P_{S^h} : H$ or $H_1 \rightarrow S^h$ and $P_{T^h} : H_2 \rightarrow T^h$. The objective is to determine $u^h \in S^h$ that satisfies one of the following problems:

$$\begin{array}{lll} \textit{variation}: & R(u^h) = \min_{v \in S^h} R(v^h), & u^h \in S^h. \\ \textit{equation}: & P_{T^h} K(u^h) = 0, & u^h \in S^h. \\ \textit{weak form}: & k^h(v^h, u^h) = 0, & \forall v^h \in T^h, u^h \in S^h. \end{array}$$

If we define the following three discrete quantities $R^h(u^h) \equiv R(P_{S^h} u^h)$, $K^h(u^h) \equiv P_{T^h} K(P_{S^h} u^h)$, and $k^h(v^h, u^h) \equiv k(P_{T^h} v^h, P_{S^h} u^h), u^h \in S^h, v^h \in T^h$, then the above discrete formulations are equivalent to the following respective formulations:

$$\begin{array}{lll} \textit{variation}: & R^h(u^h) = \min_{v^h \in T^h} R^h(v^h), & u^h \in S^h. \\ \textit{equation}: & K^h(u^h) = 0, & u^h \in S^h. \\ \textit{weak form}: & k^h(v^h, u^h) = 0, & \forall v^h \in T^h, u^h \in S^h. \end{array}$$

4. Relaxation

Loosely speaking, many conventional relaxation schemes can be cast as processes of updating the current approximation by sweeping over neighborhoods of the domain, making changes to the approximation there guided by the problem somehow localized to these neighborhoods. The current development is restricted to sequential versions of such schemes (e.g., Gauss-Seidel, not Jacobi, relaxation). Let $S_l^h \subset S^h$ and $T_l^h \subset T^h$ be subspaces, $1 \leq l \leq m$. (There may be a tendency on the part of the reader to presume that $S^h = S_1^h \oplus S_2^h \oplus \ldots \oplus S_m^h$, but it is not assumed that the S_l^h generate S^h or that the S_l^h are disjoint. Thus, it can only be said that $S^h \supset S_1^h + S_2^h + \ldots + S_m^h$.) The abstract relaxation process sweeps over

$l = 1, 2, \ldots, m$ sequentially, making the replacement $u^h \leftarrow u^h + u^h_{(l)}$, where the correction $u^h_{(l)} \in S^h_l$ is determined by one of the following projected problems ($P_{S^h_l} : S^h \rightarrow S^h_l$ and $P_{T^h_l} : T^h \rightarrow T^h_l$ denote the orthogonal projection operators onto the respective spaces S^h_l and $T^h_l, 1 \leq l \leq m$.):

$$\begin{array}{lll} variation: & R^h(u^h + u^h_{(l)}) = \min_{v^h_{(l)} \in S^h_l} R^h(u^h + v^h_{(l)}), & u^h_{(l)} \in S^h_l. \\ equation: & P_{T^h_l} K^h(u^h + u^h_{(l)}) = 0, & u^h_{(l)} \in S^h_l. \\ weak\ form: & k^h(v^h_{(l)}, u^h + u^h_{(l)}) = 0, & \forall v^h_{(l)} \in T^h_l, u^h_{(l)} \in S^h_l. \end{array}$$

5. Coarsening

Coarsening is even simpler in form, based only on subspaces $S^{2h} \subset S^h$ and $T^{2h} \subset T^h$. It consists of the replacement $u^h \leftarrow u^h + u^{2h}$, where the correction $u^{2h} \in S^{2h}$ is determined by one of the following projected problems ($P_{S^{2h}} : S^h \rightarrow S^{2h}$ and $P_{T^{2h}} : T^h \rightarrow T^{2h}$ denote the orthogonal projection operators onto the respective spaces S^{2h} and T^{2h}.):

$$\begin{array}{lll} variation: & R^h(u^h + u^{2h}) = \min_{v^{2h} \in S^{2h}} R^h(u^h + v^{2h}), & u^{2h} \in S^{2h}. \\ equation: & P_{T^{2h}} K^h(u^h + u^{2h}) = 0, & u^{2h} \in S^{2h}. \\ weak\ form: & k^h(v^{2h}, u^h + u^{2h}) = 0, & \forall v^{2h} \in T^{2h}, u^{2h} \in S^{2h}. \end{array}$$

If we define the coarse quantities $R^{2h}_{u^h}(u^{2h}) \equiv R^h(u^h + P_{S^{2h}} u^{2h})$, $K^{2h}_{u^h}(u^{2h}) \equiv P_{T^{2h}} K^h(u^h + P_{S^{2h}} u^{2h})$, and $k^{2h}_{u^h}(v^{2h}, u^{2h}) \equiv k^h(P_{T^{2h}} v^{2h}, u^h + P_{S^{2h}} u^{2h})$, then the above coarsening problems are equivalent to the following respective problems:

$$\begin{array}{lll} variation: & R^{2h}_{u^h}(u^{2h}) = \min_{v^{2h} \in S^{2h}} R^{2h}_{u^h}(v^{2h}), & u^{2h} \in S^{2h}. \\ equation: & K^{2h}_{u^h}(u^{2h}) = 0, & u^{2h} \in S^{2h}. \\ weak\ form: & k^{2h}_{u^h}(v^{2h}, u^{2h}) = 0, & \forall v^{2h} \in T^{2h}, u^{2h} \in S^{2h}. \end{array}$$

6. PML

To illustrate the basic scheme, we define a two-level PML cycle as follows for the *variation*:

Step 1. For $l = 1, 2, \ldots, m$ in turn, solve

$$R^h(u^h + u^h_{(l)}) = \min_{v^h_{(l)} \in S^h_l} R^h(u^h + v^h_{(l)}),\ u^h_{(l)} \in S^h_l,$$

and make the correction

$$u^h \leftarrow u^h + u^h_{(l)}.$$

Step 2. Solve

$$R^h(u^h + u^{2h}) = \min_{v^{2h} \in S^{2h}} R(u^h + v^{2h}), \ u^{2h} \in S^{2h},$$

and make the correction

$$u^h \leftarrow u^h + u^{2h}.$$

7. Comments

This PML formalism raises many issues, some of which are discussed in [3] and most of which are inappropriate for inclusion in this short note. However, because of their importance in practice, very brief comments will be made on the following two issues.

Nonlinearity. PML provides a direct attack on nonlinear equations (or non-quadradic functional minimization). In principle, linearization or FAS (cf. [1]) procedures are not necessary, and PML can be more efficient than the usual multigrid approaches to nonlinear problems. However, in some practical cases it may be necessary to linearize the PML steps locally, so that they can be implemented. Nevertheless, the PML principle provides a guide as an ideal that the implementation should be measured against and, in any case, whatever linearization that is necessary is also likely to be necessary in a similar way for the discretization itself.

Local refinement. PML automatically includes local refinement schemes like the fast adaptive composite grid method (FAC [2]). To see this, note that the S_l^h are not required to generate S^h. Thus, $S_0^h \equiv S_1^h + S_2^h + \ldots + S_m^h$ might correspond to a finite element space associated with a *local* grid of mesh spacing h, while S^{2h} is the finite element space associated with a *global* $2h$ grid. This means that relaxation on level h is local. Of course, it must be the case that $S^h = S^{2h} + S_0^h$ for the PML scheme even to converge in general, but this just means that S^h is the finite element space associated with the *composite* grid (i.e., the union of the global coarse and local fine grids).

Acknowledgements

This research was supported by the Air Force Office of Scientific Research under grant number AFOSR - 91-0156 and by the National Science Foundation under grant number DMS 8904404.

References

[1] Brandt, A. (1984) *Multigrid Techniques: 1984 Guide, with Applications to Fluid Dynamics,* Dept. Appl. Math, Weizmann Institute of Science, Rehovot 76100, Israel.

[2] McCormick, S. (1989) *Multilevel Adaptive Methods for Partial Differential Equations,* Frontier Series in Applied Mathematics 6, SIAM, Philadelphia.

[3] McCormick, S. (1992) *Multilevel Projection Methods for Partial Differential Equations,* CBMS-NSF Regional Conference Series in Applied Mathematics, SIAM, Philadelphia.

A PARALLEL ADAPTIVE MULTIGRID ALGORITHM FOR THE INCOMPRESSIBLE NAVIER-STOKES EQUATIONS

C. P. Thompson
Process Studies Department
AEA Petroleum Services
Harwell Laboratory
England

ABSTRACT. We describe a new algorithm for the solution of the laminar, incompressible Navier-Stokes equations for steady, two-dimensional flows. The basis of the method is a robust and efficient multigrid algorithm. A novel feature of our approach is that it dynamically adapts the grid as part of the solution process. Special interpolation operators have been used to ensure exact mass conservation. This technique can give significant savings in both CPU time and memory usage. The code has been implemented on a global shared-memory architecture, and speedups of over 13 have been observed on 20 processors. This implies that less than 2% of the code is intrinsically serial.

1. Introduction

The field of computational fluid dynamics (CFD) is extremely complex and demanding. It contains many problems of practical or fundamental interest requiring large amounts of computer power, far exceeding present resources. Moreover, considerable skill is required to develop appropriate numerical formulations because of the richness and complexity of the solutions, which typically involve wide ranges of length and time scales. Additionally, there are many numerical challenges caused by the highly nonlinear, coupled systems that arise; advanced algorithms are required to handle these in a robust and efficient manner. In this context there has been an ongoing project to tie together various recent developments in several fields. In this paper we shall present a survey of the work, draw some overall conclusions, and point out fields where we feel further work is required.

This work was initiated at the Argonne National Laboratory, and we are happy to acknowledge the contributions of our co-authors in this project: W. R. Cowell, G. K. Leaf, and J. van Rosendale. We also acknowledge the several contributions of N. S. Wilkes.

H. G. Kaper and M. Garbey (eds.),
Asymptotic and Numerical Methods for Partial Differential Equations with Critical Parameters 293–309.

Parallel computers represent the new challenge in high-performance computing. It is essential that we be able to adapt advanced algorithms to these new architectures and to obtain efficiently solutions to reasonably difficult problems. This has been a focus throughout our work, and we shall discuss the results that we have *measured*, the predicted scalability of our algorithm and in the following sections we shall make some comments on the methodology we used.

Multigrid algorithms have proved to be some of the fastest and most adaptable ways of solving partial differential equations. An overview of our basic approach is given in Section 2. Various authors [1, 10, 15] have discussed applying adaptive gridding techniques in the multigrid context, although much of the numerical work has focused on model problems. We have developed these techniques so that they can be effectively applied to CFD problems; grids are *automatically* generated that resolve shear layers and singularities without user intervention. One of the goals of this project is to implement this algorithm on parallel computers without significant loss of efficiency or robustness; a selection of our results is presented below.

1.1. TEST PROBLEMS AND GOVERNING EQUATIONS

We solve the steady laminar Navier-Stokes equations for incompressible flows in two dimensions. These form a system of differential equations that display most of the essential properties found in more complex flow regimes. The efficient solution of these equations is at the heart of many applications codes.

The governing equations are

$$-\Delta u + Re\{(u^2)_x + (vu)_y\} + p_x = S_u, \qquad (1)$$

$$-\Delta v + Re\{(uv)_x + (v^2)_y\} + p_y = S_v, \qquad (2)$$

$$u_x + v_y = 0, \qquad (3)$$

where u, v are the horizontal and vertical components of velocity; p is the pressure; and S_u, S_v are the appropriate horizontal and vertical momentum source terms.

The Reynolds number measures the ratio of inertial to viscous forces. It is defined by

$$Re = UL/\nu, \qquad (4)$$

where U and L are the characteristic velocity and length scales and ν is the laminar viscosity.

The usual way of demonstrating the capabilities of a new CFD algorithm is to apply it to a number of test cases whose results are known. We have chosen four examples (brief details are given below; for a full description see [14] and [11].

– THE DRIVEN CAVITY
  The flow domain is the unit square; on the bottom and sides of the box we define $u = 0$ and $v = 0$; the lid of the box is sliding so we have $u = 1$ and $v = 0$. There are two singularities in the top corners of the domain where the moving lid meets the stationary walls, and these cause problems for any adaptive strategy. However, the problem is geometrically simple and has been widely studied (see, for example, [18, 9, 16, 15].
– THE BACKWARD-FACING STEP
  Another widely studied problem is that of flow over a backward-facing step. It was set as a test problem for a GAMM workshop, and we have taken the finite-element benchmark solution [3] as the basis of our comparison. The geometry, computed flow fields, and automatically adapted grids are shown in Fig. 1. These results are discussed in Section 4.2.
– THE SUDDEN EXPANSION/CONTRACTION
  The domain for this problem is composed of five unit squares arranged in a cross. At the left we define a parabolic inlet, and at the right we specify fully developed flow. In the central region, where the cross is three units wide, there are two weak, recirculating flows induced by the shear forces.
– THE SUDDEN EXPANSION GEOMETRY
  This geometry is of particular interest because, for sufficiently high Reynolds number, the flow bifurcates and the stable solutions are unsymmetric. The geometry is similar to that of the backstep, except that the expansion is symmetric. This flow was studied in detail in [5].

2. An Overview of the Underlying Multigrid Solution Algorithm

In this section we give a brief description of the "standard" aspects of our algorithm. For fuller descriptions of the discretization see [13]; for an overview of multigrid techniques see [2, 8].

2.1. DISCRETIZATION

Our basic discretization uses a staggered grid: pressures are defined at the centers of control volumes, and vertical velocities are defined at the top and bottom faces; horizontal components are similarly offset (see Fig. 2). Convection terms are discretized by using hybrid differencing: if the mesh Reynolds number (uh/ν) is greater than 2, then upwind differencing is used;

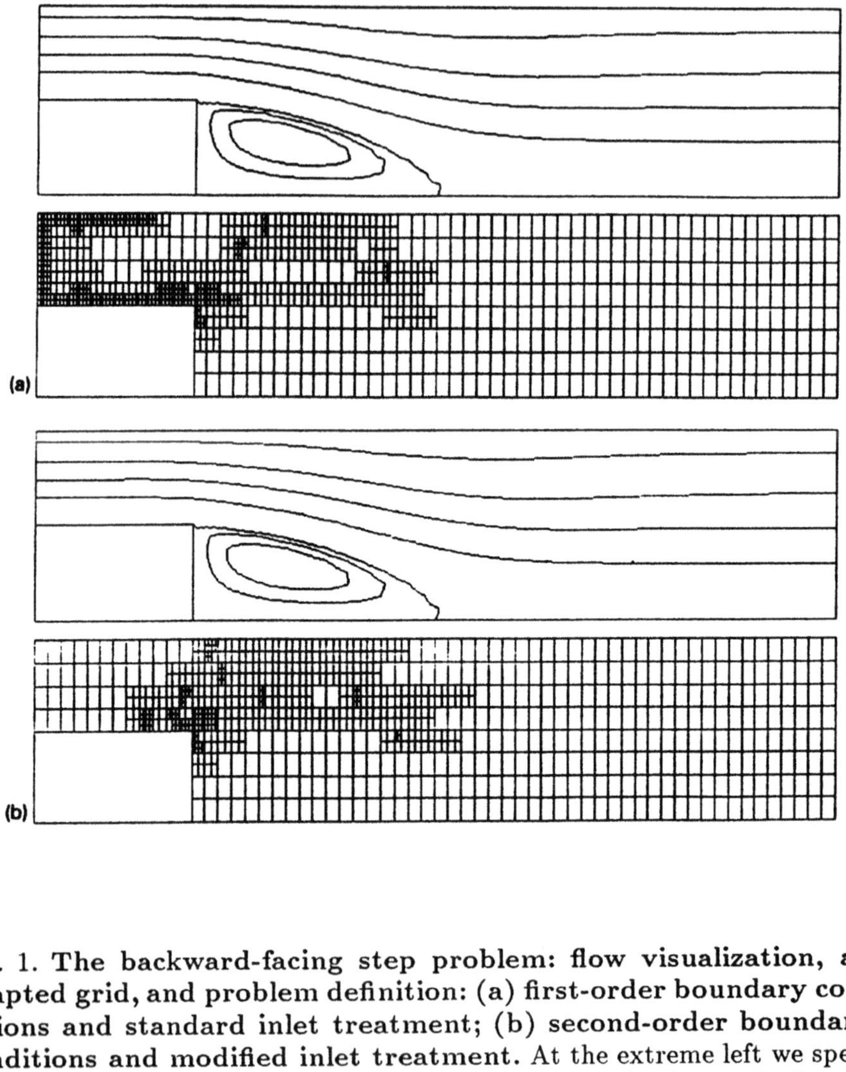

Fig. 1. **The backward-facing step problem: flow visualization, an adapted grid, and problem definition: (a) first-order boundary conditions and standard inlet treatment; (b) second-order boundary conditions and modified inlet treatment.** At the extreme left we specify a parabolic inlet profile for the horizontal velocity (and $v = 0$). On the right we specify fully developed flow conditions: $v = 0$ and $\partial u/\partial x = 0$. On all other boundaries we have $u = v = 0$. Notice that we are solving the flow *around* the step and not ignoring the singularity at the corner. This is a significantly more difficult problem than considering a rectangular domain only.

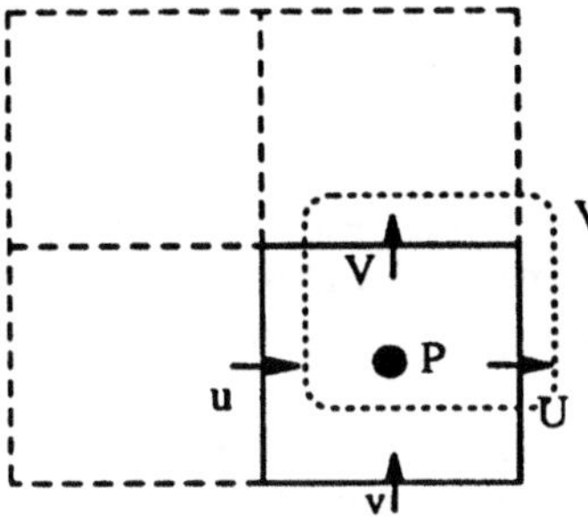

Fig. 2. **Definition of a mass control volume**

for smaller mesh Reynolds numbers central differencing is used. All other derivatives are approximated using central differences.

This approach has a number of advantages. The staggering of the grids inhibits the presence of spurious oscillatory pressure modes, provided that the momentum equations are discretized in a stable fashion. The hybrid differencing method is stable: at high mesh Reynolds numbers, there is sufficient numerical diffusion to inhibit wiggles in the solution; however, on sufficiently fine uniform grids, the whole scheme is formally second order accurate.

The properties of this approach are well understood on nonadapted grids. Convergence on coarse grids is reasonably easy because of the numerical diffusion. In contrast, on sufficiently fine grids this effect disappears, and the normal accuracy is recovered. Several groups are working with different types of discretization; however, the relative merits of these are debatable, and we wished to decouple the properties of our adaptive technique from those of our interior discretization.

2.2. THE MULTIGRID ALGORITHM

The term *multigrid* is applied to a plethora of different approaches. For definiteness we shall briefly state where our approach fits in the spectrum. The key idea is that one can obtain solutions to partial differential equations that have a convergence rate independent of grid size. This objective is achieved by defining a relaxation operator that is an effective smoother and by cycling from coarse to fine meshes in such a way that the smoother is always operating effectively.

For efficiency our algorithm is geometric (our coarse-grid operators are defined in the same way as our fine-grid operators, except that h is replaced by $2h$). Our smoother uses collective relaxation (i.e., we update all the variables associated with a control volume simultaneously). This was shown to be efficient and robust in [13], where it was applied to buoyant problems (see also [6]). The nonlinearities are represented by using full approximation storage (FAS). This means that we have approximations to all variables on all mesh levels.

Our restriction operator is simply full weighting. However, the coarse- to fine-grid mapping (prolongation) is new. We chose to avoid standard operators since they did not have all the properties we wanted on our adapted meshes (see below).

We have implemented all the common grid-cycling strategies, including adaptive cycling. Earlier work has shown that F-cycles are somewhat more efficient, and the majority of the work reported here uses this strategy.

3. Interface Conditions

Many crucial quantities in fluid flow problems satisfy conservation laws and it is important that discrete models satisfy these in some appropriate sense. For incompressible flows it is well known that solutions to the continuous problem satisfy

$$\oint_{\partial\Omega} \nabla.\mathbf{u} = 0 \tag{5}$$

around any closed path $\partial\Omega$ in the domain. The discrete analog of this expression is

$$\sum_{\partial\Omega} [u(i+1) - u(i)]/h_x + [v(j+1) - v(j)]/h_y = 0 \tag{6}$$

around any set of control volumes. There is no solution to the discrete system unless this condition is satisfied, since a singularity is induced because the pressure appears only in derivative terms. A constant pressure field can be added to any solution without altering the flow field.

Normally, this condition has implications only for the boundary conditions; however, for adaptive grids one has to pay special attention to grid interfaces.

Various workers have tried different methods to enforce exact discrete mass conservation. In [14] we describe some experiments with different formulations. The treatment that we have chosen is a compromise between robustness, accuracy, and ease of implementation. It is described below.

Our goal at interfaces is a simple one: to ensure that the mass leaving a coarse grid appears in the fine grid. Rather than attempting to allow this to drift and correct for the discrepancies, we choose to apply an interpolation operator that satisfies this condition exactly and locally.

If we consider the flux across a (horizontal) interface, then conservation requires

$$2hV(j) = hv(j) + hv(j+1), \tag{7}$$

where h is the fine mesh size and we have used the notation of Fig. 3. Our experiments show that the three-point interpolant derived by using

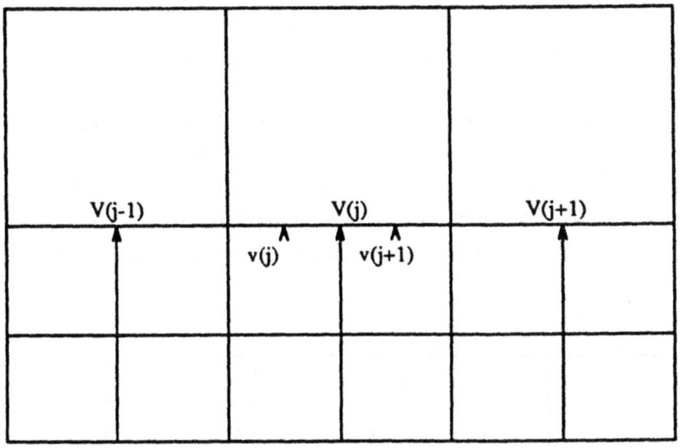

Fig. 3. **Prolongation at interfaces.** Coarse-grid unknowns are shown in upper case, fine-grid unknowns in lower case.)

this operator is a good compromise between accuracy and flexibility. For definiteness the definitions of the fine grid velocities $v(j)$ and $v(j+1)$ are given by

$$v(j+1) = V(j) + [V(j+1) - V(j-1)]/8 \tag{8}$$

and

$$v(j) = V(j) - [V(j+1) - V(j-1)]/8. \tag{9}$$

Some comparisons between first-order conservative interpolants and third-order interpolants are described in [14]. In this work it was concluded that the above expressions (Equations 8 and 9) were most appropriate for this research, and all the results reported below were obtained with these formulas.

These definitions are used as prolongation operators at all applicable cells, and there is no impact on the final solution except at interfaces. However, our interface treatment is as shown in Figs. 4 and 5. The fine-grid values in the dummy cells are defined by applying these transfer operators to the surrounding coarse-grid cells. Effectively, these values are boundary values for the fine-grid calculation, and they do have an effect on the final solution.

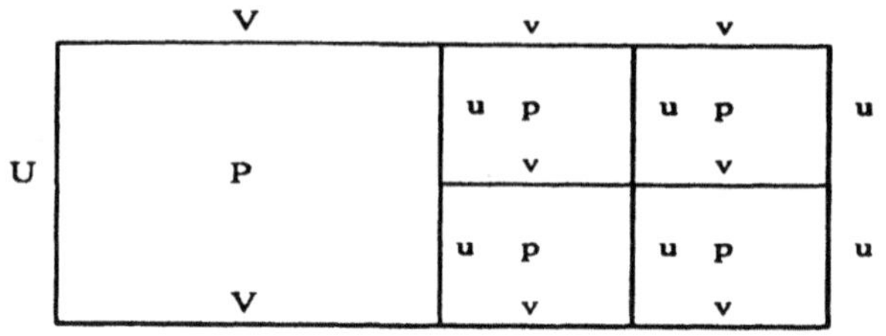

Fig. 4. **The canonic problem: one coarse-grid patch interfacing with four fine-grid patches.** This tiling is sufficient to generate all configurations of interest. We show the location of "genuine" unknowns on the compound grid. At these locations we solve approximations to the differential equations. (Coarse-grid equations are shown in upper case, fine-grid equations in lower case.)

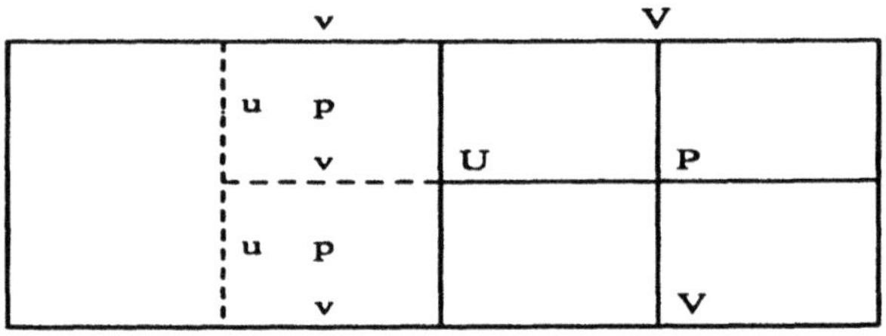

Fig. 5. **The canonic problem modified to show the dummy unknowns introduced for computational convenience.** The variables shown here are not unknowns in our system of equations but dummy variables introduced artificially. Special, additional equations are derived for them. (Coarse-grid equations are shown in upper case, fine-grid equations in lower case.)

4. Adaptivity

As we have already mentioned, we handle the nonlinearities of the system directly by using the FAS scheme. As part of the coarse-grid correction we calculate a quantity known as the defect, which is defined by

$$\tau_h^H = I_h^H \left(L^h u^h\right) - L^H \left(I_h^H u^h\right). \tag{10}$$

This may be thought of as a pointwise approximation to the local truncation error. There are many possible ways to approximate errors for adaptive schemes (use of gradients or curvature is common), and the results are qualitatively the same. We prefer to use the truncation error in that we have the following property:

$$L^H\left(e^H\right) \approx \tau_h^H. \tag{11}$$

However, it is important to note that in making this choice, we have effectively decided what we mean by the *error* of our solution. Our adaptive strategy is committed to reducing this quantity and the refinement patterns will reflect this choice. In our experience, CFD practitioners often measure errors by considering quantities that are easy to describe and measure, sometimes disregarding their intrinsic importance to the flow. These can be at variance with the above definition.

Our fundamental strategy is to refine any quadrant of a patch that fails to satisfy our error norm (see Fig. 5.1). In the work discussed here we refine any quadrant that satisfies

$$\left[\tau_h^H\right]_{quadrant} \geq \left\{\frac{1}{nVis}\left(\sum_{nVis}\left[\tau_h^H\right]^{power}\right)\right\}^{\frac{1}{power}}, \tag{12}$$

where the sum is over the visible (that is, unrefined) quadrants. The exponent (*power*) typically varies from zero to one, smaller values inducing more refinement. Additionally, we refine quadrants to ensure that the mesh ratio never exceeds two.

Subject to the above discussion, this strategy is reasonably effective; but this aspect forms a self-contained module in our program and is extremely easy to modify.

4.1. CONVERGENCE RATES

The algorithm outlined above has good convergence rates for all the cases that we have studied. Of course, the rate varies depending on the problem, but we have never observed significant degradation as the mesh becomes finer. In all cases the rates have been comparable to those of typical CFD multigrid codes.

As an example we show in Table 1 the convergence rates for the driven cavity problem for calculations on both uniform and adapted grids.

It is usual for this class of problem that, for uniform grids, the convergence rate that we observe *improves* as the mesh becomes finer. There are two reasons for this. First, the h-ellipticity of the problem increases because the convective terms become small compared to the diffusion terms. Second, on the fine grids our treatment of the convection terms becomes second order (rather than first), so the coarse-grid correction is more accurate.

This acceleration is present in some of our uniform-grid calculations, but it is masked in the adapted grid cases because there is some coarse-grid region that remains unrefined and that inhibits this effect.

Table 1. **The driven cavity problem: Rate of convergence on adaptive and nonadaptive grids, with** $Re = 400$

	Uniform Calculation				Adapted Calculation			
Grid Level	Total Pats.	Fine Pats.	No. Cycles	ρ_f	Total Pats.	Fine Pats.	No. Cycles	ρ_f
2	5	4	23	.51	5	4	23	.51
3	21	16	20	.47	21	16	20	.47
4	85	64	24	.52	43	22	19	.42
5	341	256	22	.55	101	52	17	.23
6	1365	1024	15	.31	220	102	16	.47
7	5461	4096	20	.62	462	169	15	.42
8	–	–	–	–	954	321	17	.52

4.2. ACCURACY OF THE ADAPTIVE SCHEME

It is clear from Fig. 1 that the flow field predicted by the code is qualitatively correct.

In case (a) the boundary treatments were somewhat simplistic. One-sided (first-order) boundary conditions were used at the walls, and the inlet velocity was simply set to the point value as defined in the parabolic profile. The code has automatically detected that the boundary treatments are less accurate than the internal discretization and has refined the grid at the wall above the step. (Note that the corresponding errors behind the step are smaller because the average flow velocity is less there.) The problem with the inlet specification is that, as the grid is refined, the total mass flux (integrated over the inlet) increases, even though the point values remain constant. This represents a major perturbation in the problem, and the code has refined in this region as well.

In case (b) both of these effects have been removed from the discretization. It is very pleasing to note that this situation has been detected and the grid has not been refined here. The refinement occurs only in the region of the shear layer behind the step and in the neighborhood of the step itself where there is a weak singularity in the solution. This seems to be qualitatively reasonable.

In Table 2 we show quantitative comparisons for the backstep. The various columns show results from the benchmark comparison [3], two uniform-grid calculations, and two adaptive-grid calculations. It is clear that all the results are in agreement and the calculations with two levels of adaptive refinement (shown as (2,4) in the table) use half the space of the uniform level-3 calculations for the same level of accuracy. The singularity at the re-entrant corner inhibits further improvement, but the (2,10) calculation shows that there is no degradation of the accuracy.

For this example the coarse-grid work dominates the calculations, and the adaptive strategy does not save any computing time. In other geometries, where the coarse grid is dominated by work at other levels, we observed significant savings in CPU time as well as memory usage. For the sudden expansion/contraction problem the adapted calculation uses only 60% of the CPU time and 11% of the storage required for the equivalent uniform mesh computation.

We can summarize our results by saying that our adaptive strategy made significant improvements in memory and CPU utilization in several cases when compared with equally accurate uniform-grid calculations. However, our results were dependent upon the error criterion used and the relative importance of the coarsest grids.

5. Domain Decomposition and Parallelization

"Dummy" cells were introduced at coarse-fine grid interfaces to simplify the definition of the discretization (see Fig. 5). It is natural to use this approach as the basis for our domain decomposition which is required to support our adaptive gridding strategy and our goal of parallelization. We define a group of $n \times n$ control volumes and the associated border cells as a patch (see Fig. 5.1). This forms the "atomic" data unit on which all our numerical operations are performed. All our numerical routines have been rewritten to operate upon patches with flags describing the data.

Typical lists of descriptors required to support these numerical operations are as follows:

- *Patch_Level*, defining all of the patches at a particular level. This allows operations to be performed by list traversal rather than recursive tree walk.
- *Neighbors* of each patch (simplifying the data exchange during relaxation and interpolation.
- Identifiers of the children (*Kids*) of each patch, necessary for prolongation.
- *Aunts* (neighbors of parents) of each patch, required for restriction.

It is interesting to note that this framework, which is required for the adaptive meshing, can be applied to the coarsest grid as well. This allows us to compute on "generalized simple geometries," that is, any domain defined by horizontal and vertical lines.

Table 2. **Results for the backward-facing step: Uniform and adaptive calculations and benchmark**

	Velocity				
	Cliffe	(2,2)	(3,3)	(2,4)	(2,10)
U_{max}					
1.6 step-heights	0.904	0.897	0.903	0.906	0.905
4.0 step-heights	0.725	0.725	0.722	0.726	0.725
U_{min}					
1.6 step-heights	-0.102	-0.103	-0.103	-0.102	-0.101
4.0 step-heights	-0.049	-0.036	-0.047	-0.039	-0.038
V_{max}					
1.6 step-heights	0.010	0.010	0.010	0.010	0.010
4.0 step-heights	0.0	0.0	0.0	0.0	0.0
V_{min}					
1.6 step-heights	-0.070	-0.068	-0.070	-0.071	-0.071
4.0 step-heights	-0.091	-0.077	-0.087	-0.089	-0.089
No. patches	–	180	756	391	544
Time (sec)	–	73	77	113	–
Total Time (sec)	–	73	77	278	–

Notes:
(2,2): uniform 2-level calculation with mesh size = step-height/8.
(3,3): uniform 3-level calculation with mesh size = step-height/16.
(2,N): adapted calculation with maximum mesh size = step-height/8 and minimum mesh size = step-height/$2^{(N+1)}$.

5.1. PARALLEL PERFORMANCE

The code was implemented on a global shared-memory parallel architecture (the Sequent Symmetry at the Advanced Computing Research Facility at Argonne National Laboratory). On the driven cavity problem the results were disappointing: a speedup of slightly over 3 was measured on 20 processors. This was because the refinement pattern induced by the strong singularities in the flow, which meant that no grids had large numbers of patches.

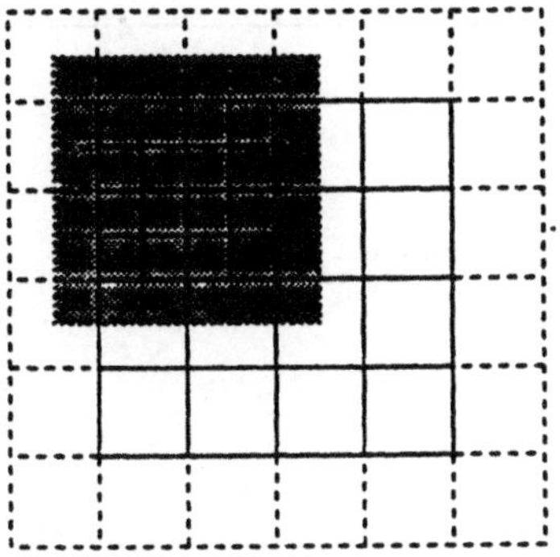

Fig. 6. **A patch (showing a refined quadrant)**

On two other geometries the results were satisfactory: a speedup of over 11 was measured on the sudden expansion/contraction, and this increased to about 13.5 for the backstep. We used a generalization of Amdahl's law to analyze our results. For this last case we estimate that less than 2% of the code is intrinsically serial. Moreover, our analysis suggests that the algorithm could perform well on larger numbers of processors.

Detailed performance of the code is given in Fig. 7. (Further details of the parallel performance is given in [12].) It should be noted that these are measured times on a complete CFD code, not a numerical kernel.

5.2. PARALLEL SOFTWARE TOOLS

It is probably appropriate to make some comments about the process of parallelization that we applied to the code. The conventional wisdom is that the numerically intensive components in multigrid (relaxation, restriction, and prolongation) dominate the work, and it is simply necessary to parallelize these. This appears to be supported by the timing breakdown given in Fig. 8. This process was reasonably easy, and we were able to use a number of software tools that we had developed to assist in the process (see [12, 17, 4, 7]).

However, when we completed this task, our speedup results were disappointing. The analysis with Amdahl's law showed that, in fact, we had to achieve a much higher proportion of parallel code before significant gains could be made. This was rather disconcerting, since the remaining routines were long and complex, involving data exchanges, etc. Unfortunately, we obtained only minimal assistance from any software tools; almost all the analysis had to be performed manually.

However, at the conclusion of this phase the parallel performance was

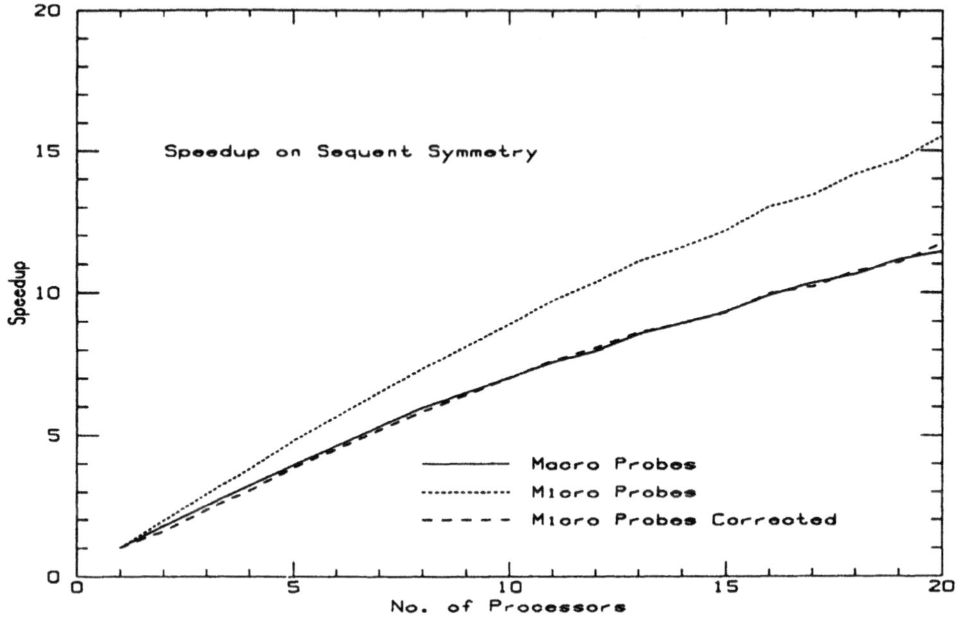

Fig. 7. **Speedups for the sudden expansion/contraction problem: measurements on a Sequent Symmetry.** Macro Probes are performance measurements using the total elapsed time for the run. The Micro Probes use performance measurements of each of the parallel loops; this information is then used in the speedup model [12].

significantly improved. Figure 9 shows the timing breakdown of the final version of the code as measured on 20 processors (note the change of scale).

6. Summary

A novel algorithm for the Navier-Stokes equations has been described. It is applicable to steady, laminar problems in two-dimensional, "generalized simple geometries." It is based on a robust, geometric multigrid solver, which uses collective relaxation as a smoother. Domain decomposition has been introduced to support dynamically adaptive gridding and parallelism.

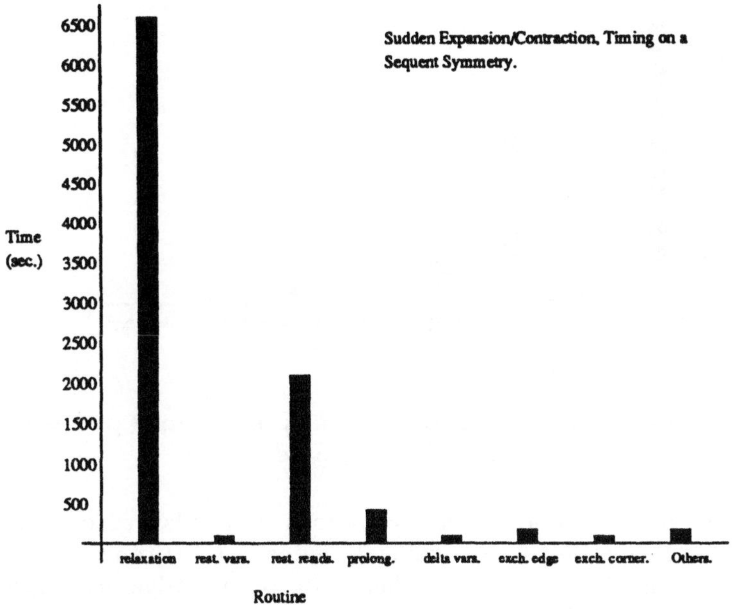

Fig. 8. **Profile of the adaptive code; timings on one processor**

Legend for Figs. 8 and 9:
(1) relax
(2) restrict variables
(3) restrict residuals
(4) prolongation
(5) calculating differences to old variable fields
(6) exchanging edges of patches
(7) exchanging corners of patches
(8) code not appearing in the above loops

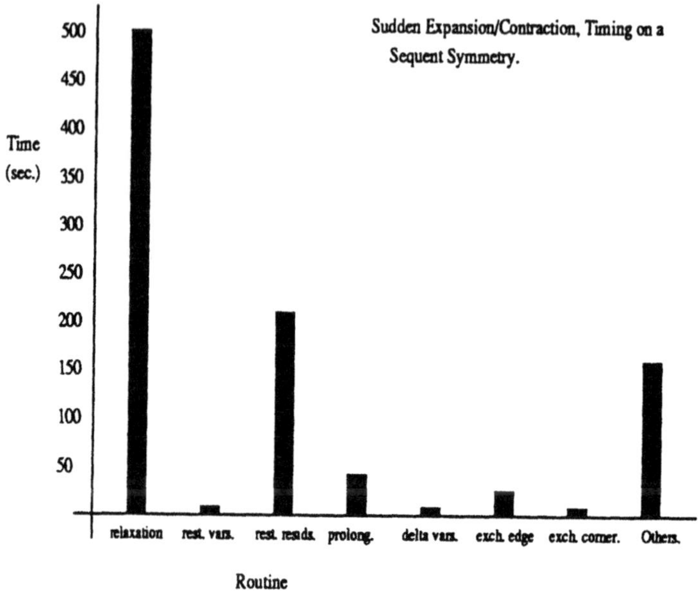

Fig. 9. **Profile of the adaptive code; timings on twenty processors.** The vertical axis is expanded for clarity (compare Fig. 8).

Comparisons of the uniform grid calculations with adapted ones show that, under certain circumstances, significant gains in CPU time and memory usage can be made. However, further work on adaptive strategies is required to consolidate this work.

The code has been implemented on various parallel architectures; speedups of over 13.5 have been measured one configuration. This implies that less than 2% of the code is intrinsically serial.

Acknowledgements

This research was supported in part by the Applied Mathematical Sciences subprogram of the Office of Energy Research, U.S. Department of Energy, under Contract W-31-109-Eng-38; in part by the National Aeronautics and Space Administration under NASA Contract No. NAS1-18605, and in part by IBM, Bergen Scientific Centre.

References

[1] Bai, D., and Brandt, A. (1987) 'Local mesh refinement multilevel techniques', *SIAM J. Sci. Stat. Comput.*, **8**, 109–134.

[2] Briggs, W. L. (1987) *A Multigrid Tutorial*, SIAM, Philadelphia.

[3] Cliffe, K. A., Jones, I. P., Porter, J., Thompson, C. P., and Wilkes, N. S. (1983) 'Laminar flow over a backward facing step: Solution to a test problem', in K. Morgan, J. Periaux, F. Thomasset (eds.), *Analysis of Laminar Flow over a Backward-Facing Step: A GAMM Workshop*, Friedr. Vieweg and Sohn, 140–161.

[4] Cowell, W. R., and Thompson, C. P. (1990) 'Tools to aid in discovering parallelism and localizing arithmetic in Fortran programs', *Software Practice and Experience*, **20**, no. 1, 25–47.

[5] Fearn, R. M., Mullin, T., and Cliffe, K. A. (1990) 'Nonlinear flow phenomena in a symmetric sudden expansion', *J. Fluid Mech.*, **211**, 595–608.

[6] Galpin, P. F., and Raithby, A. D. (1986) 'Numerical solution of problems in incompressible fluid flow: Treatment of temperature velocity coupling', *Num. Heat Transfer*, **10**, 105–129.

[7] Gjerde, K. (1991) 'Tools to parallelize Fortran 77 programs', Hovedoppgave ved Institutt for Informatikk, University of Bergen.

[8] Hackbusch, W. (1985) *Multi-Grid Methods and Applications*, Springer-Verlag, Berlin. 1985.

[9] Ku, H. C., Hirsh, R. S., and Taylor, T. D. (1987) 'A pseudospectral method for solution of the three-dimensional incompressible Navier-Stokes equations', *J. Comp. Phys.*, **70**, 439–462.

[10] McCormick, S., and Thomas, J. W. (1986) 'The fast adaptive composite grid (FAC) method for elliptic equations', *Math. Comp.*, **46**, 439–459.

[11] Thompson, C. P. (1992) 'The development and validation of a computational fluid dynamics code for global shared memory parallel architectures', preprint.

[12] Thompson, C. P., Cowell, W. R., and Leaf, G. K. (1992) 'On the parallelization of an adaptive multigrid algorithm for a class of flow problems', *Parallel Computing*, **18**, 449–466.

[13] Thompson, C. P., Leaf, G. K., and Vanka, S. P. (1988) 'Application of a method to a buoyancy-induced flow problem', in S. F. McCormick (ed.), *Multigrid Methods*, Marcel Dekker, Dordrecht, 605–629.

[14] Thompson, C. P., Leaf, G. K., and van Rosendale, J. (1992) 'A dynamically-adaptive multigrid algorithm for the incompressible Navier-Stokes equations – validation and model problems', *Applied Numerical Math.*, **9**, 511–532.

[15] Thompson, M. C., and Ferziger, J. H. (1989) 'An adaptive multigrid technique for the incompressible Navier-Stokes equations', *J. Comp. Phys.*, **82**, no. 1, 94–121.

[16] Vanka, S. P. (1986) 'Block-implicit multigrid solution of Navier-Stokes equations in primitive variables', *J. Comp. Phys.*, **65**, no. 1, 138–158.

[17] 'VecPar_77 tutorial introduction and reference manual', Numerical Algorithms Group, Inc., Downers Grove, Illinois, 1990.

[18] Winters, K. H., and Cliffe, K. A. (1979) 'A finite element study driven laminar flow in a square cavity', AERE - R 9444, AERE Harwell, UK.

NUMERICAL APPROXIMATION OF A TWO-DIMENSIONAL SINGULAR-PERTURBATION PROBLEM USING AN ASYMPTOTIC EXPANSION AND A FINITE ELEMENT METHOD

Jacques Baranger
Université Claude Bernard Lyon
69622 Villeurbanne Cedex
France

and

Mohammed El Alami El Ferricha and Harsan El Amri
ENS de Fès
BP 34 A
Fès Morocco

ABSTRACT. We consider a fourth-order operator that degenerates to a second-order operator when a small parameter ε goes to zero. For such a problem, a direct application of a finite element method gives very poor error bounds. An asymptotic expansion of the solution can be used, whose terms are approximated numerically. We propose two such methods, for which we obtain error bounds.

1. Introduction

We consider the following singular-perturbation problem:
Problem (P_ε) :

$$L_\varepsilon u_\varepsilon \equiv \varepsilon^2 \Delta^2 u_\varepsilon - \Delta u_\varepsilon = f \text{ in } \Omega \tag{1}$$

$$u_\varepsilon = 0 \text{ on } \Gamma \tag{2}$$

$$\frac{\partial u_\varepsilon}{\partial n} = 0 \text{ on } \Omega, \tag{3}$$

where Ω is a bounded convex domain in $\mathbf{R}^2$. A typical physical example is the constrained clamped plate with small flexural rigidity.

When ε tends to zero, u_ε converges in $H^1(\Omega)$ towards a solution of the formal limit problem,

H. G. Kaper and M. Garbey (eds.),
Asymptotic and Numerical Methods for Partial Differential Equations with Critical Parameters 311–319.

Problem (P_0):

$$-\Delta u_0 = f \text{ in } \Omega \tag{4}$$

$$u_0 = 0 \text{ on } \Gamma, \tag{5}$$

but there is a boundary layer, whose width is of order ε, in which $\frac{\partial u_\varepsilon}{\partial n}$ varies strongly (see [7]).

We suppose that ε is small and that, for some reason, it is not possible to look inside the boundary layer (i.e., we suppose $\varepsilon \leq h$). If a finite element method (FEM) is used to approximate u_ε, the usual error bounds are of no use. For example, for the P_3C^1 finite element of Hsieh-Clough-Tocher (HCT), we prove that

$$| u_\varepsilon - u_{\varepsilon h} |_{0,\Omega} \leq Ch^k(\varepsilon^2 + h^2)^{1/2}\varepsilon^{1/2-k-r}; \quad r = 1,2; \quad k \leq 2.$$

A natural idea is, then, to search for error estimates outside the boundary layer: $| u_\varepsilon - u_{\varepsilon h} |_{0,\Omega'}$, where Ω' is strongly included in Ω, like the one obtained in [8] for second-order problems. As far as we know, such estimates are not available for fourth-order problems.

Here we follow another method: first, we construct an asymptotic expansion (AE) of u_ε using the technique of [9] modified in [3, p. 3]; then we approximate all the terms in this expansion by a FEM. This idea has been used previously on a one-dimensional problem in [1].

The two-dimensional case studied here is more complicated because there is in some sense a contradiction: the construction of an AE requires some smoothness of the boundary Γ, whereas the use of an FEM is easier for a "polygonal" boundary.

We propose two composite AE-FEMs. The first, somewhat crude, method uses a P_1 FEM to approximate u_0 and a finite difference approximation of $\frac{\partial u_0}{\partial n}$. The second, more elaborate method uses a modification of the HCT FEM, taking into account the fact that Γ is a curve (not a polygon) following an idea of Zlamal. Denoting this last approximation by U^{0h}, we prove the estimate

$$| u_\varepsilon - U^{0h} |_{1,\Omega} \leq C(\varepsilon + h^3).$$

This paper is organized as follows: in Section 2, we prove some error estimates with negative power of ε; in Section 3, we build the AE and show its validity; in Section 4 (resp. 5), we describe the P_1 (resp. modified HCT) composite AE-FEM and give error bounds. This paper is based on [6].

2. A-Priori Error Estimates

In this section we suppose that Ω is a polygonal domain. Let T_h be a uniformly regular triangulation of Ω:

$$\bar{\Omega} = U_{K \in T_h} K.$$

We denote the diameter of K by h_K, the diameter of the largest circle inscribed in K by ρ_K, and assume that

$$\sigma_K = h_K/\rho_K \leq \sigma, \quad h_K/h \leq \mu, \ \forall K \in T_h.$$

Here, $| \ |_{m,K}$ and $\| \ \|_{m,K}$ denote Sobolev's seminorm and norm on $H^m(K)$.
We use the FEM of Hsieh-Cough-Tocher described in [4]. Each triangle K is divided arbitrarily into three triangles: K_1, K_2, K_3. The polynomial subspace is

$$P_K = \{p \in C^1(K); \quad p_{|K_i} \in P_3(K_i), \quad i = 1,2,3\},$$

and the degrees of freedom are $u(a_i), Du(a_i)$, and $\frac{\partial u}{\partial n}(b_i), i = 1,2,3$, where b_i is the midpoint of the edge opposite a_i (see Fig. 1).

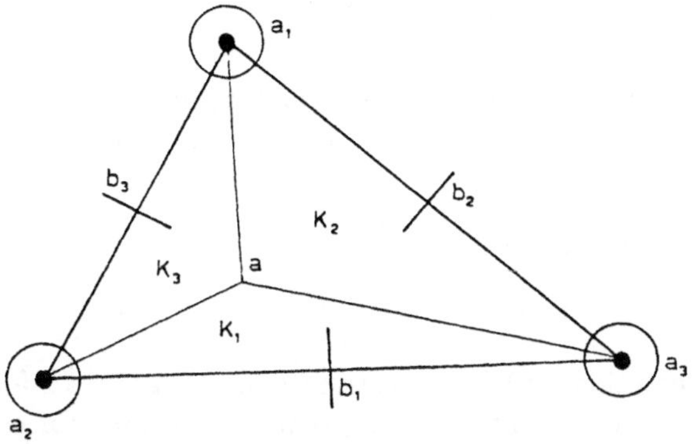

Fig. 1. **Subdivision of triangle** K

Let F_K be the affine bijective mapping from a reference element $\hat{K}$ onto K. We assume that $\{\hat{a}_K = F_K^{-1}(a_K), K \in T_h,\}$ is included in a compact set in K. We define

$$V_h = \{v_h \in C^1(\bar{\Omega} \cap H_0^2(\Omega)), \quad v_{h|K} \in P_K, \quad \forall K \in T_h\}.$$

The weak form of Problem (P_ε) is a problem for $u_\varepsilon \in H_0^2(\Omega)$:

$$a_\varepsilon(u_\varepsilon, v) = (f, v), \quad \forall v \in H_0^2(\Omega), \tag{6}$$

where $a_\varepsilon(u,v) = \varepsilon^2 \int \Delta u \Delta v + \int \nabla u \nabla v$. The approximate problem is $u_{\varepsilon h} \in V_h$:

$$a_\varepsilon(u_{\varepsilon h}, v_h) = (f, v_h), \quad \forall v_h \in V_h. \tag{7}$$

An error bound is given in the following theorem.
Theorem 1. *Suppose that* $u_\varepsilon \in H^2(\Omega)$. *Then*

$$| u_\varepsilon - u_{\varepsilon h} |_{s,\Omega} \leq C(\varepsilon^2 + h^2)^{1/2} \varepsilon^{-s+1/2}, \quad s = 1, 2. \tag{8}$$

Sketch of the proof. Let $w_h \in V_h$. From (6) and (7) we obtain

$$\varepsilon^2 | u_\varepsilon - u_{\varepsilon h} |_{2,\Omega}^2 + | u_\varepsilon - u_{\varepsilon h} |_{1,\Omega}^2 \leq C[\varepsilon^2 | u_\varepsilon - w_h |_{2,\Omega}^2 + | u_\varepsilon - w_h |_{1,\Omega}^2]. \tag{9}$$

Using the interpolation operator Π_h associated with v_h for discontinuous functions [5, 2], we set $w_h = \Pi_h u$, which implies

$$| u_\varepsilon - w_h |_{m,\Omega} \leq C h^{2-m} | u_\varepsilon |_{2,\Omega}, \quad 0 \leq m \leq 2. \tag{10}$$

Using the asymptotic expansion $u_\varepsilon = u_0 + v_\varepsilon + z_0$ of the next section, we obtain $| u_\varepsilon |_{2,\Omega} \leq C\varepsilon^{-1/2}$. Combining (9) and (10), we obtain

$$\varepsilon^2 | u - u_h |_2^2 + | u - u_h |_1^2 \leq C(\varepsilon^2 + h^2)\varepsilon^{-1}, \tag{11}$$

which gives (8).
Remark 1. The inequality (8) is an estimate in $O(h\varepsilon^{-s+\frac{1}{2}})$ for $\varepsilon \leq h$; it is, then, a bad estimate. Use of the interpolation operator for continuous functions will worsen the result; L^2 estimates obtained by duality arguments are also worse.

The estimate (11) is optimal, as can be seen from the one-dimensional example

$$\varepsilon^2 u_\varepsilon^{(4)} - u" = -2 \text{ in }]0, 1[, \tag{12}$$

with explicit solution

$$u_\varepsilon(x) = x^2 - x + \varepsilon(1 + e^{-1/\varepsilon})/(1 - e^{-1/\varepsilon}) - \varepsilon(e^{-x/\varepsilon} + e^{-(1-x)/\varepsilon})/(1 - e^{-1/\varepsilon}). \tag{13}$$

3. Asymptotic Expansion

We follow the technique of [9] modified in [3]. We look for an AE of the form

$$U_N = \sum_{i=0}^{N} \varepsilon^i u^i + \varepsilon \sum_{i=0}^{N+1} \varepsilon^i v^i. \tag{14}$$

Here $v^i = \Psi \bar{v}_i$, where Ψ is a C^∞ truncation function such that $\Psi(M) = 1$ if $d(M,\Gamma) \leq \frac{1}{3}\delta$, $\Psi(M) = 0$ if $d(M,\Gamma) \geq \frac{2}{3}\delta$. The purpose of the boundary layer functions $\bar{v}_i$ (described below) is to correct the first part, $\sum \varepsilon^i u^i$, of the AE in the boundary layer.

We assume that Γ is a regular curve parametrized by $\theta \to (x(\theta), y(\theta))$. In addition to θ, we use a second local variable, $\rho = d(M,\Gamma)$. All the functions near Γ are then considered as functions of ρ and θ. The equation (1) is written in these local coordinates:

$$\varepsilon^2 \sum_{k=1}^{4} \sum_{i+j=k, 0\leq i,j\leq k} a_{ij}(\rho,\theta) \frac{\partial^k u}{\partial \rho^i \partial \theta^j} - \sum_{k=1}^{2} \sum_{i+j=k, 0\leq i,j\leq k} b_{ij}(\rho,\theta) \frac{\partial^k u}{\partial \rho^{ij} \theta^j} = f. \tag{15}$$

We introduce the stretched variable $t = \rho/\varepsilon$ and use the Taylor formula for the a_{ij} and b_{ij}:

$$a_{ij}(\varepsilon t, \theta) = \sum_{k=0}^{N} a_{ij}^k(\theta) \varepsilon^k t^k + a_{ij}^{N+1}(\varepsilon t, \theta) \varepsilon^{N+1} t^{N+1}.$$

From (15) we can write

$$L_\varepsilon u \equiv \varepsilon^{-2} \sum_{k=0}^{N+1} \varepsilon^k M_k u,$$

where

$$M_0 u = \frac{\partial^4 u}{\partial t^4} - \frac{\partial^2 u}{\partial t^2}$$

$$M_1 u = \sum_{i=0}^{1} [(a_{3i}^0(\theta) \frac{\partial^{3+i} u}{\partial t^3 \partial \theta^i}) - b_{1i}^0(\theta) \frac{\partial^{1+i} u}{\partial t \partial \theta^i}],$$

and the $M_k, k \geq 2$, have similar expressions.

Substitution of U_N in (1) gives

$$\varepsilon^{-1}\{M_0 \bar{v}^0 + \varepsilon(M_0 \bar{v}^1 + M_1 \bar{v}^0) + \ldots\} - \Delta u^0 - \varepsilon \Delta u^1 - \ldots + \varepsilon^{N+1} R_N = f.$$

Thus, u_0 is defined as the solution of Problem (P_0) given in the introduction. To define $\bar{v}_0$, we use $M_0\bar{v}^0 = 0$ and choose the root $\lambda = -1$ in the characteristic equation $\lambda^2(\lambda^2 - 1) = 0$ associated with M_0 as being the only root giving exponential decay. From the boundary condition (3) we write at first order

$$\frac{\partial \bar{v}^0}{\partial t} = \frac{\partial u^0}{\partial n} \quad at \quad t = 0$$

and set

$$\bar{v}^0(t,\theta) = \frac{\partial u^0}{\partial n}(0,\theta)e^{-\rho/\varepsilon}.$$

Higher-order terms are built in the same way. Up to this point the AE is only formal; that is, it satisfies

$$L_\varepsilon U_N = f - \varepsilon^{N+1} R_N.$$

The validity of this expansion can be proved in Hölder norm if Γ and f are sufficiently regular. In particular, in [6] it is shown that

$$[u_\varepsilon - U_N]_j \leq C\varepsilon^{N+1-j}, \qquad j = 0 \ to \ N. \tag{16}$$

Here $[v]_j = \|v\|_{j,\infty,\Omega}$.

As $[v^i]_j \leq C\varepsilon^{-j} \qquad j = 0, \ldots, N$ we can eliminate unnecessary terms from (16) to obtain the following theorem.

Theorem 2. If Γ is sufficiently regular, then there exists, for each sufficiently regular f, a constant C independent of ε, such that

$$[u_\varepsilon - \sum_{i=0}^{N-j} \varepsilon^i u^i - \varepsilon \sum_{i=0}^{N-1} \varepsilon^i v^i]_j \leq C\varepsilon^{N+1-j}, \quad j = 0, \ldots, N. \tag{17}$$

For $j = N = 1$ we obtain

$$[u_\varepsilon - u^0 - \varepsilon v^0]_1 \leq C\varepsilon. \tag{18}$$

4. A Composite AE-FEM with P_1 Finite Elements

To simplify the description, we suppose that the triangulation T_h is such that each edge emanating from a node on Γ is normal to Γ, so that Fig. 2 is valid.

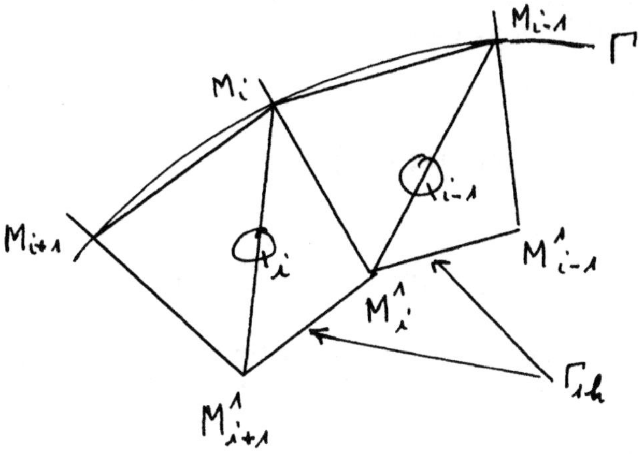

Fig. 2. **Triangularization for the composition AE-FEM**

Let V_h be the P_1 finite element space for Problem (P_0):

$$V_h = \{v_h \in C^0(\bar{\Omega}_h); v_h \mid_K \in P_1, \quad \forall K \in T_h, \quad v_h = 0 \text{ on } \Gamma_h\},$$

where

$$\bar{\Omega}_h = U_{K \in T_h}, \qquad \Gamma_h = \partial \bar{\Omega}_h.$$

The use of this FEM for Problem (P_0) gives an approximation $u^{0h} \in V_h$ of u^0.

For the approximation of v^0 we define $\Omega_{1,h} = \cup Q_i$, the first layer of triangles with interior boundary $\Gamma_{1,h}$ (see Fig. 2). Given $M \in \Omega_{1,h}$, $P(M)$ is the projection of M on Γ and M_1 the intersection of $\Gamma_{1,h}$ with $MP(M)$. We set

$$v^{0h}(M) = \frac{u^{0h}(M_1)}{P(M)M_1} \Psi(M) \exp -\frac{1}{\varepsilon} MP(M). \tag{19}$$

The truncation function Ψ is regular and such that $\Psi(M) = 1$ if $d(M,\Gamma) \leq \frac{1}{3}\mu h$ and $\Psi(M) = 0$ if $d(M,\Gamma) \geq \frac{2}{3}\mu h$.
Lemma 1. *If* $u^0 \in H_0^1(\Omega) \cap W^{2,\infty}(\Omega)$, *then*

$$\mid v^0 - v^{0,h} \mid_{1,\infty,\Omega} \leq C\varepsilon^{-1/2} h(-\log h)^{3/2}. \tag{20}$$

Proof. The proof of this lemma is based on a Taylor expansion of $w^h(M) = u^{0,h}(M_1)/P(M)M_1$ and the estimate (see [4])

$$\mid u^0 - u^{0,h} \mid_{0,\infty,\Omega_h} \leq Ch^2(-\log h)^{3/2} \|u^0\|_{2,\infty,\Omega}. \tag{21}$$

Then an explicit calculation using the coordinates (ρ, θ) gives (20).

Theorem 3. *Let Ω be such that $u^0 \in H_0^1(\Omega) \cap W^{2,\infty}(\Omega)$. Then $U_0^h = u^{0,h} + \varepsilon v^{0,h}$ is an approximation of u_ε satisfying*

$$| u_\varepsilon - U_0^h |_{1,\Omega} \leq C(\varepsilon + h) \tag{22}$$

$$| u_\varepsilon - U_0^h |_{1,\infty,\Omega_h}, \leq C[\varepsilon + h(-\log h)^{3/2}]. \tag{23}$$

Proof.

$$| u_\varepsilon - U_0^h | \leq | u_\varepsilon - U_0 | + | u^0 - u^{0,h} | + \varepsilon | v^0 - v^{0,h} | .$$

For example, to obtain (23), we use (18) for the first term, Lemma 1 for the third, and the following estimate of [4] for the second:

$$| u^0 - u^{0,h} |_{1,\infty,\Omega_h} \leq Ch(-\log h) \| u^0 \|_{2,\infty}.$$

Remark 2. The use of $u^{0,h}$ only as an approximation of u_ε gives $| u_\varepsilon - \tilde{u}^{0,h} |_{1,\Omega} \leq C(h + \varepsilon^{1/2})$ ($\tilde{u}^{0,h}$ is the extension of $u^{0,h}$ by zero in $\Omega - \Omega_h$) and $| u_\varepsilon - u^{0,h} |_{1,\infty,\Omega_h} \leq C(1+h)$. The introduction of the approximate boundary layer function $\varepsilon v^{0,h}$ corrects the normal derivative of $u^{0,h}$ in $\tilde{\Omega}_{1,h}$.

5. A Composite AE-FEM with Curved HCT Triangles

Following an idea of [10], we use a triangulation with triangles having a curved side on the boundary. The boundary triangles are described by a special mapping F_K from a reference element $\hat{K}$ with vertices $\hat{a}_1(0,0), \hat{a}_2(1,0), \hat{a}_3(0,1)$ onto K defined by

$$F_K(\xi, \eta) = (x_0(\xi, \eta) + (1 - \xi - \eta)\phi(\eta), y_0(\xi, \eta) + (1 - \xi - \eta)\Psi(\eta))$$

$$x_0(\xi, y) = x_1 + (x_2 - x_1)\xi + (x_3 - x_1)\eta$$

$$y_0(\xi, y) = y_1 + (y_2 - y_1)\xi + (y_3 - y_1)\eta$$

$$\phi(\eta) = (1 - \eta)^{-1}\{x[(\theta_1) + (\theta_3 - \theta_1)\eta] - x_1 - (x_3 - x_1)\eta\}$$

$$\Psi(\eta) = (1 - \eta)^{-1}\{y[\theta_1 + (\theta_3 - \theta_1)\eta] - y_1 - (y_3 - y_1)\eta\}.$$

Thus, $F_K(\hat{a}_1, \hat{a}_3)$ is exactly the arc $a_1 a_3$ of Γ. Let $\hat{P} = \{\hat{p} \in C^1(\hat{K}); \hat{p} |_{\hat{K}_i} \in P_3, i = 1,2,3\}$ be the space of the HCT polynomial on $\hat{K}$, as described in the introduction. We set

$$V_h = \{v_h \in C^1(\bar{\Omega}); F_K \circ v_h |_K \in \hat{P}, \quad v_h = 0 \text{ on } \Gamma\}.$$

The use of this FEM for Problem (P_0) gives an approximation $u^{0,h} \in V_h$ of u^0. We assume that the following error estimates are valid:

$$| u^0 - u^{0,h} |_{m,\infty,\Omega} \leq Ch^{4-m} \|u^0\|_{4,\infty,\Omega}. \tag{24}$$

It is easy to approximate v^0 by setting

$$v^{0,h}(M) = \frac{\partial u^{0,h}}{\partial n}(P(M))\Psi(M) \exp\left(-\frac{1}{\varepsilon} MP(M)\right), \tag{25}$$

where $P(M)$ is defined in Section 4. For this approximation, $u^{0,h} + \varepsilon v^{0,h} = U_0^h$, we can prove the following theorem.

Theorem 4. *Let Ω be such that $u^0 \in H_0^1(\Omega) \cap W^{4,\infty}(\Omega)$ and that (24) is valid. Then $U_0^h = u^{0,h} + \varepsilon v^{0,h}$ is an approximation of u_ε satisfying*

$$| u_\varepsilon - U_0^h |_{1,\infty,\Omega} \leq C(\varepsilon + h^3). \tag{26}$$

Proof. The proof of this theorem follows the same lines as the proof of Theorem 3; see [6] for details.

References

[1] Baranger, J., and El Amri, H.(1990) 'Numerical solution of spectral problem for an ODE with a small parameter using an asymptotic expansion and a finite element method', *Numer. Funct. Anal. and Optimiz.*, **11**, 621–642.

[2] Bernardi, C. (1975) 'Optimal finite interpolation on curved domains', *SIAM J. Numer. Anal.*, **26** (1989) 1212–1260.

[3] Besjes, J. G. (1975) 'Singular perturbation problems for linear differential operators of arbitrary order. I Degeneration to elliptic operators', *J. Math. Anal. Appls.*, **49**, 324–346.

[4] Ciarlet, P. G. (1978) *The Finite Element Method for Elliptic Problems. Studies in Math. and its Applications, Vol. 4*, North Holland, Amsterdam.

[5] Clément, Ph. (1978) 'Approximation by finite element functions using local regularization', *RAIRO*, **2**, 77–84.

[6] El Alami El Ferricha, M. (1992) 'Approximation numérique d'un problème de perturbations singulières en dimension deux d'espace par utilisation conjointe d'un développement asymptotique et d'une méthode d'éléments finis', Thesis, Université C. Bernard Lyon 1.

[7] Lions, J. L. (1973) *Perturbations Singulières dans les Problèmes aux Limites et en Controle Optimal, Lecture Notes in Mathematics, Vol. 323*, Springer-Verlag, Berlin.

[8] Schatz, A. H., and Wahlbin, L. (1983) 'On the finite element method for singularly perturbed reaction-diffusion problems in two and one dimension', *Math. of Comp.*, **40**, no. 161, 47–89.

[9] Vishik, N.I., and Lyusternik, L. A. (1962) 'Regular degeneration and boundary layer for linear differential equation with small parameter', *Uspechi Math. Nauk.*, **12** (*Amer. Math. Soc. Translations*, **Series 2**, 20.

[10] Zlamal, M. (1973) 'Curved elements in the finite element method', *SIAM J. Num. Anal.*, **10**, no. 1, 229–239.

INTERNAL BOUNDARY CONDITIONS AND THE FEEDBACK LOOP FOR NESTED TWO-WAY COMMUNICATION SCHEMES

A. Louise Perkins
Institute for Naval Oceanography
Bay St. Louis MS

ABSTRACT. We address the relationship between the interior boundary conditions and the feedback mechanism in nested two-way domain decomposition. In particular, we compare an averaging feedback loop with a relaxation feedback loop, using Dirichlet boundary conditions. We construct a two-way nested communication scheme that couples the internal boundary conditions with a feedback loop, maintaining consistency as defined herein. This coupled consistency requirement guarantees that the aggregate solution makes a smooth transition across an internal interface. Both feedback methods were tested by using a barotropic modon for the initial condition. For strong relaxation values combined with a limited refinement ratio, the relaxation feedback method is shown to be more accurate than the averaging feedback method.

1. Introduction

In this paper we address in detail the relationship between the interior boundary conditions and the feedback mechanism in nested two-way domain decomposition. In particular, we compare an averaging feedback loop with a relaxation feedback loop, using Dirichlet boundary conditions. In the nested two-way domain decomposition method partial differential equations are solved by overlaying a nested refined, or fine, mesh on a coarser mesh to obtain a "better" solution in the refined region. "Better" might be defined as a resolution of smaller length and time scales that resolve regional phenomena, or it might be maintaining a predefined level of accuracy throughout the domain. In such cases, and here, we assume that the refined solution is the "better" solution, and it is therefore desirable to use this "better" solution to influence the behavior in the coarser region via a feedback communication method. Two-way nested solutions use the coarser mesh to obtain boundary conditions for the fine mesh, advance the solution on the fine mesh, and then blend or feed back the fine solution to the coarser solution. This feedback loop shunts the fine resolution ("better") solution information back

H. G. Kaper and M. Garbey (eds.),
Asymptotic and Numerical Methods for Partial Differential Equations with Critical Parameters 321–334.

to the coarse-mesh solution. Because the boundary conditions for the fine region come from the coarser mesh, it is important to maintain a consistent relationship between these internal boundary conditions and the feedback loop.

We construct a two-way nested communication scheme that couples the internal boundary conditions with a feedback loop, maintaining consistency as defined herein. We then compare the accuracy of both feedback methods considered in this paper, using a numerical advection model initialized with a barotropic modon. A barotropic modon is an analytic formulation of an ocean feature that models mesoscale eddies. Such eddies require a fine mesh to adequately resolve their flow dynamics, while the overall ocean circulation flow dynamics, dominated by large-scale energetics, can be represented on much coarser meshes.

Historically, Dirichlet boundary conditions have been enforced on the fine mesh by using linear interpolation in space along the internal boundaries, combined with an averaging feedback method. We show that this combined two-way nested scheme does not, in general, lead to a smooth interface transition. As an alternative we introduce a relaxation feedback method, where we view relaxation as a weighted filter that combines the coarse- and fine-mesh results. In this scheme feedback takes the form

$$f_a = \omega f_c + (1 - \omega) f_f, \tag{1}$$

where $0 \leq \omega \leq 1$, f_a are the aggregate discrete solution values on the coarse mesh, f_c are the original discrete solution values on the coarse mesh, and f_f are the discrete solution values at the coincident fine-mesh points.

We also note that the averaging feedback method may have a wider domain of dependence than the partial differential equation. In contrast, the relaxation feedback method maintains the original domain of dependence.

2. Previous Two-Way Nesting Research

Some other methods that use domain decomposition as an integral component are the independent time-step method [6], the local uniform-mesh refinement method [1, 3, 8], and the particle-in-cell method [12]. The interested reader is referred to [3] for a brief history of adaptive methods and to [18] for a detailed study of mesh generation.

The independent time-step method, as developed by Eltgroth [5], may be characterized by the lack of a relationship in time step size between adjacent subdomains. This method was developed for the Lagrangian formulation of the inviscid Euler equations of gas dynamics on very large domains. These large domains are partitioned into many static, homogeneous subdomains. Note that this method has no corresponding spatial refinement.

Berger and Jameson [1] used a local uniform mesh refinement technique to solve inviscid hyperbolic partial differential equations. The method enforces a uniform error distribution without sacrificing local mesh uniformity, by placing uniform local fine meshes over regions of larger error. Explicit finite-difference schemes were used on the coarse mesh, with Richardson-type local truncation error estimates determining the placement of the fine meshes. In regions where the error estimates were too large, the coarse mesh was refined in time and space. These refined meshes were then recursively refined until a satisfactory error limit had been achieved. Bolstad [3] did earlier work in one spatial dimension, but decided where to refine based on a physically motivated "magnitude of gradient" criterion. Gropp [8] also discussed using a local uniform mesh refinement to follow shocks that arise in the hyperbolic two-dimensional Burgers equation. His subdomain placement is also based on a gradient threshold.

The particle-in-cell method of Mas-Gallic [12] can be viewed from two perspectives; either they partition or they decompose the molecular domain into superparticles. The Eulerian formulation is used to calculate the field information on a fixed mesh. Then particles are advanced over this mesh, based on these Eulerian forcing fields. The intended use is for problems in two or more spatial dimensions with fluid distortions, large shears, or multiple types of fluid. This method has been used by many people, including Horowitz [9].

Domain decomposition allows the mesh to evolve with the solution. It has been applied to elliptic and hyperbolic equations for several years (see [5, 4] for extensive bibliographies). But it is well known that the existing domain decomposition methods degrade when the mesh boundaries are not placed in dynamically quiescent regions. In geophysical modeling, domain decomposition has been used to solve the primitive equations of both oceanography [17] and meteorology [10], but for oceanographic modeling in particular, it is difficult (perhaps impossible) to locate relevant quiescent locations for the internal boundaries where the mesoscale features of the solution do not cross fine-mesh boundaries, and so large errors may occur about these boundaries. It is therefore necessary to develop numerical domain decomposition techniques that include flow dynamics across these internal boundaries.

3. Domain Decomposition

Formally, we describe the domain decomposition of a discrete coarse mesh, Ω^1, over the continuous Ω domain for a fixed time $t = n\Delta t$, as

$$\bigcup_{k=1}^{p(t)} \Omega^k(t) \subseteq \Omega \ , \tag{2}$$

where $(p(t)-1)$ is the number of refined subdomains used for the time step starting at time $n\Delta t$, and $\{\Omega^k\}_{k=2}^{p(t)}$ are those subdomains. We also refine in time using an integral ratio r. Let $\Delta_c t$ be the coarse-mesh time step and $\Delta_f t$ be the fine-mesh time step. Then the integral ratio of refinement relates these two times steps: $\Delta_c t = r\Delta_f t$. For the problems considered, Ω is a "four-sided" domain.

After the domain is decomposed, local uniform spatial mesh refinement, as developed in [1], is applied to the new subdomain. We use a two-level refinement scheme consisting of one coarse mesh and one overlaid refined mesh. In ocean modeling the coarse mesh must adequately represent the general circulation (larger spatial scales) in sufficient detail to provide this information to the fine mesh, while the fine meshes must accurately resolve the significant smaller-scale behaviors. Where this assumption does not hold, any domain decomposition method may fail to accurately resolve the smaller-scale behavior.

4. Internal Dirichlet Boundary Conditions with an Averaging Feedback

In this section we show that an averaging feedback loop is inconsistent with linearly interpolated Dirichlet boundary conditions on a rectangular mesh. Without loss of generality we begin by examining the simple mesh shown in Fig. 1, where a refinement value of 2 has been chosen to simplify the presentation. In Fig. 1a only the outer mesh lines are drawn, whereas in Fig. 1b both the inner and outer mesh lines are drawn. The "X" marks represent mesh points shared by both the inner and the outer mesh, while the "O" marks represent the mesh points on the inner fine mesh. All boundary values to be provided by the coarse outer mesh for the boundary conditions of the fine nested mesh have been labeled with two juxtaposed letters; the first letter represents the first dimensions relative position (L-left, C-center, R-right), while the second letter represents the second dimensions relative position (L-lower, C-center, U-upper). The fine indexing, (i,j), is used to refer to elements on either mesh.

As boundary values must be provided to the inner meshes at all fine "O" boundary points (marked with solid circles in Fig. 1b), the following question is now posed: Does interpolation from the outer "X" mesh values to the boundary "O" points produce any values that will alter the feedback average value at the "X" points? That is, at time t^{n+r}, after the outer mesh has been advanced one $\Delta_c t$ time step and the inner mesh has been advanced $r\Delta_f t$ time steps, an averaging feedback will replace the values along the internal boundary, marked with "X," with new values. At this same time t^{n+r}, the outer mesh will update these internal boundary values at the "O" locations along the internal boundary. Hence, would another feedback loop

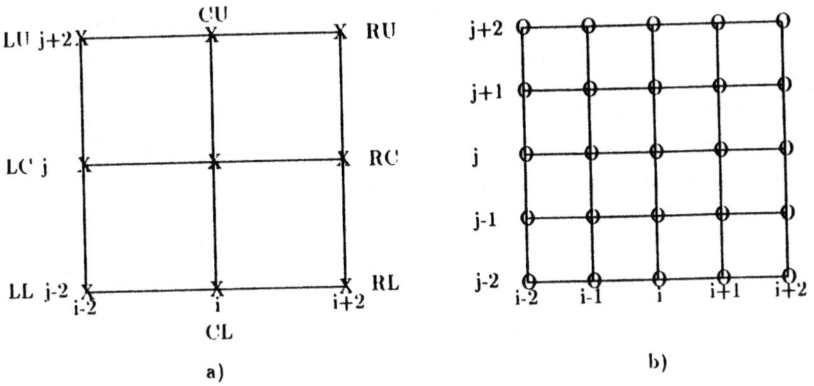

Fig. 1. **Coarse and fine mesh points**

immediately after the calculation of the prescribed boundary conditions, lead to the identical prescribed boundary conditions?

The two sequence operation of averaging feedback followed by Dirichlet boundary conditions is not nilpotent, and hence just one application of the combined operator leads to an inconsistent state between the outer and inner solutions. We believe this is the reason that some existing *ad hoc* domain decomposition methods separate the internal boundary from the feedback domain in space. Such a separation smoothes out the inconsistency by providing an interval within which the disparate solutions are smoothly joined into one aggregate solution.

To illustrate the inconsistency, we again refer to Fig. 1. Along the boundary we can identify eight classes of point, which we partition into corner classes (LL, LU, RU, and RL) and side classes (LC, RC, CU, and CL). Table 1 contains the averaging equation for each class. Notice that while the corner classes always have only one element in them (the four corner points), the side classes can contain many points.

Some of the terms in the feedback equation come from the interior of the domain, where they depend only on the prescribed boundary values at t^n and the "initial" conditions at t^n. Some terms lie on the "X" mesh points and have fixed values given by the Dirichlet boundary conditions derived from the aggregate coarse mesh solution regardless of the interpolation algorithm used for the boundary "O" values. The remaining terms are exactly the boundary "O" values that we will be interpolating to when we prescribe the boundary conditions for the next integration step. These fine boundary values had fixed (prescribed) values during the time interval $[t^n, t^{n+r}]$; but during the coupling phase that occurs after t^{n+r} but before the time interval $[t^{n+r}, t^{n+2r}]$, they are not prescribed. Indeed, this is exactly the time when we must decide how to alter them. To examine these requirements, we begin by considering the fine boundary values *for the feedback loop* as undetermined, thus providing us with degrees of freedom that allow us to couple the two operations of feedback and boundary value assignment. With these degrees of freedom we develop an adjusted set of boundary values for the fine mesh such that the coarse and fine boundary conditions can be made to be consistent along the internal interface if the two-way nested scheme allows for such consistency. When this consistency condition can be satisfied, the two-way nested scheme will make a smooth transition across the interface.

Table 1. **Averaging Equations with Refinement Ratio** $r = 2$

Class	Averaging Feedback Equation
LL	$1/4 * (U_{i-2,j-1} + U_{i-1,j-1} + U_{i-1,j-2} + U_{i-2,j-2})$
LU	$1/4 * (U_{i-2,j+1} + U_{i-1,j+1} + U_{i-1,j+2} + U_{i-2,j+2})$
RL	$1/4 * (U_{i+1,j-1} + U_{i+2,j-1} + U_{i+1,j-2} + U_{i+2,j-2})$
RU	$1/4 * (U_{i+1,j+1} + U_{i+1,j+2} + U_{i+2,j+1} + U_{i+2,j+2})$
LC (left)	$1/6 * (U_{i-2,j-1} + U_{i-2,j} + U_{i-2,j+1} + U_{i-1,j-1} + U_{i-1,j} + U_{i-1,j+1})$
RC (right)	$1/6 * (U_{i+1,j-1} + U_{i+1,j} + U_{i+1,j+1} + U_{i+2,j-1} + U_{i+2,j} + U_{i+2,j+1})$
CU (top)	$1/6 * (U_{i-1,j+1} + U_{i,j+1} + U_{i+1,j+1} + U_{i-1,j+2} + U_{i,j+2} + U_{i+1,j+2})$
CL (bottom)	$1/6 * (U_{i-1,j-2} + U_{i,j-2} + U_{i+1,j-2} + U_{i-1,j-1} + U_{i,j-1} + U_{i+1,j-1})$

We began with a simple linear interpolation algorithm from the outer mesh to these inner mesh boundary points, although we realize that linear interpolation is not always a good choice for such problems [2]. These equations are shown in Table 2, where the superscript f identifies the fine inner mesh while the superscript c identifies the coarser outer mesh. Combining the Table 2 interpolation with the consistency constraints for the feedback loop (defined in Table 1) couples these equations into eight equations in eight unknowns. At this point we now have coupled input for the interpolation calculation, with information both from the outer and the inner mesh incorporated.

Table 2. **Interior Boundary Interpolation Formulas**

Point	Linear Interpolation Equation
$U^f_{i-1,j-2}$	$U^c_{i,j-2} + 1/2 * (U^c_{i,j-2} - U^c_{i-2,j-2})$
$U^f_{i+1,j-2}$	$U^c_{i+2,j-2} + 1/2 * (U^c_{i+2,j-2} - U^c_{i,j-2})$
$U^f_{i-2,j-1}$	$U^c_{i-2,j} + 1/2 * (U^c_{i-2,j} - U^c_{i-2,j-2})$
$U^f_{i+2,j-1}$	$U^c_{i+2,j} + 1/2 * (U^c_{i+2,j} - U^c_{i+2,j-2})$
$U^f_{i-2,j+1}$	$U^c_{i-2,j+2} + 1/2 * (U^c_{i-2,j+2} - U^c_{i-2,j})$
$U^f_{i+2,j+1}$	$U^c_{i+2,j+2} + 1/2 * (U^c_{i+2,j+2} - U^c_{i+2,j})$
$U^f_{i-1,j+2}$	$U^c_{i,j+2} + 1/2 * (U^c_{i,j+2} - U^c_{i-2,j+2})$
$U^f_{i+1,j+2}$	$U^c_{i+2,j+2} + 1/2 * (U^c_{i+2,j+2} - U^c_{i,j+2})$

The resulting matrix equation can be written as

$$\begin{bmatrix} 1 & 1 & 0 & 0 & 0 & 0 & 0 & 0 \\ 0 & 1 & 1 & 0 & 0 & 0 & 0 & 0 \\ 0 & 0 & 1 & 1 & 0 & 0 & 0 & 0 \\ 0 & 0 & 0 & 1 & 1 & 0 & 0 & 0 \\ 0 & 0 & 0 & 0 & 1 & 1 & 0 & 0 \\ 0 & 0 & 0 & 0 & 0 & 1 & 1 & 0 \\ 0 & 0 & 0 & 0 & 0 & 0 & 1 & 1 \\ 1 & 0 & 0 & 0 & 0 & 0 & 0 & 1 \end{bmatrix} \begin{bmatrix} u_{i-1,j-2} \\ u_{i-2,j-1} \\ u_{i-2,j+1} \\ u_{i-1,j+2} \\ u_{i+1,j+2} \\ u_{i+2,j+1} \\ u_{i+2,j-1} \\ u_{i+1,j-2} \end{bmatrix} = \begin{bmatrix} 3u_{i-2,j-2} - u_{i-1,j-1} \\ 5u_{i-2,j} - u_{i-1,j-1} - u_{i-1,j} - u_{i-1,j+1} \\ 3u_{i-2,j+2} - u_{i-1,j+1} \\ 5u_{i,j+2} - u_{i-1,j+1} - u_{i,j+1} - u_{i+1,j+1} \\ 3u_{i+2,j+2} - u_{i+1,j+1} \\ 5u_{i+2,j} - u_{i+1,j-1} - u_{i+1,j} - u_{i+1,j+1} \\ 3u_{i+2,j-2} - u_{i+1,j+1} \\ 5u_{i,j-2} - u_{i-1,j-1} - u_{i,j-1} - u_{i+1,j-1} \end{bmatrix}, \quad (3)$$

where the elements are not presented in a natural ordering but are instead rearranged into a sequential ordering along the interior boundary curve. This ordering highlights the periodically bidiagonal nature of the resulting system: the resulting matrix is not strictly diagonally domain, nor is it symmetric. This periodic dependence leads to a nonsingular matrix, except for random fortuitous cancellations. For this combination of linear interpolation and averaging feedback, there is no general solution that will allow the inner and outer solutions to be consistent. While this does not rule out the success of different interpolation algorithms, we choose to focus here on the averaging feedback loop.

5. Averaging Feedback and Expansion of the Domain of Dependence

An averaging feedback loop introduces the possibility of violating the domain of dependence. Figure 2, where the time axis ("y ") is represented in intervals of $\Delta_r t$, illustrates this violation. As before, the coarse-mesh points are marked with an "X" and the fine-mesh points with an "O." For point $i, t + 3\Delta r$ the strict domain of dependence, on the coarse mesh *without* a feedback loop, assuming we are at the maximum CFL limit, is marked in Fig. 2a. The range of influence *after* adding an averaged feedback loop is shown in Fig. 2b. While the smoothing introduced by both averaging and

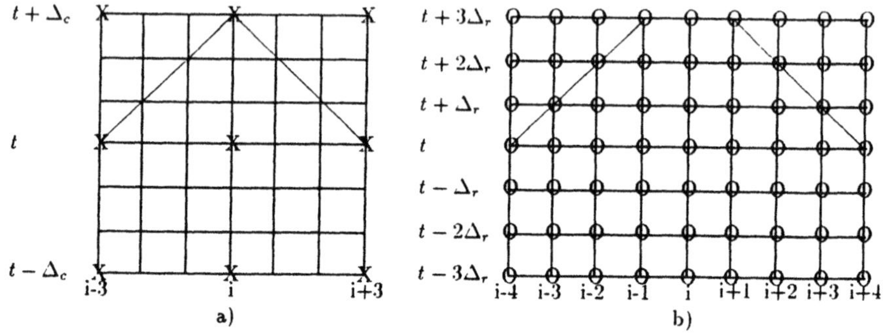

Fig. 2. **Composition of the domain**

interpolation seems, in practice, to keep the resulting solution stable, a counterexample seems at least possible to construct.

6. Error Comparison

In this section we now look at the error arising for a one-dimensional problem. Assume that the coarse-mesh solution c and the fine-mesh solution f have the same initial conditions on all coarse-mesh points ("X"). Assume also that the error on the inner fine mesh is significantly smaller than the error on the outer coarse mesh. Then our total error, ϵ, before feedback, on the coarse mesh restricted to the fine-mesh domain can be approximated as

$$\epsilon_k^{n+1} \approx f_k^{n+1} - c_k^{n+1}, \tag{4}$$

where $k = 1, ..., n$ indexes the total number of coarse-mesh points in this restricted domain. Take one time step together with the coupled averaged feedback correction. This determines the aggregate solution, C, on this restricted domain:

$$C_k^{n+1} = \frac{1}{r} \sum_{i=\lfloor\frac{r}{2}\rfloor}^{i=\lfloor\frac{r}{2}\rfloor} f_{k+i}^{n+1}, \tag{5}$$

where r is the total number of averaged fine points per coarse-mesh point. After one coupled time step the approximate error takes the form

$$\epsilon(C_k^{n+1}) = \frac{r-1}{r} f_j^{n+1} - \frac{1}{r} \sum_{i=\lfloor\frac{r}{2}\rfloor, i\neq k}^{i=\frac{r}{2}\rfloor} f_{k+i}^{n+1}, \tag{6}$$

where the behavior of the error will vary depending upon the averaged values of the inner mesh solution, and it approaches the error on the fine mesh in regions of quiescent dynamics. However, in general, the behavior of this error term can be quite varied.

In contrast, the relaxation scheme

$$C_k^{n+1} = \omega c_k^{n+1} + (1-\omega) f_k^{n+1} \tag{7}$$

has error

$$\epsilon(C_k^{n+1}) = \omega(f_k^{n+1} - c_k^{n+1}) \tag{8}$$

after one coupled time step. The right-hand side of this equation reduces to

$$\epsilon(C_k^{n+1}) = \omega(\epsilon_k^{n+1}), \tag{9}$$

where ϵ_k^{n+1} is the "total" error for a one-way nested scheme without feedback. Thus, for a relaxation feedback scheme, the error growth of the coarse aggregate solution is bounded by the error of a one-way nested scheme. It is therefore at least as accurate as the coarse-mesh solution.

7. Barotropic Modon

Another issue involved in choosing a two-way communication scheme is the absolute errors introduced. It is evident from the above error analysis that either feedback communication scheme can produce small absolute errors, depending on the behavior of the solution. Here we examine this absolute error by example, using a barotropic modon [7, 17] for our initial conditions. Such a barotropic modon represents well many of the features of oceanic mesoscale eddies, and conclusions based on its error behavior can be convincingly extrapolated to more realistic mesoscale ocean model behavior. Further, this barotropic modon has a known analytic solution, allowing us to measure error directly. Let

$$\begin{aligned} r &= 6370 \times 10^5 \\ \Omega &= \frac{\pi}{43082} \\ \beta &= 2\Omega \cos(\Theta)\frac{1}{r} \\ c &= \beta * a^2 = 17.85 \text{ cm/s}, \end{aligned}$$

where

r - radius of the earth
Ω - Rotation speed of the earth
β - Coriolis force coefficient on a beta plane
Θ - 38.5 N (latitude)
c - modon advection speed
$a = 10 \times 10^7$ cm

The barotropic modon streamfunction is then given by

$$\Psi = H_0 c \sin\Theta \left[\frac{a\beta J_1(kr)}{k^2 c J_1(ka)} - \left(1 + \frac{\beta}{k^2 c}\right) r \right], \qquad 0 \leq r \leq a, \tag{10}$$

$$\Psi = -H_0 ac \sin\Theta \left[\frac{K_1\left(\frac{r}{\sqrt{c/\beta}}\right)}{K_1\left(\frac{a}{\sqrt{c/\beta}}\right)} \right], \qquad r > a,$$

where $H_0 = 5 \times 10^5$ cm, J_1 and K_1 are Bessel functions, and the mesh size is 1540×1540 km^2 [7].

8. Test Cases

For this barotropic modon we examine the regions of (ω, r) parameter space where the relaxation error is less than the averaging error. Certainly $\omega = 0$ completely decouples the two solutions; the other extreme, $\omega = 1$, is a subsampling scheme that may introduce numerical instability.

In this experiment we use a modon as an approximation to an isolated eddy propagating through a fluid at rest. Although the modon is an exact nonlinear solution to the inviscid quasi-geostrophic equations of oceanography for a two-layer model [14], it propagates through the quiescent fluid at constant velocity. The analytic modon consists of an antisymmetric pair of nonlinear vortices that interact nonlinearly to maintain the internal organization of the modon, and the modon is strong in the sense that it transports all the fluid within the modon. The initial conditions are shown in Fig. 3.

The nonlinear nature of the modon makes it difficult to model numerically unless an extremely fine mesh is used. As the mesh size increases the modon rapidly loses the balance essential for maintaining the nonlinear interactions. Rapid degradation follows, with the modon literally breaking apart [17]. Hence the comparison between the averaging and the relaxation feedback loop is dependent upon the actual size of the fine and coarse meshes, as well as the refinement ratio between them. This requirement is related to our introductory remarks about "better" solutions.

We began with a fine mesh sufficient to maintain the structure of the modon for sufficiently long time scales. Then two test cases were chosen to allow us to partition (ω, r) parameter space into regions where the consistent

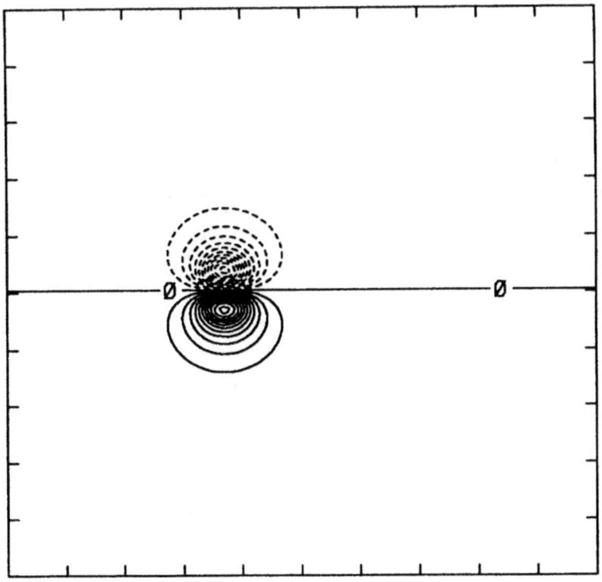

Fig. 3. **Initial barotropic modon**

relaxation feedback loop outperformed the inconsistent averaging two-way communications schemes currently in use [1, 16, 15, 17]. The test cases that define the refinement ratio r are shown in Table 3. The values used for ω were taken from 0, .25, .50, .75, .9, and 1.0. As illustrated in Fig. 4, already with reasonable values of ω, the interface consistent relaxation feedback loop is an improvement over the averaging feedback. However, as the ratio of refinement becomes larger, the weight of the relaxation scheme must be increased significantly to keep the error low. Such increases can introduce numerical instability. Hence we conclude that, for reasonable refinement ratios in the ocean, the relaxation feedback scheme is an alternative choice to the averaging feedback scheme.

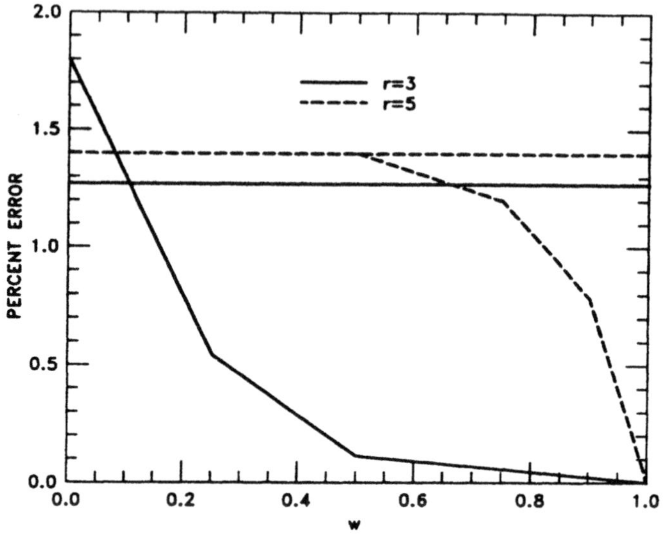

Fig. 4. **Percent error as a function of** ω **for** $r = 3.5$

Table 3. **Test cases**

Test Case 1	$\Delta t = 0.0002$	$\Delta_c x = 1/126$	$r = 1/3$
Test Case 2	$\Delta t = 0.0001$	$\Delta_c x = 1/63$	$r = 1/5$

9. Summary

We have shown that a coupled consistency requirement is sufficient to guarantee that the aggregate solution has a single-valued transition across an internal interface. We then demonstrated, by example, that an averaging feedback loop, coupled with Dirichlet boundary conditions derived by linear interpolation, cannot in general be made consistent. Both feedback methods were tested on a barotropic modon problem, where we saw that for strong enough relaxation values, combined with a limited refinement ratio, the relaxation feedback was more accurate than averaging.

Previous research concluded that when the boundary and the feedback interfaces were physically distinct in space, smoother solutions were noted across the interface. Such staggered boundary condition/feedback interfaces avoid the inconsistency problem discussed in this paper by spreading the discontinuity in space, thereby smoothing the two solutions together. We have shown why this discontinuity exists and have proposed an alternative feedback scheme that does not introduce a discontinuity. In significant regions of (ω, r) parameter space, this alternative two-way relaxation scheme with Dirichlet boundary conditions was shown to be consistent and to have less error compared to the commonly used averaging feedback loop.

There are many details still to resolve for nesting the equations of geophysics in general, and oceanography in particular. This study defined a consistent interface condition and then used that definition to show that Dirichlet boundary conditions coupled with an averaging feedback loop is not always the feedback mechanism of choice, because it is not a consistent interface condition and can have significant error.

In the future we may wish to choose which aspects of the refined solution to average onto the coarse mesh and which to relax. Clearly, more sophisticated schemes can be constructed for different needs that will greatly enhance the feedback mechanism for domain decomposition.

Acknowledgements

This work was performed under the auspices of the Institute for Naval Oceanography.

References

[1] Berger, M. J., and Jameson, A. (1983) 'Automatic adaptive grid refinement for the Euler equations', Princeton University MAE Report No. 1633.

[2] Blake, D. W. (1991) 'One-way nesting for a primitive equation ocean model', in *Proceedings for the ICASE Workshop on Heterogeneous Boundary Conditions*, ICASE Interim Report 19.

[3] Bolstad, J. H. (1982) 'An adaptive finite difference method for hyperbolic systems in one space dimension', Lawrence Berkeley Lab Report No. 13287 Rev.

[4] Chan, T. F., Glowinski, R., Periaux, J., and Widlund, O. B., eds. (1989) *Domain Decomposition Methods*, SIAM, Philadelphia.

[5] Eltgroth, P. G. (1984) 'Application of the independent time step method to multiprocessors', Lawrence Livermore National Laboratory Report UCRL-89888 Rev. 1.

[6] Eltgroth, P. G., and Porter, A. P. (1984) 'The independent time step method for hydrodynamics', Lawrence Livermore National Laboratory Report UCRL-89853, Rev. 1.

[7] Flierl, G. R., Larichev, V. D., McWilliams, J. C., and Reznik, G. M. (1980) 'The dynamics of baroclinic and barotropic solitary eddies', *Dynamics of Atmospheres and Oceans*, **5**, 1–41.

[8] Gropp, W. D. (1980) 'A test of moving mesh refinement for 2-D scalar hyperbolic problems', *SIAM J. Sci. Stat. Comput.*, **1**, 191–197.

[9] Horowitz, E. J. (1987) 'QN3D: A three dimensional quasi neutral hybrid particle in cell code with applications to the tilt mode instability in field reversed configurations', Ph.D. thesis, University of California at Davis (Lawrence Livermore National Laboratory Report UCRL-53808.

[10] Koch, S. E., and McQueen, J. T. (1987) 'A survey of nested grid techniques and their potential for use within the MASS weather prediction model', NASA Technical Memorandum 87808.

[11] Malanotte-Rizzoli, Young, R. E., and Haidvogel, D. B. (1989) 'Initialization and data assimilation experiments with a primitive equation model', *Dynamics of Atmospheres and Oceans,* **13**, 349–378.

[12] Mas-Gallic, S., and Raviart, P. A. (1987) 'A particle method for first-order symmetric systems', *Numer. Math.*, **51**, 323–352.

[13] Oliger, J., and Sundstrom, A. (1978) 'Theoretical and practical aspects of some initial boundary value problems in fluid dynamics', *SIAM J. Appl. Math.*, **35**.

[14] Pedlosky, J. (1987) *Geophysical Fluid Dynamics*, Springer-Verlag, New York.

[15] Perkins, A. L., and Rodrigue, G. (1990) 'A domain decomposition method for solving the 2-dimensional viscous Burgers' equation', *Applied Numerical Mathematics*, **6**, no. 4, 329–340.

[16] Shamarock, W., Oliger, J., and Street, R. L. (1989) 'Adaptive grid refinement for numerical weather prediction', *J. Comp. Physics*, **80**, 27–60.

[17] Spall, M. A., and Holland, W. R. (1990) 'A nested primitive equation model for oceanic applications', *J. Phys. Oceanography.*

[18] Thompson, J. F., Warsi, Z. U. A., and Mastin, C. W. (1985) *Numerical Grid Generation Foundations and Applications*, North Holland, Amsterdam.

SUBDOMAIN PREDICTOR-CORRECTOR ALGORITHMS FOR SOLVING THE INCOMPRESSIBLE NAVIER-STOKES EQUATION

P. Spiteri
IRIT-ENSEEIHT-CNRS UA1399
2 Rue Charles Camichel
31071 Toulouse
France

and

H. C. Boisson
IMFT-ENSEEIHT-CNRS UA0005
Avenue Camille Soula
31400 Toulouse
France

ABSTRACT. The discretization of the Navier-Stokes equation by an implicit method yields large-scale algebraic systems. The aim of the present paper is to model and analyze a class of subdomain algorithms without overlapping. Experimental results on a shared-memory architecture are presented for the test case of the two-dimensional driven cavity flow.

1. Introduction

Today an increasing number of industrial applications are modeled by using numerical simulations based on the Navier-Stokes equation. Among the existing methods, some are of explicit type. These can be rather easily implemented on parallel computers; however, explicit methods suffer from stability problems. The other types of method are implicit and require the solution of large-scale algebraic systems, an operation that in turn requires very large computing costs. For solving these systems as quickly as possible, new parallel algorithms must be developed.

In the present paper, we are interested in adapting subdomain algorithms to the commonly used PISO algorithm (Pressure Implicit with Split Operator) [12]. This method provides the solution of the finite-volume approximation of the linearized Navier-Stokes problem by using a predictor-corrector

H. G. Kaper and M. Garbey (eds.),
Asymptotic and Numerical Methods for Partial Differential Equations with Critical Parameters 335–347.

scheme. Among the classical subdomain algorithms, we have selected a subdomain formulation without overlapping [11, 10, 20]; more specifically, a general formulation of the relaxation algorithm is modeled [1, 2, 4]. Furthermore, the convergence analysis is made in this general context for every subdomain decomposition, and results of experimental tests on a shared-memory architecture are presented [6, 10, 15].

Previously, we worked separately on the PISO method for the Navier-Stokes equation [7] and on subdomain algorithms applied to nonlinear boundary values problems [10, 16]. The present work combines the complementary qualifications of both research groups as components of the C3NI operation [8], coordinated by CNUSC in Montpellier and IBM (France).

2. The PISO Algorithm

2.1. THE MATHEMATICAL MODEL

Let Ω be a bounded open domain; we consider here the case where $\Omega \subset R^2$ orR^3. Let $\partial\Omega$ be the boundary of Ω. We suppose that $\partial\Omega$ is sufficiently regular. Let T be a strictly positive real number. We consider the flow of an incompressible viscous fluid in the domain Ω. This flow is described by the classical Navier-Stokes equations as follows:

$$div(\overrightarrow{V}) = 0 \tag{1}$$

$$\frac{\partial \overrightarrow{V}}{\partial t} + \overrightarrow{V}\overline{\overline{grad}} + \overrightarrow{V}\overrightarrow{grad}P - \frac{1}{Re}\overline{\overline{\Delta}}\overrightarrow{V} = S \qquad \Omega \times [0,T] \tag{2}$$

$$V\mid_{\partial\Omega} = V_{\partial\Omega} \quad \forall t \tag{3}$$

$$V(x,y,z,t) = V_0(x,y,z) \quad for \quad t = 0, \tag{4}$$

where $V = (U,V,W)$ is the velocity vector, P is the pressure, Re is the constant Reynolds number, $V_{\partial\Omega}$ is the velocity vector given on $\partial\Omega$, V_0 is the initial velocity vector, and S is the source term.

2.2. THE DISCRETIZATION BY THE FINITE VOLUME METHOD

For simplicity we consider now $\Omega \subset R^2$, but the following results are still valid for $\Omega \subset R^3$. To discretize the problem (1)–(4), we consider a staggered grid arrangement sketched in Fig. 1, and we denote Ω_U, Ω_V, and Ω_P the set of points on which U, V, and P are computed, respectively.

According to Patankar [18], the time-dependent equations of the problem (1)–(4), discretized by the finite-volume method in an implicit approach, can be written as follows:

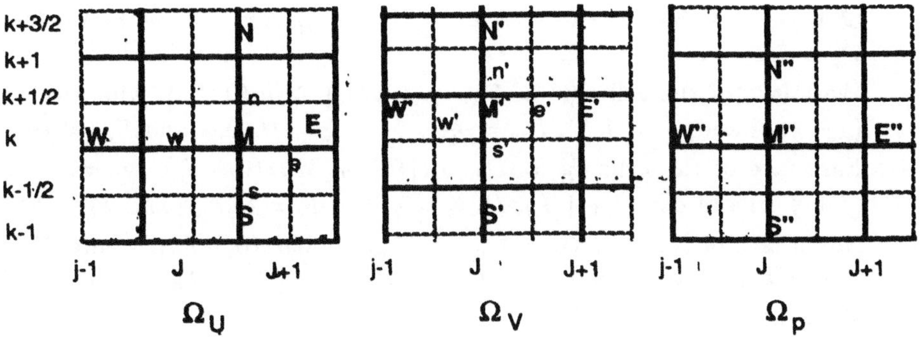

Fig. 1. The staggered grids

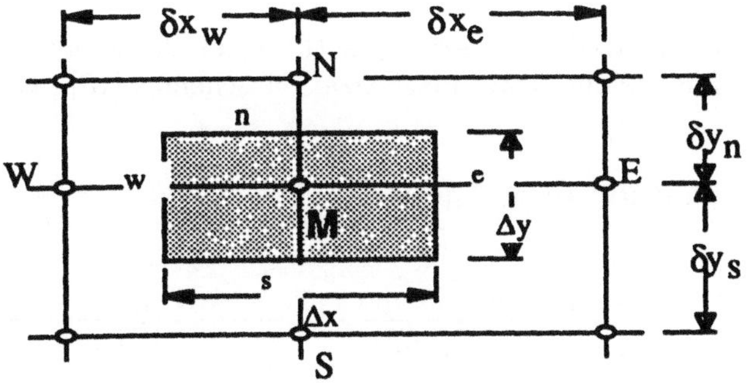

Fig. 2. The control volume

$$-b^U_S U^{(n)}_S - c^U_W U^{(n)}_W + d^U_M U^{(n)}_M - e^U_E U^{(n)}_E - f^U_N U^{(n)}_N + \Delta y \quad \delta_x(P) = S^U_M (5)$$

$$-b^V_{S'} V^{(n)}_{S'} - c^V_{W'} V^{(n)}_{W'} + d^V_{M'} V^{(n)}_{M'} - e^V_{E'} V^{(n)}_{E'} - f^V_{N'} V^{(n)}_{N'} + \Delta x \quad \delta_y(P) = S^V_{M'}, (6)$$

where the coefficient of the first (resp. second) equation is the entries of the pentadiagonal matrix A^U (resp. A^V).

Remark 1. $\delta_x()$ and $\delta_y()$ denote the first-order finite difference operator, and in Equations (5) and (6) $\delta_x(P)$ and $\delta_y(P)$ will be specified in the predictor-corrector scheme (see Section 2.3).

For the convergence analysis of the parallel algorithm, described in Section 3, the values of these coefficients are very important. We define them as follows:

1. For the U-velocity component from Equation (5):

- South face of the cell: $b_S^U = D_s \quad \alpha(|\mathcal{P}_s|) + Max(0, F_s)$, where $S = (j+\frac{1}{2}, k-1)$ and $s = (j+\frac{1}{2}, k-\frac{1}{2})$, $D_s = \frac{1}{Re}\frac{\Delta x}{(\delta y)_s}$, $F_s = V_s \Delta x$, and $\mathcal{P}_s = \frac{F_s}{D_s}$.
- West face of the cell: $c_W^U = D_w \quad \alpha(|\mathcal{P}_w|) + Max(0, F_w)$, where $W = (j-\frac{1}{2}, k)$ and $w = (j,k)$, $D_w = \frac{1}{Re}\frac{\Delta y}{(\delta x)_w}$, $F_w = U_w \Delta y$, and $\mathcal{P}_w = \frac{F_w}{D_w}$
- East face of the cell: $e_E^U = D_e \quad \alpha(|\mathcal{P}_e|) + Max(0, -F_e)$, where $E = (j+\frac{3}{2}, k)$ and $e = (j+1, k)$, $D_e = \frac{1}{Re}\frac{\Delta y}{(\delta x)_e}$, $F_e = U_e \Delta y$, and $P_e = \frac{F_e}{D_e}$,
- North face of the cell: $f_N^U = D_n \quad \alpha(|\mathcal{P}_n|) + Max(0, -F_n)$, where $N = (j+\frac{1}{2}, k+1)$ and $n = (j+\frac{1}{2}, k+\frac{1}{2})$ $D_n = \frac{1}{Re}\frac{\Delta x}{(\delta y)_n}$, $F_n = V_n \Delta x$ and $\mathcal{P}_n = \frac{F_n}{D_n}$
- and

$$d_M^U = b_S^U + c_W^U + e_E^U + f_N^U \frac{\Delta x \Delta y}{\Delta t}. \tag{7}$$

The function $\alpha(|\mathcal{P}_i|)$, where $\mathcal{P}_i$ is the local Peclet number on face i, is used to define various usual discretization schemes [18], as shown in Table 1.

Table 1. **Definition of $\alpha(|\mathcal{P}_i|)$**

Numerical Scheme	$\alpha(\|\mathcal{P}_i\|)$
Central difference	$1 - \frac{\|\mathcal{P}_i\|}{2}$
Upwind	1
Hybrid	$Max(0, 1 - \frac{\|\mathcal{P}_i\|}{2})$
Power law	$Max(0, (1 - .1\frac{\|\mathcal{P}_i\|^5}{2}))$
Exponential scheme	$\frac{\|\mathcal{P}_i\|}{\exp(\|\mathcal{P}_i\|)-1}$

For the points in the neighborhood of the boundaries, the corresponding coefficients are zero, for example:
- near the south (resp. north) boundary, b_S (resp. f_N) is null.
- near the west (resp. east) boundary, c_W (resp. e_E) is null.

The other coefficients for these cells are defined by the previous relations from Equation (5).

2. For the V-velocity component from Equation (6):

Entries of matrix A^V can be expressed in a similar way, shifting the indices in the following manner:

$$\begin{array}{llll} S & \rightarrow S' = (j, k-\frac{1}{2}) \; ; & s & \rightarrow s' = (j,k) \\ W & \rightarrow W' = (j-1, k+\frac{1}{2}) \; ; & w & \rightarrow w' = (j-\frac{1}{2}, k+\frac{1}{2}) \\ E & \rightarrow E' = (j+1, k+\frac{1}{2}) \; ; & e & \rightarrow e' = (j+\frac{1}{2}, k+\frac{1}{2}) \\ N & \rightarrow N' = (j, k+\frac{3}{2}) \; ; & n & \rightarrow n' = (j, k+1) \end{array}$$

Lemma 1. *Matrices A^U and A^V are diagonally dominant matrices.*
Proof. Evident by using the definitions of the diagonal entries of A^U and A^V.
Lemma 2. *A^U and A^V are M-matrices for all the discretization schemes of Table 1 except for the central difference scheme. For the central difference scheme and in addition, if $|P_{loc}| \leq 2$, A^U and A^V are also M-matrices.*
Proof. Indeed matrices A^U and A^V possess strictly positive diagonal entries and nonpositive off-diagonal elements. Then A^U and A^V are Z-matrices. Furthermore these matrices are irreducible, and according to the results of Lemma 1, they are M-matrices [17].
Remark 2. In practice, the central difference scheme should not be applied when the local Peclet number is such that $|P_{loc}| > 2$.

2.3. THE PREDICTOR-CORRECTOR ALGORITHM

The PISO algorithm [12] is a time-marching algorithm based upon the splitting of the solution of momentum equations and pressure equation; at each time step the computed fields are a good approximation of the exact solution of the Navier-Stokes problem.

For marching from time step $(n-1)$ to time step (n), the method consists of a predictor step followed by two correctors steps. Issa [12] has shown that two corrector steps are sufficient to obtain a suitable accuracy, compatible with the discretization scheme and the roundoff error.

2.3.1. The Predictor Step. First the momentum equations are solved by fixing the pressure at its value in the previous time step. We denote U^* and V^* as the velocity components obtained at this stage; U^* and V^* are solutions of the following systems:

$$A^U U^* = -\Delta y \quad \delta_x(P^{(n)}) + S_U \tag{8}$$

$$A^V V^* = -\Delta x \quad \delta_y(P^{(n)}) + S_V. \tag{9}$$

It can be noted that we have in this case

$$DQ^* = \Delta y \quad \delta_x(U^*) + \Delta x \quad \delta_y(V^*) \neq 0, \tag{10}$$

where DQ^* is in general nonzero; thus the continuity equation is not verified.

The systems (8), (9) are uncoupled and can be solved independently, and Equation (10) is used to compute DQ^*.

2.3.2. The Corrector Steps. In order to satisfy the continuity equation, a new velocity field denoted U^{**} and V^{**} and a new corresponding pressure field P^* are introduced in such a way that after some standard algebraic manipululations, we have to solve the following system in order to determine the pressure correction DP^*:

$$-\Delta y \quad \delta_x(\frac{1}{d_M^U}\Delta y \quad \delta_x(DP^*)) - \Delta x \delta_y(\frac{1}{d_{M'}^V}\Delta x \quad \delta_y(DP^*)) = -DQ^*. \tag{11}$$

Notation: $DP^* = P^* - P^{(n)}$ is the pressure correction; $DU^{**} = U^{**} - U^*$ and $DV^{**} = V^{**} - V^*$ are the components of the velocity correction.

Then the values of DU^{**} and DV^{**} are computed by using the following explicit relations:

$$DU_M^{**} = -\Delta y \frac{1}{d_M^U} \quad \delta_x(DP^*) \quad ; \quad DV_{M'}^{**} = -\Delta x \frac{1}{d_{M'}^V} \quad \delta_y(DP^*). \tag{12}$$

Using a finite-volume discretization, we can approximate equation (11) as follows:

$$-b_{S''}^P DP_{S''}^* - c_{W''}^P DP_{W''}^* + d_{M''}^P DP_{M''}^* - e_{E''}^P DP_{E''}^* - f_{N''}^P DP_{N''}^* = -DQ^* (13)$$

$$b_{S''}^P = \Delta y^2 \frac{1}{d_{S'}^V} \quad ; \quad f_{N''}^P = \Delta y^2 \frac{1}{d_{M'}^V}$$

$$c_{W''}^P = \Delta x^2 \frac{1}{d_W^U} \quad ; \quad e_{E''}^P = \Delta x^2 \frac{1}{d_M^U}$$

$$d_{M''}^P = b_{S''}^P + c_{W''}^P + e_{E''}^P + f_{N''}^P. \tag{14}$$

The discretized equation can be summarized by

$$A^P DP^* = -DQ^*. \tag{15}$$

Using the same procedure, one can set up a second corrector step and provide a new update of the pressure and velocity fields for the step (***), which is taken as the approximation at time step $(n+1)$ [12]. For this we have to solve the system

$$A^P DP^{**} = -DQ^{**}, \tag{16}$$

where DQ^{**} is analogous to DQ^* for the second corrector step and A^P is the same matrix as in (15).

In a similar way as Section 2.2 we can establish the following.

Lemma 3. *Matrix A^P is a diagonally dominant matrix.*

Lemma 4. *Matrix A^P is an M-matrix.*

3. The Subdomain Predictor-Corrector Algorithms

3.1. FORMULATION OF THE PARALLEL ALGORITHMS

In the preceding section, we showed that it was necessary to solve systems (8), (9), (15), and (16); these four systems are large-scale algebraic linear systems that must be solved as quickly as possible. Thus these systems will solved by using synchronous or asynchronous subdomain relaxation algorithms without overlapping.

Let N be the set of integers and I denote the set $\{1, 2, ..., \nu\}$, where $\nu \in N$; let us consider the system

$$AX = B, \tag{17}$$

where $\beta \in N$, $\beta = dim(A)$, A is a $\beta \times \beta$ matrix, $X \in R^{\beta}$, and $B \in R^{\beta}$.

We consider now a block decomposition of the problem (17) into ν blocks, which can be written

$$A_{ll}\tilde{x}_l + \sum_{k=1}^{\nu} A_{lk}x_k = b_l \quad \forall l \in I, \tag{18}$$

where A_{lk} is a $\beta_l \times \beta_k$ submatrix, $x_k \in R^{\beta_k}$, $b_l \in R^{\beta_l}$, and $\beta_k \in N$ for every $l, k \in I$, $\beta = \sum_{k=1}^{\nu} \beta_k$ and

$$X = \{x_1, ..., x_l, ..., x_\nu\} \tag{19}$$

$$B = \{b_1, ..., b_l, ..., b_\nu\}. \tag{20}$$

We associate with this problem the following fixed-point problem:

$$X = F(X). \tag{21}$$

According to the preceding block decomposition, we can also split the fixed-point mapping as follows:

$$F(X) = \{F_1(X), ..., F_l(X), ..., F_\nu(X)\}. \tag{22}$$

The parallel asynchronous relaxation algorithms executed with ν processors may be defined as follows.

Let $X^0 \in R^{\beta}$ be given, and for every integer m, suppose that we could get X^1,...,X^m ; then X^{m+1} is recursively defined by

$$\forall m \in N, \quad \forall l \in I, \quad x_l^{m+1} = \left\{ \begin{array}{ll} x_l^m & if\ l \notin \mathcal{S}(m) \\ F_l(W) & if\ l \in \mathcal{S}(m) \end{array} \right\}, \tag{23}$$

where

- $\mathcal{S}(m)$ is a sequence of nonvoid subsets of the set I, and each subset $\mathcal{S}(m)$ contains the number of components updated at step number m by each processor in a parallel way; furthermore, for every $l \in I$ the set $\{m \in N | l \in \mathcal{S}(m)\}$ is infinite, which means that the evaluation of every component is never left out during the calculation.
- For every integer m

$$W = \{x_1^{m-r_1(m)}, ..., x_k^{m-r_k(m)}, ..., x_\nu^{m-r_\nu(m)}\}, \tag{24}$$

where $r(m)$ is a ν component vector

$$r(m) = \{r_1(m), ..., r_\nu(m)\}. \tag{25}$$

In fact, $r(m)$ is a sequence of delays that describe the asynchronism between the ν processors; moreover, the mappings $r_k(m)$ verify the following assumptions:

$$m - r_k(m) \geq 0, \forall k \in I \tag{26}$$

$$r_l(m) = 0, \forall l \in I \tag{27}$$

$$\lim_{m \to \infty} (m - r_k(m)) = +\infty \tag{28}$$

Remark 3. In practice, the choice of the components may be guided by any criterion. In particular, a natural criterion is to pick up the most recently available values of the components.
Remark 4. If for every m, $r(m)$ is null, then the algorithm (23) describes a parallel synchronous relaxation algorithm [19]. Moreover, for the particular choice:

$$\mathcal{S}(m) = I \qquad (resp. \quad \mathcal{S}(m) = m \quad mod(\nu) + 1) \tag{29}$$

the preceding algorithm describes the block Jacobi algorithm (resp. block Gauss-Seidel algorithm).
Remark 5. The block decomposition of (17) can be the natural block decomposition of the problem; but in practice we have not as many processors as blocks, so that we consider a splitting of the problem into "superblocks" which in fact corresponds to a subdomain decomposition without overlapping. In this context, each "superblock" is formed by several adjacent blocks of the natural decomposition.

3.2. CONVERGENCE ANALYSIS FOR LEXICOGRAPHICAL ORDERING

We consider now that the grid points defining Ω_U, Ω_V, and Ω_P are numbered by using a lexicographical ordering of the grid points.
Theorem 1. *If the assumptions of Lemma 2 are valid, and if the grid points of Ω_U, Ω_V, and Ω_P are numbered by using a lexicographical ordering, then*

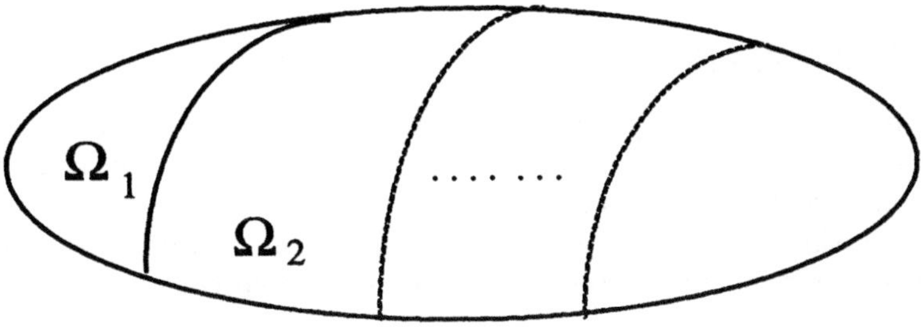

Fig. 3. **Subdomain decomposition**

the synchronous and asynchronous parallel subdomain relaxation algorithms applied to the resolution of problems (8), (9), (15), and (16) converge to the unique solution of these problems for every subdomain decomposition considered.

Proof. In the following, let us consider the system (17) and consider now a point-by-point decomposition of the previous system. After some simple manipulations we can write

$$x_l^{(m+1)} - x_l^* = -\sum_k \neq l \frac{a_{lk}}{a_{ll}}(w_k - x_k^*), \tag{30}$$

where a_{lk} denotes the entries of matrix A, X^* is the exact solution of system (17), x_l^{m+1} is given by the algorithm (23) for the point decomposition considered, and W is the vector of components $x_k^{m-r_k(m)}$. Then the off-diagonal entries of the M-matrix A are nonpositive and

$$|x_l^{(m+1)} - x_l^*| \leq \sum_{k \neq l} \frac{|a_{lk}|}{a_{ll}} |w_k - x_k^*| \quad , \quad \forall l. \tag{31}$$

Let us set

$$J_{lk} = \left\{ \begin{array}{ll} \frac{|a_{lk}|}{a_{ll}} & for\ k \neq l \\ 0 & for\ l = k \end{array} \right\}. \tag{32}$$

Then matrix J is the Jacobi matrix associated to matrix A. By applying Lemmas 2 and 4, A is in general an M-matrix; then

$$\rho(J) < 1, \tag{33}$$

where $\rho(J)$ denotes the spectral radius of matrix J. Furthermore, since J is a nonnegative matrix, we can write

$$y(X^{(m+1)} - X^*) \leq Jy(W - X^*) \quad , \quad \forall W, \tag{34}$$

where y is a vectorial norm defined by

$$y(X) = \{|x_1|, |x_2|, ..., |x_l|, ..., |x_\beta|\} \quad , \quad x_l \in R \quad \forall l. \tag{35}$$

Then. applying a result proved by Miellou [13], we conclude that the considered algorithm associated with the sharpest decomposition converges. Furthermore, if we consider now a coarser subdomain decomposition, we can apply a result shown by Miellou and Spiteri [16] and Spiteri [20], and we conclude also that the subdomain algorithms considered converges for every decomposition.

3.3. THE CONVERGENCE FOR ZEBRA-LINE ORDERING

We consider now the case when the grid points defining Ω_U, Ω_V, and Ω_P are numbered by using a zebra-line ordering. Then, using the same argument, we can establish in a similar way the following results.

Theorem 2. *If the assumptions of Lemma 2 are valid and if the grid points of Ω_U, Ω_V, and Ω_P are numbered by using a zebra-line ordering, then the synchronous and asynchronous parallel subdomain relaxation algorithms applied to the resolution of problems (8), (9), (15), and (16) converge to the unique solution of these problems.*

Remark 6. For both orderings considered, the matrix arising in systems (8), (9), (15), and (16) are M-matrices (cf. Lemmas 2 and 4). Then the convergence of asynchronous parallel algorithms can be also proved by using the result shown by Miellou [14], a result founded on the property of monotone convergence of the iterative processes. It is worth noting also that in this last context the communications between the processors may be very flexible; indeed, in this case, a processor can use iterates just computed by the other processors.

4. The Numerical Results

The preceding methods have been tested in the case of the bidimensional driven square cavity flow. The flow inside the cavity is imposed by the constant velocity of the upper boundary which is fixed to one, and it is governed by the constant Reynolds number. This test case has been studied by various authors such as de Vahl Davies [5], and the sequential version of this program has been fully tested by Estivalezes et al. [7]. In the following, the Reynolds number is fixed at 100.

The numerical experiments have been conducted on the IBM 3090/VF6 of CNUSC in Montpellier (6 vector processors), and the results are presented in details in [3].

It has been observed that solving the linear algebraic systems (8), (9), (15), and (16) is the operation for which the computational costs are the highest. More than 90% of the CPU time is actually spent in the corresponding subroutine (that is, approximately 10% for the predictor step of the velocity and 80% for the computation of the pressure corrections). Thus this phase has been parallelized first.

The algorithms used here for solving the linear systems are overrelaxation ones; preliminary tests on the sequential implementation have shown that an optimal value of the overrelaxation parameter of 1.5 is suitable in the present case. This value is taken for all the subsequent tests.

The time step is chosen as 0.025 for comparing different grid size corresponding to different number of unknowns and different values of T, which are presented in Table 2.

Table 2. **Time step and number of unknowns for different grids**

Grid Size	Number of Unknowns	Time Interval
64×64	12160	1.000
128×128	48896	0.100
256×256	196096	0.025

The comparison of the parallel implementations of the algorithms with lexicographical and zebra-line ordering is presented for 1, 2, 4, and 6 processors respectively in Tables 3–5, corresponding respectively to the elapsed time, the speedup, and the efficiency.

Table 3. **Ellapsed time for both orderings**

Grid size	64×64		128×128		256×256	
Method	Lexico	Zebra	Lexico	Zebra	Lexico	Zebra
1 proc.	281	892	548	571	613	698
2 proc.	181	495	292	301	320	363
4 proc.	113	367	160	170	169	192
6 proc.	113	349	159	179	189	196

Table 4. **Speedup for both orderings**

Rank	64×64		128×128		256×256	
Method	Lexico	Zebra	Lexico	Zebra	Lexico	Zebra
2 proc.	1.55	1.80	1.88	1.90	1.92	1.92
4 proc.	2.49	2.43	3.42	3.36	3.63	3.63
6 proc.	2.49	2.56	3.45	3.19	3.24	3.56

Table 5. **Efficiency for both orderings**

Rank	64×64		128×128		256×256	
Method	Lexico	Zebra	Lexico	Zebra	Lexico	Zebra
2 proc.	0.78	0.90	0.94	0.95	0.96	0.96
4 proc.	0.62	0.61	0.86	0.84	0.91	0.91
6 proc.	0.41	0.43	0.57	0.53	0.54	0.59

For this complex problem the behavior that was already observed for the Poisson equation and various nonlinear boundary value problems is confirmed [10], [11]. Thus, to reduce the costs, we have limited the scope of this experimental study to the synchronous case.

In this case, for small grid sizes, the lexicographical ordering provides better performances. However, when the grid size increases, the differences in performances between both orderings vanish. On the other hand, when the number of processors is increased, the efficiency decreases, but it increases conversely when the grid size is increased.

5. Conclusion

As a conclusion, we can state that the parallel implementation of the resolution of the previous linear systems yields a substantial improvement in the elapsed time necessary to solve the discretized Navier-Stokes problem, especially for large grids. The present implementation can be easily and safely extended to an asynchronous communication, as far as the theoretical convergence is proved for every decomposition. On a practical point of view this last remark is very important, as it can be used to reduce the communication overheads for these algorithms. Furthermore, in tests we have conducted and presented elsewhere [3], these robust implicit algorithms have been shown to be more efficient for sufficiently high Reynolds numbers than some semi-implicit algorithms.

Acknowledgements

We thank Miss Isabelle d'Ast and Mr. Luc Giraud for their technical assistance. We also acknowledge the help of Gérad Leblanc and Benoît Delhaye and support of the C3NI group of CNUSC and IBM France in Montpellier, especially Georges Urbach, Michèle Battle, Sylvain Carta, and Jean Loic Delhaye.

References

[1] Baudet, G. M. (1978) 'Asynchronous iterative methods for multi-processor', *J. ACM*, **25**, 226–244.

[2] Bertsekas, D., and Tsitsiklis, J. (1989) *Parallel and Distributed Computation. Numerical Methods*, Prentice Hall, Englewood Cliffs, N.J.

[3] Boisson, H. C., d'Ast, I., and Spiteri, P. (1992) 'Some subdomain algorithms and their parallel implementation for solving incompressible Navier-Stokes equations', in Ch. Hirsch (ed.), *Computational Fluid Dynamics*, Elsevier North Holland, Amsterdam.

[4] Chazan, D., and Mirkanher, M. (1969) 'Chaotic relaxation', *Linear Algebra and Its Appl.*, **2**, 199–222.

[5] de Vahl Davies, G. (1983) 'Natural convection of air in a square cavity: A benchmark numerical solution', *Int. J. Numer. Methods in Fluids*, **3**, 249–264.

[6] El Tarazi, M. N. (1982) 'Some convergence results for asynchronous algorithms', *Num. Math.*, **39**, 325–340.

[7] Estivalezes, J. L., Boisson, H.C., Kourta, A., Chassaing, P., and Ha Minh, H. (1990) *Performances of the PISO Algorithm Applied to Natural Oscillating Convection in Low Pr Fluids, Series on Numerical Methods in Fluid Dynamics*, Vieweg Verlag, Braunschweig.

[8] Estivalezes, J. L., Boisson, H. C., and Kourta, A. (1989) 'Simulation d'écoulements internes: Comparaison de deux méthodes et application à la convection naturelle oscillatoire', CNUSC-C3NI.

[9] Euvrar, D. (1987) 'Résolution numérique des équations aux dérivées partielles de la physique de la mécanique et des sciences de l'ingénieur', Masson.

[10] Giraud, L., and Spiteri, P. (1991) Résolution de problèmes aux limites non-linéaires', *Mathematical Modelling and Numerical Analysis*, **2**, no. 515, 579–606.

[11] Giraud, L., and Spiteri, P. (1992) 'Implementation of parallel solutions for nonlinear boundary value problems', in D. J. Evans (ed.), *Advances in Parallel Computing*, North-Holland, New York, to appear.

[12] Issa, R. I. (1986) 'Solution of the implicitly discretised fluid flow equations by operator splitting', *J. Comp. Physics*, **62**, 40–65.

[13] Miellou, J. C. (1975) Algorithmes de relaxation chaotiques à retards', *R.A.I.R.O.*, **R1**, 55–82.

[14] Miellou, J. C. (1986) 'Asynchronous iterations in order intervals', in M. Cosnard et al.(eds.), *Parallel Algorithms and Architecture*, North Holland, New York, 85–96.

[15] Miellou, J. C., Cortey-Dumont, P., and Boulbrachene, M. (1990) 'Perturbation of fixed point iterative methods', *Advances in Parallel Processing*, **1**, 81–122.

[16] Miellou, J. C., and Spiteri, P. (1985) 'Un critère de convergence pour des méthodes générales de point fixe', *Mathematical Modelling and Num. Anal.*, **19**, no. 4, 645–669.

[17] Ortega, J. M., and Rheinbold, W. C. (1970) *Iterative Solution of Non Linear Equation in Several Variables*, Academic Press, New York.

[18] Patankar, S. V. (1980) *Numerical Heat Transfer and Fluid Flow*, McGraw-Hill, New York.

[19] Robert, F., Charnay, M., and Musy, F. (1975) 'Iterations chaotiques série parallèles pour des equations nonlinéaires de point fixe', *Appl. Math.*, **20**, 1–38.

[20] Spiteri, P. (1986) 'Parallel asynchronous algorithms for solving boundary value problems', in Cosnard, M., et al. (eds.), *Parallel Algorithms and Architecture*, North Holland, New York, 73–84.

PARALLEL BLAS AND BLACS FOR NUMERICAL ALGORITHMS ON A RECONFIGURABLE NETWORK

F. Desprez and B. Tourancheau
LIP/IMAG
CNRS
ENS Lyon
69364 Lyon Cedex 07
France

ABSTRACT. We study basic parallel linear algebra algorithms: matrix product, matrix transposition, broadcast and scattering operations, and their chaining (rank-2k updates). For all those basic routines, we compare one method using a fixed 2D-torus topology and another one using the capabilities of reconfiguration of the interconnection network. Generally, distributed-memory machines use fixed networks such as multidimensional tori or hypercubes. Today, machines that can be reconfigured during program execution are available. Thus, a large number of possibilities are available to the programmer, who can adapt his configuration during runtime to suit both best algorithm and data distribution. Obviously, this dynamical reconfiguration introduces an overhead through the setting of the network switch. This overhead must be taken into account in the cost of the whole computation. Using complexity analysis, we show the advantages and the drawbacks of the switching solution.

1. Introduction

Programming today's new supercomputers is often a problem, because of the lack of tools and libraries. Some people have tried to develop linear algebra libraries that can be easily used on a broad range of supercomputers. This is the goal of the BLAS [11] and LAPACK which is built on top of it [6, 5]. These libraries first addressed sequential and shared-memory machines. Today, with the development of distributed-memory machines, similar libraries must be provided to enable programmers to get better efficiency without having to redevelop the same low-level routines. In the case of distributed-memory multicomputers, existing numerical routines [11] are not sufficient; communication routines such as the BLACS [12] have to be added for data movement (distributing and storing data across the machine).

Those basic routines are involved in other linear algebra programs, for example, in the LAPACK library. The performance of such high-level routines

H. G. Kaper and M. Garbey (eds.),
Asymptotic and Numerical Methods for Partial Differential Equations with Critical Parameters 349–366.

is then directly associated with the efficient implementation of the basic kernels involved in their construction.

In this paper, we show the major benefits and limitations of communication network reconfigurability for distributed-memory multicomputers. For each of the basic operations studied, we compare one algorithm on a classical fixed topology and one algorithm exploiting the reconfiguration of the network for its execution. The chosen fixed topology is a two-dimensional torus, which gives good average results and thus allows one to chain successive operations efficiently. If it is possible to configure the network before the execution of one routine, we try to determine the best topology. We note that the program topologies can change from one topology for the whole computation, to one topology for each computation or communication subroutine. This is called a quasi-dynamic reconfiguration.

After modeling the different costs involved in the execution of the parallel algorithms, we study the following basic operations: broadcast, scattering, matrix product, and matrix transposition. We then analyze the succession of the four operations: one scattering to load matrices on the network, one matrix product, one transposition of the result matrix, one other matrix product, and the gathering of the results. This succession is a solution for computing the rank-2k updates of a symmetric matrix, which is another basic routine of the Level 3 BLAS library. The solution validates our approach on this set of basic numerical routines. We conclude with a brief discussion of future research.

2. Model of Analysis

The target machine is a reconfigurable distributed-memory parallel multicomputer (DMPC) and a DMPC with a 2D-torus topology. We assume that its N processors $(0..N-1)$ are identical and that the same code runs at the same time on each individual processor. Each processor has independent units for communication (managed by direct memory access, or DMA) and computation (CPU). Thus, on the same node it is possible to perform, in parallel, bidirectional data transfers on each link (full duplex and k-port assumptions) and arithmetic operations. Each node is assumed to have k links $(0..k-1)$ to connect with other nodes through the switching network. This assumption allows us to realize every topology with a degree less than or equal to k. We assume that $k \ll N$. Note that the 2-D torus topology is available on most parallel machines (e.g., hypercubes) and on reconfigurable machines with degree k greater than or equal to 4.

The communication protocol is a "point-to-point rendezvous." Outside of the communication handshake, all the processors are independent and run asynchronously.

For arithmetic operations, the ALU of the CPU is supposed to be scalar.

We always use real, single precision. We call τ_{add} the cost of one addition, τ_{mul} the cost of one multiplication. For the matrix product algorithm, we always use SAXPY (multiply and add on vectors). Thus, we take τ_a for the average time between + and *; a SAXPY then costs $2\tau_a$ [4]. The cost of one valuation equals τ_{aff}. The cost for one memory exchange using three valuations is $\tau_e = 3\tau_{aff}$.

For the communication cost, the time needed to communicate a message of size L between two neighbor processors is modeled as a start-up time β plus the length L times the throughput τ of the communication channel [20]; see Table 1. The total throughput τ_b is not two times τ, and $\beta_{i(b)}$ usually increases with i [8].

Table 1. **Communication costs**

No. of Links in Parallel	Unidirectional	Bidirectional
1	$\beta + L\tau$	$\beta_b + L\tau_b$
i	$\beta_i + L\tau$	$\beta_{ib} + L\tau_b$

In the following sections, we conduct a complexity analysis using parameter values corresponding to the Transputer microprocessor [8]. The reconfiguration method we use is the quasi-dynamic reconfiguration, according to the classification given in [2] and [10]. The analysis is divided into phases. Each phase uses a different topology. Reconfiguration occurs before the start of the current phase of the program; the topology remains fixed within the phases (cf. the case of the Telmat Tnode machine with the C-net environment [1]). Specifically, we study quasi-dynamic topologies versus static ones.

The cost of the switching operation is modeled just as is the communication cost. If we assume that there is only one switch, our model is linear in the number of links in set C, with a start-up time β_r:

$$\beta_r + L\tau_r. \tag{1}$$

From the viewpoint of hardware, this is simply the cost of the load of the switch memory command. The configuration itself loses only a few nanoseconds. However, in this paper, we call the reconfiguration cost the cost of the whole operation. If one can load the switch's memory in parallel, the cost of the reconfiguration of a bigger hierarchical network equals the cost of the reconfiguration of one switch. Moreover, one can imagine an overlap between the load of a new configuration, computations, and communications, thus decreasing the overhead cost.

3. Broadcast Algorithms

The broadcast operation is one of the basic operations of many parallel algorithms. It is used, for example, to load the same code or data or to communicate results on a network of processors. This operation has been studied for various distributed-memory machines such as rings, hypercubes, tori [18, 21, 15, 20], and De Bruijn networks [15] and for different models of communication [16].

3.1. FIXED TOPOLOGY

In this case, we use a two-dimensional torus (2DT). The 2DT topology is a grid with wraparound links. This topology is very suitable for a number of linear algebra operations such as matrix multiplication [14, 4] and Gaussian elimination [22]. It can embedded on a hypercube of dimension greater than or equal to 4. Its diameter is equal to $2\lfloor\frac{\sqrt{N}}{2}\rfloor$. Because of the maximum degree of our nodes, one must keep at most one link connecting our torus to the host machine. In the following, we call this topology the host-connected 2DT (HC-2DT). The diameter of the HC-2DT is $2\lfloor\frac{\sqrt{N}}{2}\rfloor + 1$.

To have an efficient broadcast, we must find a spanning tree of a minimal depth. The construction of such a tree is quite simple [10]. Its depth is $2\lfloor\frac{\sqrt{N}}{2}\rfloor$. With a HC-2DT, the cut link might increase the depth of 1, but it is always possible to find a spanning tree with this link placed in a shorter path, preserving the diameter of the topology.

If we have a short message, the broadcast operation can be done in

$$\beta_4 + (\lfloor\frac{\sqrt{N}}{2}\rfloor - 1)\beta_2 + \lfloor\frac{\sqrt{N}}{2}\rfloor\beta + 2\lfloor\frac{\sqrt{N}}{2}\rfloor L\tau. \tag{2}$$

To improve this operation, one can use different spanning trees with disjoint edges (each broadcasting a part of the data). Trees such as the previous one cannot be used because of links contention. We must therefore use trees of depth equal to $\sqrt{N} + 1$ (in case of a square torus). Those trees and the broadcasting method on a 2DT have been described in [19]. Using a rotation of those trees in the four directions, we can then have four trees with disjoint edges.

With long messages, with a pipeline method [20, 19], the four messages are sent in

$$(\sqrt{\frac{L}{4}\tau_b} + \sqrt{(\sqrt{N} + 2)\beta_4})^2. \tag{3}$$

3.2. RECONFIGURABLE NETWORK

3.2.1. One Reconfiguration Algorithm. The first idea is to use the topology for which the diameter is the smallest for this operation. On our target machine with k links per nodes, the smallest diameter topology is the $k(k-1)$-ary tree. The $k(k-1)$-ary tree is a tree in which the root has k sons and every node has a maximum of $k-1$ sons (degree k for the associated graph). The number of links created for this topology equals $N-1$.

We can send the message in one piece in parallel on every k links of the root and so on, until every node gets the message. This is a valid strategy for short messages. We obtain

$$T_{b\ ar} = T_{reconf} + \lceil \log_{k-1}(\frac{(k-2)N+2}{k})\rceil(\beta_k + L\tau). \quad (4)$$

One optimization is to pipeline the broadcast operation in the tree. Because of the k-port assumption, this operation is equivalent to one pipeline on a linear array whose length equals the depth of the tree [20], which is $\lceil \log_{k-1}(\frac{(k-2)N+2}{k})\rceil$:

$$\nu_{opt} = \sqrt{\frac{(\lceil \log_{k-1}(\frac{(k-2)N+2}{k})\rceil - 1)L\tau}{\beta_k}}$$

$$T_{opt\ b\ ar} = T_{reconf} + (\sqrt{L\tau} + \sqrt{(\lceil \log_{k-1}(\frac{(k-2)N+2}{k})\rceil - 1)\beta_k})^2. \quad (5)$$

3.2.2. Multiple Reconfigurations Algorithm. In the case of multiple reconfiguration, the data is broadcast by reconfiguring the network each time a processor obtains it.

In the first step, the root possessing the data connects itself to k processors. Thus the root can broadcast the data to those processors. In the next step, every processor having the data behaves like the root in the first step. This step is iterated until all the processors involved in the broadcast operation are reached (see the algorithm in [10]). (We cannot use a pipeline method here because the length of the communication paths is one.)

The number of new processors reached every step is given by the following sequence:

$$\begin{aligned}
X_0 &= k \\
X_1 &= k + k^2 = (k+1)k = (k+1)X_0 \\
X_2 &= (k+1)(k+k^2) = (k+1)X_1 \\
&\vdots \\
X_i &= (k+1)X_{i-1}
\end{aligned}$$

Let $h = \lceil \log_{k+1}(N) \rceil$ be the number of steps to reach every node. Thus, we have $\lceil \log_{k+1}(N) \rceil$ reconfiguration steps. Therefore, we have $X_{h-1} = k(k+1)^{h-1}$

The total number of processors is given by the sum of every X_i plus one (the root):

$$\sum_{j=0}^{h-1} X_j + 1 = (k+1)^h.$$

The total number of modified links is the total number of processors reached minus one.

With our model, we have a reconfiguration cost given by

$$T_{reconf} = h\beta_r + ((k+1)^h - 1)\tau_r = \lceil \log_{k+1}(N) \rceil \beta_r + (N-1)\tau_r.$$

The broadcast time is then given by the number of steps of the algorithm. At every step, each root processor sends its data in parallel on its k links. That is, at every step we have $(\beta_k + L\tau)$. Thus, we have a total communication time of

$$T_{com} = h(\beta_k + L\tau) = \lceil \log_{k+1}(N) \rceil (\beta_k + L\tau). \tag{6}$$

The broadcast cost with such a network equals the sum of the reconfiguration and communication costs, giving

$$T_{b\ rec} = T_{reconf} + \lceil \log_{k+1}(N) \rceil (\beta_k + L\tau). \tag{7}$$

3.3. FIXED TOPOLOGY VERSUS RECONFIGURABLE NETWORK

With our model and parameters, it is easy to determine which algorithm is the best for a given message length and a given machine size. With 25 nodes and for messages under 80 bytes long, the best solution is the multiple reconfiguration algorithm regardless of the reconfiguration cost; after 80 bytes the fixed-topology pipeline algorithm is better. With 81 nodes, the fixed-topology pipeline algorithm starts to be better after a message length of 210 bytes.

4. Scattering Algorithms

The scattering operation is a communication routine that consists of distributing different data from one node to a set of nodes. It is often used to distribute the data to be computed on a distributed-memory machine. The inverse operation (gathering) is used, for example, to gather the results from the processors at the end of the execution of a parallel program.

4.1. FIXED TOPOLOGY

On a fixed topology, the scattering operation has to be correctly implemented to give the best results for the complete problem. It is of no use to make an efficient parallel computation algorithm if the network load has a prohibitive cost.

In this section, we consider the scattering operation from the root node to every other nodes. This is important because for the fixed topology case, the HC-2DT is not symmetric and one can have different complexity results if the root node of the scattering operation is changed.

Fraignaud [15] gives an efficient method to distribute a personal message from one node to all other nodes on a 2DT. The 2DT is viewed as two series of rings (horizontal and vertical ones). The distribution operation is directly based on the ring method proposed in [17]. We obtain the total cost formula

$$T_{D\ HC-2DT} = \lfloor \frac{2\sqrt{N}+1}{2} \rfloor \beta_2 + \frac{N+\sqrt{N}-1}{4} L\tau. \tag{8}$$

4.2. RECONFIGURABLE NETWORK

4.2.1. One Reconfiguration Algorithm. On a $k(k-1)$-ary tree, we use the following strategy: First, the root sends the data to the leaves of the tree (i.e., nodes) at depth h. This step is done in parallel on k links, giving a cost $\beta_k + (k-1)^{h-1} L\tau$.

Second, the root sends the data to the nodes at depth $h-1$. At the same time, its sons (the nodes at depth 1), while receiving data from the root, send to their respective sons the data received at the previous step, giving a cost $\beta_k + (k-1)^{h-2} L\tau$. The process continues until the root sends to its sons their own data at the hth step. At this step, every node has received its own data. The cost of the last step equals $\beta_k + L\tau$. Thus, the total communication cost formula is

$$T_{com\ s\ a} = \sum_{i=0}^{h-1} (\beta_k + (k-1)^i L\tau) = h\beta_k + \frac{(k-1)^h - 1}{k-2} L\tau.$$

We replace h in the preceding relation, and we add the reconfiguration cost, giving the total cost formula

$$T_{s\ a} = T_{reconf} + \lceil \log_{k-1}(\frac{(k-2)N+2}{k}) \rceil \beta_k + \lceil \frac{N-1}{k} \rceil L\tau. \tag{9}$$

4.2.2. Multiple Reconfigurations Algorithm. In the case where the network is completely reconfigurable, we propose the following algorithm to implement the scattering operation.

At the beginning, the root possesses all the messages, corresponding to NL bytes. In order to have a good load balancing of the work, the root divides the message to be scattered into $k+1$ equal parts. One of the resultant packets stays on the root (it contains messages for the next steps). The k others are sent to the k nodes that are connected at this step. Just as for the broadcast operation, the root makes a connection to k other nodes, giving a reconfiguration cost equal to $\beta_r + k\tau_r$.

After the (re)configuration step, the root sends in parallel the k messages on its k links, with a communication cost equal to $\beta_k + \frac{NL}{k+1}\tau$.

During the next step, every node reached at the previous step starts the same operation. They can now have the same behavior as the previous root. In this step we have $k+1$ "roots" possessing $\frac{NL}{k+1}$ bytes. At this time of the scattering, the original root and every node reached at the previous step connect themselves to k sons and send one piece of the message, giving the cost formula $\beta_k + \frac{NL}{(k+1)^2}\tau$.

This process stops when the message on every node is of length L (the size of the personal message). At this step, every node is reached. The number of steps ($\lceil \log_{k+1}(N) \rceil$) and the total reconfiguration cost are the same as the multiple reconfigurations broadcast.

The total cost of this algorithm equals the sum of the costs for each step, giving

$$
\begin{aligned}
T_{s\ rec} &= T_{reconf} + T_{com} \\
&= T_{reconf} + \sum_{i=1}^{\lceil \log_{k+1}(N) \rceil} (\beta_k + \frac{NL}{(k+1)^i}\tau) \\
&= T_{reconf} + \lceil \log_{k+1}(N) \rceil \beta_k + (\frac{(k+1)^{\lceil \log_{k+1}(N) \rceil} - 1}{k(k+1)^{\lceil \log_{k+1}(N) \rceil}}) NL\tau \\
&= T_{reconf} + \lceil \log_{k+1}(N) \rceil \beta_k + \lceil \frac{N-1}{k} \rceil L\tau. \qquad (10)
\end{aligned}
$$

4.3. FIXED TOPOLOGY VERSUS RECONFIGURABLE NETWORK

When we consider the complexity analysis of the three solutions of the scattering problem, the two reconfiguration solutions are almost equal (they differ only in the startup time). The reconfiguration solution is interesting only in the case of the scattering from the host with big messages.

5. Matrix Product Algorithm

Matrix multiply is one of the basic linear algebra operations used to solve a broad range of problems. It can be implemented very efficiently on parallel machines [4, 14, 9]. We use the matrix product routine GEMM, from the well-known Level 3 BLAS [11] (GEMM is used as a basic routine of matrix inversion and as a benchmark for sequential or parallel computers).

For this operation, and for the sake of simplicity, we first study the case with a reconfigurable network.

5.1. RECONFIGURABLE NETWORK

5.1.1. Data Distribution. In our study, we use square matrices of size M. The amount of computation is the same for the calculation of each element. The algorithms proposed act on subblocks distributed on a 2DT. The number of blocks in a matrix is equal to the number of processors. In this paper, we use a matrix of size M, where $M \bmod \sqrt{N} = 0$ and where $\sqrt{N}$ is an integer. The blocks are of equal size $m = \frac{M}{\sqrt{N}}$ to ensure load balancing.

5.1.2. Parallel Algorithm. Let A, B, and C be three matrices of size M. The multiplication $C = AB$ has a sequential complexity of $2n^3\tau_a$.

Usually the parallelization is made on a 2D torus of $\sqrt{N}$x$\sqrt{N}$ processors, but quite good results have been obtained on a ring topology as well [7, 3]. In [4] the classical parallel implementation on a 2D torus is described. The data distribution by square subblocks fits the topology very well. Let C_{ij} be the subblock ij of the matrix. We have

$$C_{hk} = C_{hk} + \sum A_{hn}B_{nk} \quad (0 \leq n < \frac{M}{\sqrt{N}}).$$

The algorithm is shown in Fig. 1.

for $i = 1$ to $\sqrt{N}$ do
  inparallel do
    $C_{ij} = C_{ij} + A_{ij} * B_{ij}$ *submatrices multiplication*
    Rolling of A horizontally
    Rolling of B vertically
  enddo
endfor

Fig. 1. **Parallel matrix-multiply algorithm**

Using our complexity models (i.e., assuming communications on different links can be done in parallel), we obtain an overlapping of the two rollings and the product of two submatrices, leading to the following [9]:

$$\max(2\frac{M^3}{N}\tau_a, \frac{M^2}{\sqrt{N}}\tau + \sqrt{N}\beta_4). \tag{11}$$

We are able to calculate, for a fixed number of processors N, the value of m for which the computations and the communications totally overlap [9]. We call it m_c; values of $m \geq m_c$ will lead to an optimal efficiency of the parallel algorithm.

5.2. FIXED TOPOLOGY

In this case communication between the extreme nodes of the first row of processors occurs in two parts because of the presence of the host on the link between the two nodes. With this additional cost, the total cost of the matrix product becomes

$$\max(2\frac{M^3}{N}\tau_a, 2(\frac{M^2}{\sqrt{N}}\tau + \sqrt{N}\beta_4)). \tag{12}$$

Applying the same method used for the previous topology, we compute the value of m for which we have a total overlap between communications and computations.

5.3. CONCLUSION

As we have seen before with the computation of the optimal size of the submatrices, the total overlap between computations and communications starts for very small matrices. Thus we conclude that, for the matrix product algorithm and our parameters, the reconfiguration is not useful.

6. Matrix Transposition Algorithms

The matrix transposition operation is used in a broad range of linear algebra algorithms on distributed-memory multicomputers in order to modify the data distribution of submatrices on the network and hence to perform operations with less communication.

Given a matrix $A = (a_{ij})$ and its transpose $A^T = (a'_{ij}) = (a_{ji})$, the sequential algorithm costs $\frac{m(m-1)}{2}\tau_e$. In the following, we denote $T_t = \frac{m(m-1)}{2}\tau_e$.

The matrix is distributed by square subblocks on the network, and we wish to obtain the transpose of the matrix distributed by the same manner. Notice that it is not the same operation as in [20] and [15] where the transposition operation is equivalent to one broadcast from every node (for example, the transposition of a matrix distributed by rows on the network). Here we have exchanges only between nodes containing the M_{ij} and M_{ji} parts of the matrix (for $i \neq j$).

6.1. FIXED TOPOLOGY

Using the two-dimensional torus with a cut link, we can compute the transpose of a matrix simply by taking this topology as a grid. We therefore have a symmetric topology with a diameter of $2(\sqrt{N} - 1)$.

A simple strategy can be used. The submatrix is broken in two parts of the same size and sent in two directions. The two messages travel by a stair path until reaching their destination. The total cost of the algorithm is given by the longest path of the grid, giving the cost formula

$$T_{tr\ grid} = 2(\sqrt{N} - 1)(\beta_2 + \frac{L}{2}\tau_b). \tag{13}$$

Using a 2DT, we can apply the same strategy. In this case, however, we break the matrix in four parts and send them in four directions using wrapped-around links:

$$T_{tr\ 2DT} = 2(\sqrt{N} - 1)(\beta_4 + \frac{L}{4}\tau_b). \tag{14}$$

One improvement to these solutions is to use a pipeline method on the different communication paths and an overlap of the memory management by the communications. Notice that in this case there is a problem of contention on the links. By grouping the different messages in one unique message, however, we can reduce the communication overhead.

6.2. RECONFIGURABLE NETWORK

The topology used for the transposition is the graph in which pairs of processors that have to exchange data are connected by at least one link (diagonally opposed processors on a square grid). In this way a matrix can be transposed directly by using neighbor-to-neighbor communication. The transposition takes place in two steps. First, the subblocks are transmitted to the corresponding nodes. Second, each node computes the transpose of its subblock. By storing the subblocks in a contiguous way in the memory, the exchange of a subblock requires only one message.

The number of links required to connect the diagonally opposed processors with one bidirectional link is $\frac{\sqrt{N}(\sqrt{N}-1)}{2}$.

The lowest cost of transposition is a function of the number of links available to connect the two diagonally opposed processors. The number of links connecting two processors can be varied from 1 to k, the maximal number of links available on each node. But as the communication cost decreases, the reconfiguration cost increases. The optimal interconnection graph depends on the physical characteristics of the machine.

One possible optimization is to overlap the internal transposition with the communications. We transform this algorithm in a *macro-pipeline* where while one packet of data is transferred, and at the same time the previous one is transposed (Fig. 2).

```
inparallel do
  receive sub-block
  send sub-block
enddo
for i = 1 to n - 1 do
  inparallel do
    receive sub-block
    send sub-block
    transpose preceding sub-block
  enddo
endfor
transpose last sub-block
```

Fig. 2. **Algorithm with** *macro-pipeline*

The data is divided into ν packets of size $\frac{m^2}{\nu}$. After receiving the first subblock, exchanges and communications take place in parallel, using the same method as for pipelining.

With the formula (1) of our model, the reconfiguration cost is

$$T_r = \beta_r + \frac{\sqrt{N}(\sqrt{N}-1)}{2}\tau_r.$$

and the cost of the parallel transposition is

$$T_{tr} = \beta_b + \frac{m^2}{\nu}\tau_b + (\nu - 1)\max(\beta_b + \frac{m^2}{\nu}\tau_b, T_{it}) + T_{it},$$

where $T_{it} = \frac{m(m-1)}{2\nu}\tau_e$ = cost of the internal rearrangement of one packet. Here, the overlap of the memory management is possible only if the internal transposition time of one packet of data is smaller than the communication time of the next packet, giving the inequality

$$\frac{m(m-1)}{2\nu}\tau_e \leq \beta_b + \frac{m^2}{\nu}\tau_b,$$

which leads to

$$\frac{(\tau_e - 2\tau_b)m^2 - m\tau_e}{2\beta_b} \le \nu.$$

In the other case, the communications are overlapped by the memory management. Roughly, with our parameters, the right part of the inequality is negative as soon as $\tau_e \le 2\tau_b$; that is, the communication cost is more than half the exchange cost. We can now calculate the optimal size of the packets (in the case of one link between nodes):

$$\nu_{opt} = \sqrt{\frac{T_t}{\beta_b}} \text{ where } T_t = \nu T_{it}$$

We replace ν and obtain the following:

$$T_{tr} = \beta_r + \frac{\sqrt{N}(\sqrt{N}-1)}{2}\tau_r + 2\sqrt{T_t\beta_b} + m^2\tau_b. \tag{15}$$

Analog formulas can be derived for 2 and k links in parallel [9].

To reduce the internal transposition cost (and if there is enough memory space), we no longer need to transpose the received packets but to store their elements so that the resulting matrix is the transposed matrix. In this case we have the following cost:

$$T_{tr2} = \beta_b + \frac{m^2}{\nu}\tau_b + (\nu - 1)\max(\beta_b + \frac{m^2}{\nu}\tau_b, T_{irgt}) + T_{irgt},$$

where $T_{irgt} = \frac{m^2}{\nu}\tau_{aff}$ =cost of the store of one packet, with a total overlap of the memory management. The optimal size of the packets becomes smaller:

$$\nu_{opt} = \sqrt{\frac{T_{rgt}}{\beta_b}} = m\sqrt{\frac{\tau_{aff}}{\beta_b}} \text{ where } T_{rgt} = \nu T_{irgt}.$$

We replace ν and obtain the following:

$$T_{tr2} = \beta_r + \frac{\sqrt{N}(\sqrt{N}-1)}{2}\tau_r + 2m\sqrt{\tau_{aff}\beta_b} + m^2\tau_b. \tag{16}$$

6.3. COMPARISON OF THE TWO SOLUTIONS

The reconfiguration solution remains more interesting than the fixed-topology solution and leads to realistic reconfiguration costs. But the pipeline improvement on the torus solution can reduce the difference between the three solutions.

7. Rank-2k Update Algorithms

In this part, we consider the routine SYR2K [11] from the Level 3 BLAS. This routine performs the rank-2k updates of a symmetric matrix C:

$$C = \alpha AB^T + \alpha BA^T + \beta C$$
$$C = \alpha A^T B + \alpha B^T A + \beta C$$

The updating operation is an interesting example of the use of all the previously studied routines with the reconfiguration in parallel linear algebra.

We have here only matrix products of the form $A^T B$ or AB^T. Because $AB^T = (BA^T)^T$, we need only one matrix product, one transposition of the result, and one addition (with scaling with α or β if needed) [13].

Figure 3 describes the general algorithm for computing the rank-2k updates $C = \alpha AB^T + \alpha BA^T + \beta C$ on a parallel distributed-memory machine.

begin
(I) load matrix A, B^T and C
(II) compute $T = \alpha AB^T$
(III) compute $C = \beta C + T$
(IV) transpose T
(V) compute $C = C + T^T$
(VI) unload matrix C
end

Fig. 3. General algorithm for rank-2k updates of a symmetric matrix

The submatrix B_{ij} must be loaded on the processors whose coordinates in the torus are ji, so that we do not have to compute the transposition of the initial matrix. We store its elements with an implicit transposition.

7.1. FIXED TOPOLOGY

We use the algorithms described earlier when we computer the rank-2k updates of a symmetric matrix on a fixed 2DT topology. One might, of course, be able to find other algorithms for such a topology.

7.2. COMPLEXITY ANALYSIS

We break the algorithm into five parts.

Part 1 corresponds to (I).

We load the network with the 3 matrices A, B^T and C.

$$T_{ph1} = \lfloor \frac{2\sqrt{N}+1}{2} \rfloor \beta_2 + \frac{N+\sqrt{N}-1}{4} (\frac{3M^2}{N}) \tau \tag{17}$$

Part 2 corresponds to (II).

It is the matrix product $T = \alpha A B^T$.

$$T_{ph2} = 2\frac{M^3}{N}\tau_a \tag{18}$$

Part 3 corresponds to (III) and (IV).

We overlap the transposition of matrix T and its addition with C scaled with β.

$$T_{ph3} = \max(2(\sqrt{N}-1)(\beta_4 + \frac{M^2}{4N}\tau_b), \frac{2M^2}{N}\tau_{add}) \tag{19}$$

Part 4 corresponds to (V).

It is the addition of the transposed matrix T with C.

$$T_{ph4} = \frac{M^2}{N}\tau_{add} \tag{20}$$

Part 5 corresponds to (VI).

The last phase consists of the gathering of the matrix C on the host.

$$T_{ph5} = \lfloor \frac{2\sqrt{N}+1}{2} \rfloor \beta_2 + \frac{N+\sqrt{N}-1}{4} (\frac{M^2}{N}) \tau \tag{21}$$

7.3. RECONFIGURABLE NETWORK

On a reconfigurable network, we decompose the algorithm into four phases (Fig. 4), each one corresponding to an algorithmic strategy: two multiple reconfigurations, one 2DT, and one direct network. Each phase corresponds to one of the algorithms described in the preceding sections. The two differences with the fixed topology case are that we join the matrix product and the following addition in one phase (same topology), and we overlap the transposition and the second addition using a macro-pipeline method.

First, the matrices are loaded by using the multiple reconfiguration scattering algorithm. Then $T = \alpha A B^T$ and $C = \beta C + T$ are computed on a 2DT. After that, the topology is set to a direct network for the transposition. Then T is cut into packets of the optimal size, and in parallel we compute the sum $C = C + T^T$ while exchanging the packets. The last operation consists of downloading the resulting matrix C to the host, using the multiple reconfiguration gathering algorithm.

Phase 1: [Multiple rec. algorithm]
    Scattering of the matrices A, B^T and C
Phase 2: [Topology = 2DT]
    Compute $T = \alpha AB^T$ using matrix product algorithm
    Compute $C = \beta \dot{C} + T$
Phase 3: [Topology = direct network]
    inparallel do
      exchange submatrices T
      compute $C = C + T^T$
    done
Phase 4:
    Gathering of matrix C

Fig. 4. **Rank-2k updates on a reconfigurable network**

The following presents the complexity analysis of each phase. In Phases 1, 2, and 4, we use the preceding algorithm complexities (scattering and matrix product). The sum of two matrices distributed by subblocks is straightforward.

Phase 1

$$T_{ph1} = \lceil \log_{k+1}(N+1) \rceil \beta_k + \lceil \frac{N}{k} \rceil (3\frac{M^2}{N})\tau \tag{22}$$

Phase 2

$$T_{ph2} = 2\frac{M^3}{N}\tau_a + \frac{M^2}{N}\tau_{add} \tag{23}$$

Phase 3

For this phase, we use a variation of the macro-pipeline described for the matrix transposition algorithm.

For the operation $C = C + T^T$, we have the total computation cost $m^2\tau_{add}$. We break T in ν packets of size $\frac{m^2}{\nu}$:

$$T_{ph\ 3} = \beta_b + \frac{m^2}{\nu}\tau_b + (\nu - 1)\max(\beta_b + \frac{m^2}{\nu}\tau_b, \frac{m^2}{\nu}\tau_{add}) + \frac{m^2}{\nu}\tau_{add}.$$

Now we determine, with our parameters, the maximum cost between the communication and the computation of one packet of size $\frac{m^2}{\nu}$:

$$\beta_b + \frac{m^2}{\nu}\tau_b \geq \frac{m^2}{\nu}\tau_{add} \Rightarrow \nu \geq \frac{m^2(\tau_{add} - \tau_b)}{\beta_b}.$$

Thus we have a total overlap of the computation by the communication, giving the cost formula

$$T_{ph\ 3} = \nu\beta_b + m^2\tau_b + \frac{m^2}{\nu}\tau_{add}.$$

We have $\nu_{opt} = \sqrt{\frac{m^2\tau_{add}}{\tau_b}}$, giving the following cost formula for Phase 3:

$$T_{ph\ 3} = 2m\sqrt{\tau_{add}\beta_b} + m^2\tau_b. \tag{24}$$

Phase 4

Note that the multiple reconfiguration gathering algorithm is the inverse of the scattering:

$$T_{ph1} = \lceil \log_{k+1}(N+1)\rceil\beta_k + \lceil\frac{N}{k}\rceil(\frac{M^2}{N})\tau. \tag{25}$$

7.4. CONCLUSION

The modularity of this algorithm leads directly to a phase decomposition that fits very well to the reconfiguration solution on a quasi-dynamic reconfigurable network.

Notice that if we are able to overlap the reconfiguration of the network and computations/communications on the nodes, the reconfiguration overhead should be at least partially hidden in Phase 2.

8. Conclusion and Future Work

In this paper we have shown that with our model and parameters, the reconfiguration solution must be of interest for some algorithms if its cost is not too important and especially if one can overlap the reconfiguration with the computations/communications of the algorithm. This capability of the switching architecture must be taken into account in further research.

However, we pointed out that good parallel algorithmic solutions on the HC-2DT are better than any others in some cases when the problem size is not very small and/or when pipeline or macro-pipeline solutions are possible. This demonstrates that research in this area is not obsolete and will still be used in message-passing DMPCs.

Our future research will try to generalize the macro-pipeline and overlap approaches, to be able to provide libraries and tools to help the programmers of such applications.

References

[1] Adamo, J. M., et al. (1991) 'Developing a high level programming environment for supernode', in *Proceedings of Transputer Application*. IOS Press, Glasgow, 632–637.

[2] Bauch, A., Braam, R., and Maehle, E. (1991) 'Damp – a dynamic reconfigurable multiprocessor system with a distributed switching network', in *Proceedings of the 2nd European Distributed Memory Computing Conference, Lecture Notes in Computer Science*, Springer-Verlag, Berlin, 495–504.

[3] Berger, P., Dayde, M., Lacoste, R., and Morere, P. (1990) 'Implementation of Level 3 BLAS kernel on a transputer network', in *Workshop on Software Standard for MIMD Machines*, Abdington, UK.

[4] Bernsen, J. (1989) 'Communication efficient matrix multiplication on hypercubes', *Parallel Computing*, **12**, 335–342.

[5] Bischof, C. (1988) 'Fundamental linear algebra computations on high-performance computers', Preprint MCS-P150-0490, Mathematics and Computer Science Division, Argonne National Laboratory, Argonne, Illinois.

[6] Bischof, C., Demmel, J., Dongarra, J., DuCroz, J., Greenbaum, A., Hammarling, S. H., and Sorensen, D. (1989) 'LAPACK working note #5: Provisional contents', Argonne National Laboratory Technical Report ANL-88-38, Argonne, Illinois.

[7] Desprez, F. (1990) 'Algèbre linéaire sur tnode', Master's thesis, LIP/ENS-Lyon.

[8] Desprez, F., and Tourancheau, B. (1990) 'Modélisation des performances de communication sur le Tnode avec le logical system transputer toolset', *La lettre du Transputer et des calculateurs distribués*, no. 7, 65–72.

[9] Desprez, F., and Tourancheau, B. (1991) 'Re-configuration versus fixed topology', in *Proceedings of the EPCA Bohn Workshop*, 1–10.

[10] Desprez, F., and Tourancheau, B. (1992) 'Communication and linear algebra subroutines on a quasi-dynamic network', Technical Report, Laboratoire LIP-IMAG.

[11] Dongarra, J., du Croz, J., Hammarling, S., and Duff, I. (1990) 'A set of level 3 basic linear algebra subprograms', *ACM Trans. Math. Software*, **16**, no. 1, 1–17.

[12] Dongarra, J., and de Geijn, R. (1991) 'LAPACK working note: Two dimensional basic linear algebra communication subprograms, University of Tennessee Technical Report 37.

[13] Elster, A. C. (1990) 'Basic matrix subprograms for distributed memory systems', in *Proceedings of DMCC5*. IEEE Computer Society Press, Charleston, 311–316,

[14] Fox, G., Johnson, M., Lycenza, G., Otto, S., Salmon, J., and Walker, D. (1988) *Solving Problems on Concurrent Processors*, Vol. 1, Chap. 10, 167–185.

[15] Fraignaud, P. (1990) 'Communications Intensives dans les Architectures à Mémoire Distribuée et Algorithmes Parallèles pour la recherche de racines de polynômes', Ph.D. thesis, ENS Lyon - Université Claude Bernard de Lyon I.

[16] Fraignaud, P., and Lazard, E. (1991) 'Methods and problems of communication in usual networks', Technical Report 33, Laboratoire LIP-IMAG.

[17] Fraignaud, P., Miguet, S., and Robert, Y. (1990) 'Scattering on a ring of processors', *Parallel Computing*, **13**, 377–384.

[18] Johnsson, S. L., and Ho, C. T. (1989) 'Optimum broadcasting and personalized communication in hypercubes', *IEEE Transactions on Computers*, **38**, no. 9, 1249–1268.

[19] Michallon, P., Trystram, D., and Villard, G. (1992) 'Optimal broadcasting algorithms on torus', Technical Report 872, Laboratoire LMC-IMAG.

[20] Saad, Y., and Schultz, M. H. (1988) 'Data communication in parallel architectures', *Parallel Computing*, **11**, 131–150.

[21] Stout, Q. F., and Wagar, B. (1987) 'Intensive hypercube communication', Technical Report, University of Michigan Computing Research Laboratory.

[22] Tourancheau, B. (1989) 'Algorithmique Parallèle pour les Machines à Mémoires Distribuées (application aux algorithmes matriciels)', Ph.D. thesis, INP Grenoble.

Index

GPSR Compliance

The European Union's (EU) General Product Safety Regulation (GPSR) is a set of rules that requires consumer products to be safe and our obligations to ensure this.

If you have any concerns about our products, you can contact us on ProductSafety@springernature.com

In case Publisher is established outside the EU, the EU authorized representative is:

Springer Nature Customer Service Center GmbH
Europaplatz 3
69115 Heidelberg, Germany

Batch number: 10816188

Printed by Printforce, the Netherlands